Charles Darwin

On the origin of natural species

Charles Darwin

On the origin of natural species

ISBN/EAN: 9783742842442

Manufactured in Europe, USA, Canada, Australia, Japa

Cover: Foto ©berggeist007 / pixelio.de

Manufactured and distributed by brebook publishing software
(www.brebook.com)

Charles Darwin

On the origin of natural species

ON

THE ORIGIN OF SPECIES

BY MEANS OF NATURAL SELECTION,

OR THE

PRESERVATION OF FAVOURED RACES IN THE STRUGGLE FOR LIFE.

By CHARLES DARWIN, M.A.,

FELLOW OF THE ROYAL, GEOLOGICAL, LINNEAN, ETC., SOCIETIES;
AUTHOR OF 'JOURNAL OF RESEARCHES DURING H. M. S. BEAGLE'S VOYAGE ROUND THE WORLD.'

THIRD EDITION, WITH ADDITIONS AND CORRECTIONS.

(SEVENTH THOUSAND.)

LONDON:
JOHN MURRAY, ALBEMARLE STREET.
1861.

LONDON: PRINTED BY W. CLOWES AND SONS, STAMFORD STREET AND CHARING CROSS.

CONTENTS.

CHAPTER I.

VARIATION UNDER DOMESTICATION.

CHAPTER II.

VARIATION UNDER NATURE.

CHAPTER III.

STRUGGLE FOR EXISTENCE.

CHAPTER IV.

NATURAL SELECTION.

CHAPTER V.

LAWS OF VARIATION.

CHAPTER VI.

DIFFICULTIES ON THEORY.

CHAPTER VII.

INSTINCT.

CHAPTER VIII.

HYBRIDISM.

CHAPTER IX.

ON THE IMPERFECTION OF THE GEOLOGICAL RECORD.

CHAPTER X.

ON THE GEOLOGICAL SUCCESSION OF ORGANIC BEINGS.

CHAPTER XI.

GEOGRAPHICAL DISTRIBUTION.

CHAPTER XII.

Geographical Distribution—*continued.*

CHAPTER XIII.

Mutual Affinities of Organic Beings: Morphology: Embryology: Rudimentary Organs.

CHAPTER XIV.

Recapitulation and Conclusion.

INSTRUCTION TO BINDER.

The Diagram to front page 123, and to face the latter part of the Volume.

ADDITIONS AND CORRECTIONS,

TO THE SECOND AND THIRD EDITIONS.

— — — —

[The more important Additions and Corrections alone are here
tabulated for the convenience of those interested in the subject,
who possess only the earlier editions. Most of these additions
have appeared in the German and in the Second American
editions. The paging of the first and second English editions is
nearly the same.]

Second Edition.	Third Edition.	Additions and Corrections.
Page	Page	
12	12	Case of correlation of colour and liability to poison given.
45	46	On the improbability of the occurrence of sudden and great modifications of form.
49	51	On the doctrine recently advanced by some naturalists that animals present no variations.
54	56	The meaning of the term "dominant species" made clearer.
73 } 95 }	76 } 100 }	The dependence of the fertility of clover on the visits of insects amplified.
81	84	Various objections to the term "Natural Selection" answered.
84	89	On the two assumptions of an indefinite and definite amount of variation.
102	107	Power of reversion exaggerated.
105	110	Mistaken idea of the efficiency of time alone in producing change.
112	117	On the accidental accumulation of similar variations effecting little without the aid of selection.
126	133 to 143	On the tendency of organisation to advance—On the present coexistence of high and lowly organised beings—Difficulty of conceiving how at the dawn of life the simplest beings were modified — Various objections considered—On convergence—On the indefinite multiplication of specific forms.
135	152	On mutilations being inherited.
137	154 } to 157 }	The blindness of cave-animals amplified and corrected.
184	202	Case of ground-woodpecker amplified.
186 } 187 }	205 } 207 }	On the modification of the eye amplified.
192	211	On the ovigerous frena of cirripedes.
198	218	On the effect of little exercise and abundant food.
212	232	On variations in the habits of the hive-bee.
231	252	Thickness of walls of cell of hive-bee corrected.
245	256	On related variations in instinct and structure not necessarily simultaneous.

POSTSCRIPT.

An admirable, and, to a certain extent, favourable Review of this work, including an able discussion on the Theological bearing of the belief in the descent of species, has now been separately published by Professor Asa Gray as a pamphlet, about 60 pages in length. It is entitled, 'Natural Selection not inconsistent with Natural Theology. A Free Examination of Darwin's Treatise on the Origin of Species, and of its American Reviewers. By Asa Gray, M.D., Fisher Professor of Natural History in Harvard University. Reprinted from the Atlantic Monthly for July, August, and October, 1860. London: Trübner and Co., 60, Paternoster-row. Boston: Ticknor and Fields. 1861.' (Price 1s. 6d.)

An Historical Sketch of the recent Progress of Opinion on the Origin of Species.

I will here attempt to give a brief, but imperfect sketch of the progress of opinion on the Origin of Species. The great majority of naturalists believe that species are immutable productions, and have been separately created. This view has been ably maintained by many authors. Some few naturalists, on the other hand, believe that species undergo modification, and that the existing forms of life have descended by true generation from pre-existing forms. Passing over authors from the classical period to that of Buffon, with whose writings I am not familiar, Lamarck was the first man whose conclusions on this subject excited much attention. This justly-celebrated naturalist first published his views in 1801, and he much enlarged them in 1809 in his ' Philosophie Zoologique,' and subsequently, in 1815, in his Introduction to his ' Hist, Nat. des Animaux sans Vertèbres.' In these works he upholds the doctrine that all species, including man, are descended from other species. He first did the eminent service of arousing attention to the probability of all change in the organic as well as in the inorganic world being the result of law, and not of miraculous interposition. Lamarck seems to have been chiefly led to his conclusion on the gradual change of species, by the difficulty of distinguishing species and varieties, by the almost perfect gradation of forms in certain organic groups, and by the analogy of domestic productions. With respect to the means of modification, he attributed something to the direct action of the physical conditions of life, something to the crossing of already existing forms, and much to use and disuse, that is, to the effects of habit. To this latter agency he seems to attribute all the beautiful adaptations in nature;—such as the long neck of the giraffe for browsing on the branches of trees. But he likewise believed in a law of progressive development; and as all the forms of life thus tended to progress, in order to account for the existence at the present day of very simple productions, he maintained that such forms were now spontaneously generated.*

* I have taken the date of the first publication of Lamarck from Isid. Geoffroy Saint Hilaire's (' Hist. Nat. Generale,' tom. ii. p. 405, 1859, ex-

Geoffroy Saint Hilaire, as is stated in his 'Life,' written by his son, suspected, as early as 1795, that what we call species are various degenerations of the same type. It was not until 1828 that he published his conviction that the same forms have not been perpetuated since the origin of all things. Geoffroy seems to have relied chiefly on the conditions of life, or the "monde ambiant," as the cause of change. He was cautious in drawing conclusions, and did not believe that existing species are now undergoing modification; and, as his son adds, "C'est donc un problème à réserver entièrement à l'avenir, supposé même que l'avenir doive avoir prise sur lui."

In England the Hon. and Rev. W. Herbert, afterwards Dean of Manchester, in the fourth volume of the 'Horticultural Transactions,' 1822, and in his work on the 'Amaryllidaceæ' (1837, p. 19, 339), declares that "horticultural experiments have established, beyond the possibility of refutation, that botanical species are only a higher and more permanent class of varieties." He extends the same view to animals. The Dean believes that single species of each genus were created in an originally highly plastic condition, and that these have produced, chiefly by intercrossing, but likewise by variation, all our existing species.

In 1826, Professor Grant, in the concluding paragraph in his well-known paper (Edinburgh Philosophical Journal, vol. xiv. p. 283) on the Spongilla, clearly declares his belief that species are descended from other species, and that they become improved in the course of modification. This same view was given in his 55th Lecture, published in the 'Lancet' in 1834.

In 1831 Mr. Patrick Matthew published his work on 'Naval Timber and Arboriculture,' in which he gives precisely the same view on the origin of species as that (presently to be alluded to) propounded by Mr. Wallace and myself in the 'Linnean Journal,' and as that enlarged on in the present volume. Unfortunately the view was given by Mr. Matthew very briefly in scattered passages in an Appendix to a work on a different subject, so that it remained unnoticed until Mr. Matthew himself drew attention to it in the

cellent history of opinion on this subject. In this work a full account is given of Buffon's fluctuating conclusions on the same subject. It is curious how largely my grandfather, Dr. Erasmus Darwin, anticipated the erroneous grounds of opinion, and the views of Lamarck, in his 'Zoonomia' (vol. i. p. 500-510), published in 1794. According to Isid. Geoffroy there is no doubt that Goethe was an extreme partisan of similar views, as shown in the Introduction to a work written in 1794 and 1795, but not published till long afterwards. It is rather a singular instance of the manner in which similar views arise at about the same period, that Goethe in Germany, Dr. Darwin in England, and Geoffroy Saint Hilaire (as we shall immediately see) in France, came to the same conclusion on the origin of species, in the years 1794-5.

'Gardener's Chronicle,' on April 7th, 1860. The differences of Mr. Matthew's view from mine are not of much importance: he seems to consider that the world was nearly depopulated at successive periods, and then re-stocked; and he gives, as an alternative, that new forms may be generated "without the presence of any mould or germ of former aggregates." I am not sure that I understand some passages; but it seems that he attributes much influence to the direct action of the conditions of life. He clearly saw, however, the full force of the principle of natural selection. In answer to a letter of mine (published in Gard. Chron., April 13th), fully acknowledging that Mr. Matthew had anticipated me, he with generous candour wrote a letter (Gard. Chron. May 12th) containing the following passage:—"To me the conception of this law of Nature came intuitively as a self-evident fact, almost without an effort of concentrated thought. Mr. Darwin here seems to have more merit in the discovery than I have had; to me it did not appear a discovery. He seems to have worked it out by inductive reason, slowly and with due caution to have made his way synthetically from fact to fact onwards; while with me it was by a general glance at the schemo of Nature that I estimated this select production of species as an à priori recognisable fact—an axiom requiring only to be pointed out to be admitted by unprejudiced minds of sufficient grasp."

Rafinesque, in his 'New Flora of North America,' published in 1836, wrote (p. 6) as follows:—"All species might have been varieties once, and many varieties are gradually becoming species by assuming constant and peculiar characters:" but farther on (p. 18) he adds, "except the original types or ancestors of the genus."

In 1843-44 Professor Haldeman (Boston Journal of Nat. Hist. U. States, vol. iv. p. 468) has ably given the arguments for and against the hypothesis of the development and modification of species : he seems to lean towards the side of change.

The 'Vestiges of Creation' appeared in 1844. In the tenth and much improved edition (1853) the anonymous author says (p. 155):—"The proposition determined on after much consideration is, that the several series of animated beings, from the simplest and oldest up to the highest and most recent, are, under the providence of God, the results, *first*, of an impulse which has been imparted to the forms of life, advancing them, in definite times, by generation, through grades of organisation terminating in the highest dicotyledons and vertebrata, these grades being few in number, and generally marked by intervals of organic character, which we find to be

a practical difficulty in ascertaining affinities; second, of another
impulse connected with the vital forces, tending, in the course of
generations, to modify organic structures in accordance with external
circumstances, as food, the nature of the habitat, and the meteoric
agencies, these being the 'adaptations' of the natural theologian.'
The author apparently believes that organisation progresses by
sudden leaps, but that the effects produced by the conditions of life
are gradual. He argues with much force on general grounds that
species are not immutable productions. But I cannot see how the
two supposed "impulses" account in a scientific sense for the
numerous and beautiful co-adaptations which we see throughout
nature; I cannot see that we thus gain any insight how, for instance, a woodpecker has become adapted to its peculiar habits of
life. The work, from its powerful and brilliant style, though displaying in the earlier editions little accurate knowledge and a great
want of scientific caution, immediately had a very wide circulation.
In my opinion it has done excellent service in calling in this country
attention to the subject, in removing prejudice, and in thus preparing the ground for the reception of analogous views.

In 1846 the veteran geologist M. J. d'Omalius d'Halloy published
in an excellent, though short paper (' Bulletins de l'Acad. Roy.
Bruxelles,' tom. xiii. p. 581), his opinion that it is more probable
that new species have been produced by descent with modification,
than that they have been separately created: the author first promulgated this opinion in 1831.

Professor Owen, in 1849 ('Nature of Limbs,' p. 86), wrote as
follows :—" The archetypal idea was manifested in the flesh under
diverse such modifications, upon this planet, long prior to the existence of those animal species that actually exemplify it. To what
natural laws or secondary causes the orderly succession and progres-
sion of such organic phenomena may have been committed, we, as yet,
are ignorant." In his Address to the British Association, in 1858, he
speaks (p. li.) of " the axiom of the continuous operation of creative
power, or of the ordained becoming of living things." Farther on
(p. xc.), after referring to geographical distribution, he adds, " These
phenomena shake our confidence in the conclusion that the Apteryx
of New Zealand and the Red Grouse of England were distinct creations in and for those islands respectively. Always, also, it may be
well to bear in mind that by the word ' creation ' the zoologist means
' a process he knows not what.' " He amplifies this idea by adding,
that when such cases as that of the Red Grouse are " enumerated by
the zoologist as evidence of distinct creation of the bird in and for
such islands, he chiefly expresses that he knows not how the Red

Grouse came to be there, and there exclusively ; signifying also by this mode of expressing such ignorance his belief, that both the bird and the islands owed their origin to a great first Creative Cause."

M. Isidore Geoffroy Saint Hilaire, in his Lectures delivered in 1850 (of which a Résumé appeared in the ' Revue et Mag. de Zoolog.,' Jan. 1851), briefly gives his reason for believing that specific characters "sont fixes, pour chaque espèce, tant qu'elle se perpétue au milieu des mêmes circonstances : ils se modifient, si les circonstances ambiantes viennent à changer." " En résumé, *l'observation* des animaux sauvages démontre déjà la variabilité *limitée* des espèces. Les *expériences* sur les animaux sauvages devenus domestiques, et sur les animaux domestiques redevenus sauvages, la démontrent plus clairement encore. Ces mêmes expériences prouvent, de plus, que les différences produites peuvent être de *valeur générique*." In his ' Hist. Nat. Générale ' (tom. ii. p. 430, 1859) he amplifies analogous conclusions.

From a circular lately issued it appears that Dr. Freke, in 1851 (' Dublin Medical Press,' p. 322), propounded the doctrine that all organic beings have descended from one primordial form. His grounds of belief and treatment of the subject are wholly different from mine ; but as Dr. Freke has now (1861) published his Essay on ' the Origin of Species by means of Organic Affinity,' the difficult attempt to give any idea of his views would be superfluous on my part.

Mr. Herbert Spencer, in an Essay (originally published in the ' Leader,' March 1852, and republished in his ' Essays ' in 1858), has contrasted the theories of the Creation and the Development of organic beings with remarkable skill and force. He argues from the analogy of domestic productions, from the changes which the embryos of many species undergo, from the difficulty of distinguishing species and varieties, and from the principle of general gradation, that species have been modified ; and he attributes the modification to the change of circumstances. The author (1855) has also treated Psychology on the principle of the necessary acquirement of each mental power and capacity by gradation.

In 1852 (' Revue Horticole,' p. 102) M. Naudin, a distinguished botanist, has expressly stated his belief that species are formed in an analogous manner as varieties are under cultivation ; and the latter process he attributes to man's power of selection. But he does not show how selection acts under nature. He believes, like Dean Herbert, that species, when nascent, were more plastic than at present. He lays weight on what he calls the principle of finality, " puissance mystérieuse, indéterminée ; fatalité pour les uns ; pour les autres, volonté providentielle, dont l'action incessante sur les êtres vivants détermine, à toutes les époques de l'existence du monde, la forme, le volume, et la durée de chacun d'eux, en raison do sa destinée dans l'ordre de choses dont il fait partie. C'est cette puissance qui harmonise chaque membre à l'ensemble en l'appropriant à la fonction qu'il doit remplir dans l'organisme général de la nature, fonction qui est pour lui sa raison d'être." *

In 1853, a celebrated geologist, Count Keyserling (' Bulletin de la Soc. Géolog.,' 2nd Ser., tom. x. p. 357), suggested that as new diseases, supposed to have been caused by some miasma, have arisen and spread over the world, so at certain periods the germs of existing species may have been chemically affected by circumambient molecules of a particular nature, and thus have given rise to new forms.

In this same year, 1853, Dr. Schaaffhausen published an excellent pamphlet (' Verhand. des Naturhist. Vereins der Preuss. Rheinlands,' &c.), in which he maintains the progressive development of organic forms on the earth. He infers that many species have kept true for long periods, whereas a few have become modified. The distinction of species he explains by the destruction of intermediate graduated forms. " Thus living plants and animals are not separated from the extinct by new creations, but are to be regarded as their descendants through continued reproduction."

The ' Philosophy of Creation' has been treated in a masterly manner by the Rev. Baden Powell, in his ' Essays on the Unity of Worlds,' 1855. Nothing can be more striking than the manner in which he shows that the introduction of new species is " a regular, not a casual phenomenon," or, as Sir John Herschel expresses it, " a natural in contra-distinction to a miraculous process."

The third volume of the ' Journal of the Linnean Society' contains papers, read July 1st, 1858, by Mr. Wallace and myself, in which, as stated in the introductory remarks to this volume, the

* From references in Bronn's ' Untersuchungen über die Entwickelungs-Gesetze' it appears that the celebrated botanist and palæontologist Unger published, in 1852, his belief that species undergo development and modification. D'Alton, likewise, in Pander and d'Alton's work on Fossil Sloths, expressed, in 1821, a similar belief. Similar views have, as is well known, been maintained by Oken in his mystical ' Natur-Philosophie.' From other references in Godron's work ' Sur l'Espèce,' it seems that Bory St. Vincent, Burdach, Poiret, and Fries, have all admitted that new species are continually being produced.

I may add, that of the thirty authors named in this Historical Sketch, who believe in the modification of species, or at least disbelieve in separate acts of creation, twenty-five have written on special branches of natural history: of these only three are simple geologists, nine are botanists, and thirteen zoologists; but several of the botanists and zoologists have written on palæontology or on geology.

theory of Natural Selection is promulgated by Mr. Wallace with admirable force and clearness.

In June, 1859, Professor Huxley gave a lecture before the Royal Institution on the 'Persistent Types of Animal Life.' Referring to such cases, he remarks, " It is difficult to comprehend the meaning of such facts as these, if we suppose that each species of animal and plant, or each great type of organisation, was formed and placed upon the surface of the globe at long intervals by a distinct act of creative power; and it is well to recollect that such an assumption is as unsupported by tradition or revelation as it is opposed to the general analogy of nature. If, on the other hand, we view 'Persistent Types' in relation to that hypothesis which supposes the species living at any time to be the result of the gradual modification of pre-existing species—a hypothesis which, though unproven, and sadly damaged by some of its supporters, is yet the only one to which physiology lends any countenance; their existence would seem to show that the amount of modification which living beings have undergone during geological time is but very small in relation to the whole series of changes which they have suffered."

In December, 1859, Dr. Hooker published his 'Introduction to the Australian Flora.' In the first part of this great work he admits the truth of the descent and modification of species, and supports this doctrine by many original observations.

The first edition of this work was published on November 24th, 1859, and the second edition on January 7th, 1860.

ON THE ORIGIN OF SPECIES.

INTRODUCTION.

WHEN on board H.M.S. 'Beagle,' as naturalist, I was much struck with certain facts in the distribution of the organic beings inhabiting South America, and in the geological relations of the present to the past inhabitants of that continent. These facts, as will be seen in the latter chapters of this volume, seemed to throw some light on the origin of species—that mystery of mysteries, as it has been called by one of our greatest philosophers. On my return home, it occurred to. me, in 1837, that something might perhaps be made out on this question by patiently accumulating and reflecting on all sorts of facts which could possibly have any bearing on it. After five years' work I allowed myself to speculate on the subject, and drew up some short notes : these I enlarged in 1844 into a sketch of the conclusions, which then seemed to me probable: from that period to the present day I have steadily pursued the same object. I hope that I may be excused for entering on these personal details, as I give them to show that I have not been hasty in coming to a decision.

My work is now nearly finished; but as it will take me two or three more years to complete it, and as my health is far from strong, I have been urged to publish this Abstract. I have more especially been induced

B

to do this, as Mr. Wallace, who is now studying the natural history of the Malay archipelago, has arrived at almost exactly the same general conclusions that I have on the origin of species. In 1858 he sent me a memoir on this subject, with a request that I would forward it to Sir Charles Lyell, who sent it to the Linnean Society, and it is published in the third volume of the Journal of that Society. Sir C. Lyell and Dr. Hooker, who both knew of my work—the latter having read my sketch of 1844—honoured me by thinking it advisable to publish, with Mr. Wallace's excellent memoir, some brief extracts from my manuscripts.

This Abstract, which I now publish, must necessarily be imperfect. I cannot here give references and authorities for my several statements; and I must trust to the reader reposing some confidence in my accuracy. No doubt errors will have crept in, though I hope I have always been cautious in trusting to good authorities alone. I can here give only the general conclusions at which I have arrived, with a few facts in illustration, but which, I hope, in most cases will suffice. No one can feel more sensible than I do of the necessity of hereafter publishing in detail all the facts, with references, on which my conclusions have been grounded; and I hope in a future work to do this. For I am well aware that scarcely a single point is discussed in this volume on which facts cannot be adduced, often apparently leading to conclusions directly opposite to those at which I have arrived. A fair result can be obtained only by fully stating and balancing the facts and arguments on both sides of each question; and this cannot possibly be here done.

I much regret that want of space prevents my having the satisfaction of acknowledging the generous assistance which I have received from very many naturalists, some of them personally unknown to me. I cannot, however, let this opportunity pass without expressing my deep obligations to Dr. Hooker, who for the last fifteen years has aided me in every possible way by his large stores of knowledge and his excellent judgment.

In considering the Origin of Species, it is quite conceivable that a naturalist, reflecting on the mutual affinities of organic beings, on their embryological relations, their geographical distribution, geological succession, and other such facts, might come to the conclusion that each species had not been independently created, but had descended, like varieties, from other species. Nevertheless, such a conclusion, even if well founded, would be unsatisfactory, until it could be shown how the innumerable species inhabiting this world have been modified, so as to acquire that perfection of structure and coadaptation which justly excites our admiration. Naturalists continually refer to external conditions, such as climate, food, &c., as the only possible cause of variation. In one very limited sense, as we shall hereafter see, this may be true; but it is preposterous to attribute to mere external conditions, the structure, for instance, of the woodpecker, with its feet, tail, beak, and tongue, so admirably adapted to catch insects under the bark of trees. In the case of the mistletoe, which draws its nourishment from certain trees, which has seeds that must be transported by certain birds, and which has flowers with separate sexes absolutely requiring the agency of certain insects to bring pollen from one flower to the other, it is equally preposterous to account for the structure of this parasite, with its relations to several distinct organic beings, by the effects of external conditions, or of habit, or of the volition of the plant itself.

It is, therefore, of the highest importance to gain a

clear insight into the means of modification and coadaptation. At the commencement of my observations
it seemed to me probable that a careful study of domesticated animals and of cultivated plants would offer the
best chance of making out this obscure problem. Nor
have I been disappointed; in this and in all other
perplexing cases I have invariably found that our
knowledge, imperfect though it be, of variation under
domestication, afforded the best and safest clue. I may
venture to express my conviction of the high value of
such studies, although they have been very commonly
neglected by naturalists.   ▪

From these considerations, I shall devote the first
chapter of this Abstract to Variation under Domestication.
We shall thus see that a large amount of hereditary
modification is at least possible; and, what is equally or
more important, we shall see how great is the power of
man in accumulating by his Selection successive slight
variations. I will then pass on to the variability of
species in a state of nature; but I shall, unfortunately,
be compelled to treat this subject far too briefly, as it
can be treated properly only by giving long catalogues
of facts. We shall, however, be enabled to discuss
what circumstances are most favourable to variation.
In the next chapter the Struggle for Existence amongst
all organic beings throughout the world, which inevitably follows from the high geometrical ratio of their
increase, will be treated of. This is the doctrine of
Malthus, applied to the whole animal and vegetable
kingdoms. As many more individuals of each species
are born than can possibly survive; and as, consequently, there is a frequently recurring struggle for
existence, it follows that any being, if it vary however
slightly in any manner profitable to itself, under the
complex and sometimes varying conditions of life, will

have a better chance of surviving, and thus be *naturally selected*. From the strong principle of inheritance, any selected variety will tend to propagate its new and modified form.

This fundamental subject of Natural Selection will be treated at some length in the fourth chapter; and we shall then see how Natural Selection almost inevitably causes much Extinction of the less improved forms of life, and leads to what I have called Divergence of Character. In the next chapter I shall discuss the complex and little known laws of variation and of correlation of growth. In the four succeeding chapters, the most apparent and gravest difficulties on the theory will be given: namely, first, the difficulties of transitions, or in understanding how a simple being or a simple organ can be changed and perfected into a highly developed being or elaborately constructed organ; secondly, the subject of Instinct, or the mental powers of animals; thirdly, Hybridism, or the infertility of species and the fertility of varieties when intercrossed; and fourthly, the imperfection of the Geological Record. In the next chapter I shall consider the geological succession of organic beings throughout time; in the eleventh and twelfth, their geographical distribution throughout space; in the thirteenth, their classification or mutual affinities, both when mature and in an embryonic condition. In the last chapter I shall give a brief recapitulation of the whole work, and a few concluding remarks.

No one ought to feel surprise at much remaining as yet unexplained in regard to the origin of species and varieties, if he make due allowance for our profound ignorance in regard to the mutual relations of all the beings which live around us. Who can explain why one species ranges widely and is very numerous, and why

another allied species has a narrow range and is rare?
Yet these relations are of the highest importance, for
they determine the present welfare, and, as I believe,
the future success and modification of every inhabitant
of this world. Still less do we know of the mutual
relations of the innumerable inhabitants of the world
during the many past geological epochs in its history.
Although much remains obscure, and will long remain
obscure, I can entertain no doubt, after the most de-
liberate study and dispassionate judgment of which I
am capable, that the view which most naturalists entertain, and which I formerly entertained—namely, that
each species has been independently created—is erroneous. I am fully convinced that species are not immutable; but that those belonging to what are called
the same genera are lineal descendants of some other
and generally extinct species, in the same manner as
the acknowledged varieties of any one species are the
descendants of that species. Furthermore, I am convinced that Natural Selection has been the main but
not exclusive means of modification.

CHAPTER I.

VARIATION UNDER DOMESTICATION.

Causes of Variability — Effects of Habit — Correlation of Growth — Inheritance — Character of Domestic Varieties — Difficulty of distinguishing between Varieties and Species — Origin of Domestic Varieties from one or more Species — Domestic Pigeons, their Differences and Origin — Principle of Selection anciently followed, its Effects — Methodical and Unconscious Selection — Unknown Origin of our Domestic Productions — Circumstances favourable to Man's power of Selection.

WHEN we look to the individuals of the same variety or sub-variety of our older cultivated plants and animals, one of the first points which strikes us, is, that they generally differ more from each other than do the individuals of any one species or variety in a state of nature. When we reflect on the vast diversity of the plants and animals which have been cultivated, and which have varied during all ages under the most different climates and treatment, I think we are driven to conclude that this great variability is simply due to our domestic productions having been raised under conditions of life not so uniform as, and somewhat different from, those to which the parent-species have been exposed under nature. There is also, I think, some probability in the view propounded by Andrew Knight, that this variability may be partly connected with excess of food. It seems pretty clear that organic beings must be exposed during several generations to the new conditions of life to cause any appreciable amount of variation; and that when the organisation has once begun to vary, it generally continues to vary for many generations.

No case is on record of a variable being ceasing to be variable under cultivation. Our oldest cultivated plants, such as wheat, still often yield new varieties : our oldest domesticated animals are still capable of rapid improvement or modification.

It has been disputed at what period of life the causes of variability, whatever they may be, generally act ; whether during the early or late period of development of the embryo, or at the instant of conception. Geoffroy St. Hilaire's experiments show that unnatural treatment of the embryo causes monstrosities ; and monstrosities cannot be separated by any clear line of distinction from mere variations. But I am strongly inclined to suspect that the most frequent cause of variability may be attribnted to the male and female reproductive elements having been affected prior to the act of conception. Several reasons make me believe in this ; but the chief one is the remarkable effect which confinement or cultivation has on the function of the reproductive system : this system appearing to be far more susceptible than any other part of the organisation, to the action of any change in the conditions of life. Nothing is more easy than to tame an animal, and few things more difficult than to get it to breed freely under confinement, even in the many cases when the male and female unite. How many animals there are which will not breed, though living long under not very close confinement in their native country! This is generally attributed to vitiated instincts; but how many cultivated plants display the utmost vigour, and yet rarely or never seed! In some few such cases it has been discovered that very trifling changes, such as a little more or less water at some particular period of growth, will determine whether or not the plant sets a seed. I cannot here enter on the copious details which I have collected on

this curious subject ; but to show how singular the laws
are which determine the reproduction of animals under
confinement, I may just mention that carnivorous animals, even from the tropics, breed in this country pretty
freely under confinement, with the exception of the plan-
tigrades or bear family, which seldom produce young ;
whereas carnivorous birds, with the rarest exceptions,
hardly ever lay fertile eggs. Many exotic plants have
pollen utterly worthless, in the same exact condition as
in the most sterile hybrids. When, on the one hand,
we see domesticated animals and plants, though often
weak and sickly, yet breeding quite freely under confinement ; and when, on the other hand, we see individuals, though taken young from a state of nature, perfectly tamed, long-lived, and healthy (of which I could
give numerous instances), yet having their reproductive
system so seriously affected by unperceived causes as to
fail in acting, we need not be surprised at this system,
when it does act under confinement, acting not quite
regularly, and producing offspring not perfectly like
their parents.

Sterility has been said to be the bane of horticulture ;
but on this view we owe variability to the same cause
which produces sterility ; and variability is the source
of all the choicest productions of the garden. I may
add, that as some organisms will breed freely under
the most unnatural conditions (for instance, the rabbit
and ferret kept in hutches), showing that their reproductive system has not been thus affected ; so will some
animals and plants withstand domestication or cultivation, and vary very slightly—perhaps hardly more than
in a state of nature.

A long list could easily be given of "sporting plants ;"
by this term gardeners mean a single bud or offset,
which suddenly assumes a new and sometimes very

different character from that of the rest of tho plant.
Such buds can bo propagated by grafting, &c., and
sometimes by seed. These "sports" are extremely
rare under nature, but far from rare under cultivation ;
and in this case we see that the treatment of the parent
has affected a bud or offset, and not the ovules or pollen.
But it is the opinion of most physiologists that there is no
essential difference between a bud and an ovule in their
earliest stages of formation ; so that, in fact, "sports"
support the view, that variability may be largely attributed to tho ovules or pollen, or to both, having been
affected by the treatment of the parent prior to the act
of conception. These cases anyhow show that variation
is not necessarily connected, as some authors have supposed, with the act of generation.

Seedlings from the same fruit, and the young of the
same litter, sometimes differ considerably from each
other, though both the young and the parents, as Müller
has remarked, have apparently been exposed to exactly
the same conditions of life ; and this shows how unimportant the direct effects of the conditions of life are
in comparison with the laws of reproduction, of growth,
and of inheritance ; for had the action of the conditions
been direct, if any of the young had varied, all would
probably have varied in the same manner. To judge how
much, in the case of any variation, we should attribute
to the direct action of heat, moisture, light, food, &c.,
is most difficult : my impression is, that with animals
such agencies have produced very little direct effect,
though apparently more in the case of plants. Under
this point of view, Mr. Buckman's recent experiments
on plants are extremely valuable. When all or nearly
all the individuals exposed to certain conditions are
affected in the same way, the change at first appears to
be directly due to such conditions ; but in some cases it

can be shown that quite opposite conditions produce similar changes of structure. Nevertheless some slight amount of change may, I think, be attributed to the direct action of the conditions of life—as, in some cases, increased size from amount of food, colour from particular kinds of food, and perhaps the thickness of fur from climate.

Habit also has a decided influence, as in the period of flowering with plants when trausported from one climate to another. In animals it has a more marked effect; for instance, I find in the domestic duck that the bones of the wing weigh less and the bones of the leg more, in proportion to the whole skeleton, than do the same bones in the wild-duck; and I presume that this change may be safely attributed to the domestic duck flying much less, and walking more, than its wild parent. The great and inherited development of the udders in cows and goats in countries where they are habitually milked, in comparison with the state of these organs in other countries, is another instance of the effect of use. Not a single domestic animal can be named which has not in some country drooping ears; and the view suggested by some authors, that the drooping is due to the disuse of the muscles of the ear, from the animals not being much alarmed by danger, seems probable.

There are many laws regulating variation, some few of which can be dimly seen, and will be hereafter briefly mentioned. I will here only allude to what may be called correlation of growth. Any change in the embryo or larva will almost certainly entail changes in the mature animal. In monstrosities, the correlations between quite distinct parts are very curious; and many instances are given in Isidore Geoffroy St. Hilairo's great work on this subject. Breeders believe that long limbs are almost always accompanied by an elongated head.

Some instances of correlation are quite whimsical : thus cats with blue eyes are generally deaf. Colour and constitutional peculiarities go together, of which many remarkable cases could be given amongst animals and plants. From the facts collected by Heusinger, it appears that white sheep and pigs are differently affected from coloured individuals by certain vegetable poisons : Professor Wyman has recently communicated to me a good illustration of this fact; on asking some farmers in Florida how it was that all their pigs were black, they informed him that the pigs ate the paint-root (Lachnanthes), which coloured their bones pink, and which caused the hoofs of all but the black varieties to drop off; and one of the "crackers" (i.e. Florida squatters) added, " we select the black members of a litter for raising, as they alone have a good chance of living." Hairless dogs have imperfect teeth : long-haired and coarse-haired animals are apt to have, as is asserted, long or many horns ; pigeons with feathered feet have skin between their outer toes ; pigeons with short beaks have small feet, and those with long beaks large feet. Hence if man goes on selecting, and thus augmenting, any peculiarity, he will almost certainly unintentionally modify other parts of the structure, owing to the mysterious laws of the correlation of growth.

The result of the various, quite unknown, or dimly seen laws of variation is infinitely complex and diversified. It is well worth while carefully to study the several treatises published on some of our old cultivated plants, as on the hyacinth, potato, even the dahlia, &c. ; and it is really surprising to note the endless points in structure and constitution in which the varieties and sub-varieties differ slightly from each other. The whole organisation seems to have become plastic, and tends to depart in some small degree from that of the parental type.

Any variation which is not inherited is unimportant for us. But the number and diversity of inheritable deviations of structure, both those of slight and those of considerable physiological importance, are endless. Dr. Prosper Lucas's treatise, in two large volumes, is the fullest and the best on this subject. No breeder doubts how strong is the tendency to inheritance: like produces like is his fundamental belief: doubts have been thrown on this principle by theoretical writers alone. When any deviation of structure often appears, and we see it in the father and child, we cannot tell whether it may not be due to the same cause having acted on both; but when amongst individuals, apparently exposed to the same conditions, any very rare deviation, due to some extraordinary combination of circumstances, appears in the parent—say, once amongst several million individuals—and it reappears in the child, the mere doctrine of chances almost compels us to attribute its reappearance to inheritance. Every one must have heard of cases of albinism, prickly skin, hairy bodies, &c., appearing in several members of the same family. If strange and rare deviations of structure are truly inherited, less strange and commoner deviations may be freely admitted to be inheritable. Perhaps the correct way of viewing the whole subject, would be, to look at the inheritance of every character whatever as the rule, and non-inheritance as the anomaly.

The laws governing inheritance are quite unknown; no one can say why a peculiarity in different individuals of the same species, or in individuals of different species, is sometimes inherited and sometimes not so; why the child often reverts in certain characters to its grandfather or grandmother or other more remote ancestor; why a peculiarity is often transmitted from one sex to both sexes, or to one sex alone, more commonly but not exclusively to the like sex. It is a fact of some little importance to us, that peculiarities appearing in the males of our domestic breeds are often transmitted either exclusively, or in a much greater degree, to males alone. A much more important rule, which I think may be trusted, is that, at whatever period of life a peculiarity first appears, it tends to appear in the offspring at a corresponding age, though sometimes earlier. In many cases this could not be otherwise: thus the inherited peculiarities in the horns of cattle could appear only in the offspring when nearly mature; peculiarities in the silkworm are known to appear at the corresponding caterpillar or cocoon stage. But hereditary diseases and some other facts make me believe that the rule has a wider extension, and that when there is no apparent reason why a peculiarity should appear at any particular age, yet that it does tend to appear in the offspring at the same period at which it first appeared in the parent. I believe this rule to be of the highest importance in explaining the laws of embryology. These remarks are of course confined to the first *appearance* of the peculiarity, and not to its primary cause, which may have acted on the ovules or on the male element; in nearly the same manner as in the offspring from a .short-horned cow by a long-horned bull, the greater length of horn, though appearing late in life, is clearly due to the male element.

Having alluded to the subject of reversion, I may here refer to a statement often made by naturalists— namely, that our domestic varieties, when run wild, gradually but certainly revert in character to their aboriginal stocks. Hence it has been argued that no deductions can be drawn from domestic races to species in a state of nature. I have in vain endeavoured to discover on what decisive facts the above statement has so often and so boldly been made. There would be great difficulty in proving its truth: we may safely conclude that very many of the most strongly-marked domestic varieties could not possibly live in a wild state. In many cases we do not know what the aboriginal stock was, and so could not tell whether or not nearly perfect reversion had ensued. It would be quite necessary, in order to prevent the effects of intercrossing, that only a single variety should be turned loose in its new home. Nevertheless, as our varieties certainly do occasionally revert in some of their characters to ancestral forms, it seems to me not improbable, that if we could succeed in naturalising, or were to cultivate, during many generations, the several races, for instance, of the cabbage, in very poor soil (in which case, however, some effect would have to be attributed to the direct action of the poor soil), that they would to a large extent, or even wholly, revert to the wild aboriginal stock. Whether or not the experiment would succeed, is not of great importance for our line of argument; for by the experiment itself the conditions of life are changed. If it could be shown that our domestic varieties manifested a strong tendency to reversion,—that is, to lose their acquired characters, whilst kept under the same conditions, and whilst kept in a considerable body, so that free intercrossing might check, by blending together, any slight deviations in their structure, in such case, I grant that we could deduce nothing from domestic varieties in regard to species. But there is not a shadow of evidence in favour of this view: to assert that we could not breed our cart and race-horses, long and short-horned cattle, and poultry of various breeds, and esculent vegetables, for an almost infinite number of generations, would be opposed to all experience. I may add, that when under nature the conditions of life do change, variations and reversions of character probably do occur; but natural selection, as will hereafter be explained, will determine how far the new characters thus arising shall be preserved.

When we look to the hereditary varieties or races of our domestic animals and plants, and compare them with closely allied species, we generally perceive in each domestic race, as already remarked, less uniformity of character than in true species. Domestic races of the same species, also, often have a somewhat monstrous character; by which I mean, that, although differing from each other, and from other species of the same genus, in several trifling respects, they often differ in an extreme degree in some one part, both when compared one with another, and more especially when compared with all the species in nature to which they are nearest allied. With these exceptions (and with that of the perfect fertility of varieties when crossed,—a subject hereafter to be discussed), domestic races of the same species differ from each other in the same manner as, only in most cases in a lesser degree than, do closely-allied species of the same genus in a state of nature. I think that this must be admitted, when we find that there are hardly any domestic races, either amongst animals or plants, which have not been ranked by competent judges as the descendants of aboriginally distinct species, and by other competent judges as mere varieties. If any marked distinction existed between domestic races and species, this source of doubt could not so perpetually recur. It has often been stated that domestic races do not differ from each other in characters of generic value. I think it can be shown that this statement is hardly correct; but naturalists differ widely in determining what characters are of generic value; all such valuations being at present empirical. Moreover, on the view of the origin of genera which I shall presently give, we have no right to expect often to meet with generic differences in our domesticated productions.

When we attempt to estimate the amount of structural difference between the domestic races of the same species, we are soon involved in doubt, from not knowing whether they have descended from one or several parent-species. This point, if it could be cleared up, would be interesting; if, for instance, it could be shown that the greyhound, bloodhound, terrier, spaniel, and bull-dog, which we all know propagate their kind so truly, were the offspring of any single species, then such facts would have great weight in making us doubt about the immutability of the many very closely allied natural species —for instance, of the many foxes—inhabiting different quarters of the world. I do not believe, as we shall presently see, that the whole amount of difference between the several breeds of the dog has been produced under domestication; I believe that some small part of the difference is due to their having descended from distinct species. In the case of some other domesticated species, there is presumptive, or even strong evidence, that all the breeds have descended from a single wild stock.

It has often been assumed that man has chosen for domestication animals and plants having an extraordinary inherent tendency to vary, and likewise to withstand diverse climates. I do not dispute that these capacities have added largely to the value of most of our domesticated productions; but how could a savage possibly know, when he first tamed an animal, whether it would vary in succeeding generations, and whether it would endure other climates? Has the little variability of the ass or guinea-fowl, or the small power of endurance of warmth by the reindeer, or of cold by the common camel prevented their domestication? I cannot

donbt that if other animals and plants, equal in number
to our domesticated productions, and belonging to equally
diverse classes and countries, were taken from a state
of nature, and could be made to breed for an equal
number of generations under domestication, they
would vary on an average as largely as the parent
species of our existing domesticated productions have
varied.

In the case of most of our anciently domesticated
animals and plants, it is not possible to come to any
definite conclusion, whether they have descended from
one or several wild species. The argument mainly
relied on by those who believe in the multiple origin
of our domestic animals is, that we find in the most
ancient records, more especially on the monuments of
Egypt, much diversity in the breeds; and that some of
the breeds closely resemble, perhaps are identical with,
those still existing. Even if this latter fact were found
more strictly and generally true than seems to me to
be the case, what does it show, but that some of our
breeds originated there, four or five thousand years
ago? Since the recent discoveries of flint tools or celts
in the superficial deposits of France and England, few
geologists will doubt that man, in a sufficiently civilized
state to have manufactured weapons, existed at a
period extremely remote as measured by years; and
we know that at the present day there is hardly a tribe
so barbarous as not to have domesticated at least the
dog.

The origin of most of our domestic animals will probably for ever remain vague. But I may here state,
that looking to the domestic dogs of the whole world, I
have, after a laborious collection of all known facts,
come to the conclusion that several wild species of
Canidæ have been tamed, and that their blood, more or

less mingled, flows in the veins of our many domestic
breeds. In regard to sheep and goats I can form no
opinion. From facts communicated to me by Mr. Blyth,
on the habits, voice, and constitution, &c., of the humped
Indian cattle, it is probable that these descended from
a different aboriginal stock from our European cattle;
and several competent judges believe that these latter
have had more than one wild parent. With respect to
horses, from reasons which I cannot here give, I am
with much doubt inclined to believe, in opposition to
several authors, that all the races have descended from
one wild stock. Mr. Blyth, whose opinion, from his
large and varied stores of knowledge, I should value
more than that of almost any one, thinks that all the
breeds of poultry have proceeded from the common
wild Indian fowl (Gallus bankiva). In regard to ducks
and rabbits, the breeds of which differ considerably from
each other in structure, the evidence preponderates in
favour of their having all descended from the common
wild duck and rabbit.

The doctrine of the origin of our several domestic
races from several aboriginal stocks, has been carried to
an absurd extreme by some authors. They believe that
every race which breeds true, let the distinctive characters be ever so slight, has had its wild prototype.
At this rate there must have existed at least a score of
species of wild cattle, as many sheep, and several goats
in Europe alone, and several even within Great Britain.
One author believes that there formerly existed in
Great Britain eleven wild species of sheep peculiar to
it! When we bear in mind that Britain has now hardly
one peculiar mammal, and France but few distinct from
those of Germany and conversely, and so with Hungary,
Spain, &c., but that each of these kingdoms possesses
several peculiar breeds of cattle, sheep, &c., we must

admit that many domestic breeds have originated in
Europe; for whence could they have been derived, as
these several countries do not possess a number of
peculiar species as distinct parent-stocks? So it is in
India. Even in the case of the domestic dogs of the
whole world, which I admit to have descended from
several wild species, it cannot be doubted that there has
been an immense amount of inherited variation; for
who will believe that animals closely resembling the
Italian greyhound, the bloodhound, the bull-dog, pug-
dog, or Blenheim spaniel, &c.—so unlike all wild Canidæ
—ever existed freely in a state of nature? It has often
been loosely said that all our races of dogs have been
produced by the crossing of a few aboriginal species; .
but by crossing we can only get forms in some degree
intermediate between their parents; and if we account
for our several domestic races by this process, we must
admit the former existence of the most extreme forms, .
as the Italian greyhound, bloodhound, bull-dog, &c., in
the wild state. Moreover, the possibility of making distinct races by crossing has been greatly exaggerated.
Many cases are on record, showing that a race may be
modified by occasional crosses, if aided by the careful
selection of those individual mongrels which present
any desired character; but that a race could be obtained
nearly intermediate between two extremely different
races or species, I can hardly believe. Sir J. Sebright
expressly experimented for this object, and failed. The
offspring from the first cross between two pure breeds
is tolerably and sometimes (as I have found with
pigeons) extremely uniform, and everything seems simple
enough; but when these mongrels are crossed one with
another for several generations, hardly two of them will
be alike; and then the extreme difficulty, or rather
utter hopelessness, of the task becomes apparent. Cer-

tainly, a breed intermediate between *two very distinct*
breeds could not be got without extreme care and long-
continued selection; nor can I find a single case on
record of a permanent race having been thus formed.

On the Breeds of the Domestic Pigeon.—Believing that
it is always best to study some special group, I have,
after deliberation, taken up domestic pigeons. I have
kept every breed which I could purchase or obtain, and
have been most kindly favoured with skins from several
quarters of the world, more especially by the Hon. W.
Elliot from India, and by the Hon. C. Murray from
Persia. Many treatises in different languages have been
published on pigeons, and some of them are very important, as being of considerable antiquity. I have
associated with several eminent fanciers, and have been
permitted to join two of the London Pigeon Clubs.
The diversity of the breeds is something astonishing.
Compare the English carrier and the short-faced
tumbler, and see the wonderful difference in their beaks,
entailing corresponding differences in their skulls. The
carrier, more especially the male bird, is also remarkable from the wonderful development of the caruncu-
lated skin about the head; and this is accompanied by
greatly elongated eyelids, very large external orifices to
the nostrils, and a wide gape of mouth. The short-faced
tumbler has a beak in outline almost like that of a
finch; and the common tumbler has the singular inherited habit of flying at a great height in a compact
flock, and tumbling in the air head over heels. The
runt is a bird of great size, with long, massive beak and
large feet; some of the sub-breeds of runts have very
long necks, others very long wings and tails, others
singularly short tails. The barb is allied to the carrier,
but, instead of a very long beak, has a very short and

very broad one. The pouter has a much elongated body, wings, and legs; and its enormously developed crop, which it glories in inflating, may well excite astonishment and even laughter. The turbit has a very short and conical beak, with a line of reversed feathers down the breast; and it has the habit of continually expanding slightly the upper part of the œsophagus. The Jacobin has the feathers so much reversed along the back of the neck that they form a hood; and it has, proportionally to its size, much elongated wing and tail feathers. The trumpeter and laugher, as their names express, utter a very different coo from the other breeds. The fantail has thirty or even forty tail-feathers, instead of twelve or fourteen—the normal number in all members of the great pigeon family; and these feathers are kept expanded, and are carried so erect, that in good birds the head and tail touch: the oil-gland is quite aborted. Several other less distinct breeds might be specified.

In the skeletons of the several breeds, the development of the bones of the face in length and breadth and curvature differs enormously. The shape, as well as the breadth and length of the ramus of the lower jaw, varies in a highly remarkable manner. The caudal and sacral vertebræ vary in number; as does the number of the ribs, together with their relative breadth and the presence of processes. The size and shape of the apertures in the sternum are highly variable; so is the degree of divergence and relative size of the two arms of the furcula. The proportional width of the gape of mouth, the proportional length of the eyelids, of the orifice of the nostrils, of the tongue (not always in strict correlation with the length of beak), the size of the crop and of the upper part of the œsophagus; the development and abortion of the oil-gland; the number of the primary wing and caudal feathers; the relative length of wing and tail to each other and to the body; the relative length of leg and of the feet; the number of scutellæ on the toes, the development of skin between the toes, are all points of structure which are variable. The period at which the perfect plumage is acquired varies, as does the state of the down with which the nestling birds are clothed when hatched. The shape and size of the eggs vary. The manner of flight, and in some breeds the voice and disposition, differ remarkably. Lastly, in certain breeds, the males and females have come to differ to a slight degree from each other.

Altogether at least a score of pigeons might be chosen, which, if shown to an ornithologist, and he were told that they were wild birds, would certainly be ranked by him as well-defined species. Moreover, I do not believe that any ornithologist would place the English carrier, the short-faced tumbler, the runt, the barb, pouter, and fantail in the same genus; more especially as in each of these breeds several truly-inherited sub-breeds, or species as he might have called them, could be shown him.

Great as the differences are between the breeds of pigeons, I am fully convinced that the common opinion of naturalists is correct, namely, that all have descended from the rock-pigeon (Columba livia), including under this term several geographical races or sub-species, which differ from each other in the most trifling respects. As several of the reasons which have led me to this belief are in some degree applicable in other cases, I will here briefly give them. If the several breeds are not varieties, and have not proceeded from the rock-pigeon, they must have descended from at least seven or eight aboriginal stocks; for it is impossible to make the present domestic

breeds by the crossing of any lesser number : how, for
instance, could a pouter be produced by crossing two
breeds unless one of the parent-stocks possessed the
characteristic enormous crop? The supposed aboriginal
stocks must all have been rock-pigeons, that is, not
breeding or willingly perching on trees. But besides
C. livia, with its geographical sub-species, only two or
three other species of rock-pigeons are known ; and these
have not any of the characters of the domestic breeds.
Hence the supposed aboriginal stocks must either still
exist in the countries where they were originally domesticated, and yet be unknown to ornithologists ; and this,
considering their size, habits, and remarkable characters,
seems very improbable ; or they must have become
extinct in the wild state. But birds breeding on preci-
pices, and good fliers, are unlikely to be exterminated ;
and the common rock-pigeon, which has the same habits
with the domestic breeds, has not been exterminated
even on several of the smaller British islets, or on the
shores of the Mediterranean. Hence the supposed extermination of so many species having similar habits
with the rock-pigeon seems to me a very rash assumption. Moreover, the several above-named domesticated
breeds have been transported to all parts of the world,
and, therefore, some of them must have been carried
back again into their native country ; but not one has
ever become wild or feral, though the dovecot-pigeon,
which is the rock-pigeon in a very slightly altered state,
has become feral in several places. Again, all recent
experience shows that it is difficult to get any wild
animal to breed freely under domestication ; yet, on the
hypothesis of the multiple origin of our pigeons, it must
be assumed that at least seven or eight species were so
thoroughly domesticated in ancient times by half-civilized man, as to be quite prolific under confinement.

An argument, as it seems to me, of great weight, and
applicable in several other cases, is, that the above-
specified breeds, though agreeing generally in constitution, habits, voice, colouring, and in most parts of their
structure, with the wild rock-pigeon, yet are certainly
highly abnormal in other parts of their structure; we
may look in vain throughout the whole great family of
Columbidæ for a beak like that of the English carrier,
or that of the short-faced tumbler, or barb; for reversed
feathers like those of the Jacobin; for a crop like that
of the pouter; for tail-feathers like those of the fantail.
Hence it must be assumed not only that half-civilized
man succeeded in thoroughly domesticating several
species, but that he intentionally or by chance picked
out extraordinarily abnormal species; and further, that
these very species have since all become extinct or unknown. So many strange contingencies are improbable
in the highest degree.

Some facts in regard to the colouring of pigeons well
deserve consideration. The rock-pigeon is of a slaty-
blue, and has a white croup (the Indian sub-species,
C. intermedia of Strickland, having it bluish); the tail
has a terminal dark bar, with the bases of the outer
feathers externally edged with white; the wings have
two black bars: some semi-domestic breeds, and some
apparently truly wild breeds, have, besides the two
black bars, the wings chequered with black. These
several marks do not occur together in any other species
of the whole family. Now, in every one of the domestic
breeds, taking thoroughly well-bred birds, all the above
marks, even to the white edging of the outer tail-
feathers, sometimes concur perfectly developed. Moreover, when birds belonging to two or more distinct
breeds are crossed, none of which are blue or have any
of the above-specified marks, the mongrel offspring are

c

very apt suddenly to acquire these characters. To give
one instance out of several which I have observed:—I
crossed some white fantails, which breed very true, with
some black barbs—and it so happens that blue varieties
of barbs are so rare that I never heard of an instance in
England; and the mongrels were black, brown, and
mottled. I also crossed a barb with a spot, which is a
white bird with a red tail and red spot on the forehead,
and which notoriously breeds very true; the mongrels
were dusky and mottled. I then crossed one of the
mongrel barb-fantails with a mongrel barb-spot, and
they produced a bird of as beautiful a blue colour, with
the white croup, double black wing-bar, and barred and
white-edged tail-feathers, as any wild rock-pigeon! We
can understand these facts, on the well-known principle
of reversion to ancestral characters (confined, as far as I
have seen, to colour alone), if all the domestic breeds
have descended from the rock-pigeon. But if we deny
this, we must make one of the two following highly
improbable suppositions. Either, firstly, that all the
several imagined aboriginal stocks were coloured and
marked like the rock-pigeon, although no other existing
species is thus coloured and marked, so that in each
separate breed there might be a tendency to revert to
the very same colours and markings. Or, secondly,
that each breed, even the purest, has within a dozen, or
at most within a score, of generations, been crossed by
the rock-pigeon: I say within a dozen or twenty generations, for we know of no fact countenancing the belief
that the child ever reverts to some one ancestor, removed
by a greater number of generations. In a breed which
has been crossed only once with some distinct breed, the
tendency to reversion to any character derived from
such cross will naturally become less and less, as in each
succeeding generation there will be less of the foreign

blood; but when there has been no cross with a distinct breed, and there is a tendency in both parents to revert to a character which has been lost during some former generation, this tendency, for all that we can see to the contrary, may be transmitted undiminished for an indefinite number of generations. These two quite distinct cases are often confounded by those who have written on inheritance.

Lastly, the hybrids or mongrels from between all the domestic breeds of pigeons are perfectly fertile. I can state this from my own observations, purposely made, on the most distinct breeds. Now, it is difficult, perhaps impossible, to bring forward one case of the hybrid offspring of two animals *clearly distinct* being themselves perfectly fertile. Some authors believe that long-continued domestication eliminates this strong tendency to sterility: from the history of the dog I think there is some probability in this hypothesis, if applied to species closely related together, though it is unsupported by a single experiment. But to extend the hypothesis so far as to suppose that species, aboriginally as distinct as carriers, tumblers, pouters, and fantails now are, should yield offspring perfectly fertile *inter se*, seems to me rash in the extreme.

From these several reasons, namely, the improbability of man having formerly got seven or eight supposed species of pigeons to breed freely under domestication; these supposed species being quite unknown in a wild state, and their becoming nowhere feral; these species having very abnormal characters in certain respects, as compared with all other Columbidæ, though so like in most other respects to the rock-pigeon; the blue colour and various black marks occasionally appearing in all the breeds, both when kept pure and when crossed; the mongrel offspring being perfectly fertile;—from

c 2

these several reasons, taken together, we may safely
conclude that all our domestic breeds have descended
from the Columba livia with its geographical sub-
species.

In favour of this view, I may add, firstly, that C. livia,
or the rock-pigeon, has been found capable of domestication in Europe and in India; and that it agrees in habits
and in a great number of points of structure with all
the domestic breeds. Secondly, although an English
carrier or short-faced tumbler differs immensely in certain characters from the rock-pigeon, yet by comparing
the several sub-breeds of these varieties, more especially
those brought from distant countries, we can make an
almost perfect series between the extremes of structure.
Thirdly, those characters which are mainly distinctive of
each breed, for instance the wattle and length of beak
of the carrier, the shortness of that of the tumbler, and
the number of tail-feathers in the fantail, are in each
breed eminently variable; and the explanation of this
fact will be obvious when we come to treat of Selection.
Fourthly, pigeons have been watched and tended with
the utmost care, and loved by many people. They have
been domesticated for thousands of years in several quarters of the world; the earliest known record of pigeons
is in the fifth Ægyptian dynasty, about 3000 B.C.,
as was pointed out to me by Professor Lepsius; but
Mr. Birch informs me that pigeons are given in a bill
of fare in the previous dynasty. In the time of the
Romans, as we hear from Pliny, immense prices were
given for pigeons; "nay, they are come to this pass, that
they can reckon up their pedigree and race." Pigeons
were much valued by Akber Khan in India, about the
year 1600; never less than 20,000 pigeons were taken
with the court. "The monarchs of Iran and Turan sent
him some very rare birds;" and, continues the courtly

historian, "His Majesty by crossing the breeds, which method was never practised before, has improved them astonishingly." ; About this same period the Dutch were as eager about pigeons as were the old Romans. The paramount importance of these considerations in explaining the immense amount of variation which pigeons have undergone, will be obvious when we treat of Selection. We shall then, also, see how it is that the several breeds so often have a somewhat monstrous character. It is also a most favourable circumstance for the production of distinct breeds, that male and female pigeons can be easily mated for life; and thus different breeds can be kept together in the same aviary.

I have discussed the probable origin of domestic pigeons at some, yet quite insufficient, length; because when I first kept pigeons and watched the several kinds, knowing well how true they bred, I felt fully as much difficulty in believing that since they were domesticated they could all have descended from a common parent, as any naturalist could in coming to a similar conclusion in regard to the many species of finches, or other large groups of birds, in nature. One circumstance has struck me much; namely, that nearly all the breeders of the various domestic animals and the cultivators of plants, with whom I have conversed, or whose treatises I have read, are firmly convinced that the several breeds to which each has attended, are descended from so many aboriginally distinct species. Ask, as I have asked, a celebrated raiser of Hereford cattle, whether his cattle might not have descended from long-horns, and he will laugh you to scorn. I have never met a pigeon, or poultry, or duck, or rabbit fancier, who was not fully convinced that each main breed was descended from a distinct species. Van Mons, in his treatise on pears and apples, shows how utterly he disbelieves that the several sorts, for instance a Ribston-pippin or Codlin-apple, could ever have proceeded from the seeds of the same tree. Innumerable other examples could be given. The explanation, I think, is simple: from long-continued study they are strongly impressed with the differences between the several races; and though they well know that each race varies slightly, for they win their prizes by selecting such slight differences, yet they ignore all general arguments, and refuse to sum up in their minds slight differences accumulated during many successive generations. May not those naturalists who, knowing far less of the laws of inheritance than does the breeder, and knowing no more than he does of the intermediate links in the long lines of descent, yet admit that many of our domestic races have descended from the same parents—may they not learn a lesson of caution, when they deride the idea of species in a state of nature being lineal descendants of other species?

Selection.—Let us now briefly consider the steps by which domestic races have been produced, either from one or from several allied species. Some little effect may be attributed to the direct action of the external conditions of life, and some little to habit; but he would be a bold man who would account by such agencies for the differences of a dray and race horse, a greyhound and bloodhound, a carrier and tumbler pigeon. One of the most remarkable features in our domesticated races is that we see in them adaptation, not indeed to the animal's or plant's own good, but to man's use or fancy. Some variations useful to him have probably arisen suddenly, or by one step; many botanists, for instance, believe that the fuller's teazle, with its hooks, which cannot be rivalled by any mechanical contrivance, is only a variety of the wild Dipsacus; and this amount of change may have suddenly arisen in a seedling. So it has probably been with the turnspit dog; and this is known to have been the case with the ancon sheep. But when we compare the dray-horse and race-horse, the dromedary and camel, the various breeds of sheep fitted either for cultivated land or mountain pasture, with the wool of one breed good for one purpose, and that of another breed for another purpose; when we compare the many breeds of dogs, each good for man in very different ways; when we compare the game-cock, so pertinacious in battle, with other breeds so little quarrelsome, with "everlasting layers" which never desire to sit, and with the bantam so small and elegant; when we compare the host of agricultural, culinary, orchard, and flower-garden races of plants, most useful to man at different seasons and for different purposes, or so beautiful in his eyes, we must, I think, look further than to mere variability. We cannot suppose that all the breeds were suddenly produced as perfect and as useful as we now see them; indeed, in several cases, we know that this has not been their history. The key is man's power of accumulative selection: nature gives successive variations; man adds them up in certain directions useful to him. In this sense he may be said to make for himself useful breeds.

The great power of this principle of selection is not hypothetical. It is certain that several of our eminent breeders have, even within a single lifetime, modified to a large extent some breeds of cattle and sheep. In order fully to realise what they have done, it is almost necessary to read several of the many treatises devoted to this subject, and to inspect the animals. Breeders habitually speak of an animal's organisation as something quite plastic, which they can model almost as they please. If I had space I could quote numerous passages to this effect from highly competent authorities. Youatt,

who was probably better acquainted with the works of
agriculturists than almost any other individual, and who
was himself a very good judge of an animal, speaks of
the principle of selection' as "that which enables the
agriculturist, not only to modify the character of his
flock, but to change it altogether. It is the magician's
wand, by means of which he may summon into life
whatever form and mould he pleases." Lord Somerville,
speaking of what breeders have done for sheep, says:—
" It would seem as if they had chalked out upon a wall
a form perfect in itself, and then had given it existence."
That most skilful breeder, Sir John Sebright, used to
say, with respect to pigeons, that "he would produce
any given feather in three years, but it would take him
six years to obtain head and beak." In Saxony the importance of the principle of selection in regard to merino
sheep is so fully recognised, that men follow it as a
trade : the sheep are placed on a table and are studied,
like a picture by a connoisseur ; this is done three times
at intervals of months, and the sheep are each time
marked and classed, so that the very best may ultimately
be selected for breeding.

  What English breeders have actually effected is
proved by the enormous prices given for animals with a
good pedigree ; and these have now been exported to
almost every quarter of the world. The improvement is
by no means generally due to crossing different breeds ;
all the best breeders are strongly opposed to this practice, except sometimes amongst closely allied sub-breeds.
And when a cross has been made, the closest selection is
far more indispensable even than in ordinary cases. If
selection consisted merely in separating some very distinct variety, and breeding from it, the principle would
be so obvious as hardly to be worth notice ; but its importance consists in the great effect produced by the

accumulation in one direction, during successive generations, of differences absolutely inappreciable by an uneducated eye—differences which I for one have vainly attempted to appreciate. Not one man in a thousand has accuracy of eye and judgment sufficient to become an eminent breeder. If gifted with these qualities, and he studies his subject for years, and devotes his lifetime to it with indomitable perseverance, he will succeed, and may make great improvements; if he wants any of these qualities, he will assuredly fail. Few would readily believe in the natural capacity and years of practice requisite to become even a skilful pigeon-fancier.

The same principles are followed by horticulturists; but the variations are here often more abrupt. No one supposes that our choicest productions have been produced by a single variation from the aboriginal stock. We have proofs that this is not so in some cases in which exact records have been kept; thus, to give a very trifling instance, the steadily-increasing size of the common gooseberry may be quoted. We see an astonishing improvement in many florists' flowers, when the flowers of the present day are compared with drawings made only twenty or thirty years ago. When a race of plants is once pretty well established, the seed-raisers do not pick out the best plants, but merely go over their seed-beds, and pull up the "rogues," as they call the plants that deviate from the proper standard. With animals this kind of selection is, in fact, also followed; for hardly any one is so careless as to allow his worst animals to breed.

In regard to plants, there is another means of observing the accumulated effects of selection—namely, by comparing the diversity of flowers in the different varieties of the same species in the flower-garden; the diversity of leaves, pods, or tubers, or whatever part is valued, in the kitchen-garden, in comparison with the flowers of the same varieties; and the diversity of fruit of the same species in the orchard, in comparison with the leaves and flowers of the same set of varieties. See how different the leaves of the cabbage are, and how extremely alike the flowers; how unlike the flowers of the heartsease are, and how alike the leaves; how much the fruit of the different kinds of gooseberries differ in size, colour, shape, and hairiness, and yet the flowers present very slight differences. It is not that the varieties which differ largely in some one point do not differ at all in other points; this is hardly ever, I speak after careful observation, perhaps never, the case. The laws of correlation of growth, the importance of which should never be overlooked, will ensure some differences; but, as a general rule, I cannot doubt that the continued selection of slight variations, either in the leaves, the flowers, or the fruit, will produce races differing from each other chiefly in these characters.

It may be objected that the principle of selection has been reduced to methodical practice for scarcely more than three-quarters of a century; it has certainly been more attended to of late years, and many treatises have been published on the subject; and the result has been, in a corresponding degree, rapid and important. But it is very far from true that the principle is a modern discovery. I could give several references to the full acknowledgment of the importance of the principle in works of high antiquity. In rude and barbarous periods of English history choice animals were often imported, and laws were passed to prevent their exportation: the destruction of horses under a certain size was ordered, and this may be compared to the "roguing" of plants by nurserymen. The principle of selection I find distinctly given in an ancient Chinese encyclopædia. Ex-

plicit rules are laid down by some of the Roman classical
writers. From passages in Genesis, it is clear that the
colour of domestic animals was at that early period
attended to. Savages now sometimes cross their dogs
with wild canine animals, to improve the breed, and they
formerly did so, as is attested by passages in Pliny. The
savages in South Africa match their draught cattle by
colour, as do some of the Esquimaux their teams of dogs.
Livingstone shows how much good domestic breeds are
valued by the negroes of the interior of Africa who have
not associated with Europeans. Some of these facts do
not show actual selection, but they show that the breeding of domestic animals was carefully attended to in
ancient times, and is now attended to by the lowest
savages. It would, indeed, have been a strange fact, had
attention not been paid to breeding, for the inheritance
of good and bad qualities is so obvious.

At the present time, eminent breeders try by methodical selection, with a distinct object in view, to make
a new strain or sub-breed, superior to anything existing
in the country. But, for our purpose, a kind of Selection,
which may be called Unconscious, and which results
from every one trying to possess and breed from the best
individual animals, is more important. Thus, a man
who intends keeping pointers naturally tries to get as
good dogs as he can, and afterwards breeds from his
own best dogs, but he has no wish or expectation of
permanently altering the breed. Nevertheless we
may infer that this process, continued during centuries,
would improve and modify any breed, in the same way
as Bakewell, Collins, &c., by this very same process,
only carried on more methodically, did greatly modify,
even during their own lifetimes, the forms and qualities
of their cattle. Slow and insensible changes of this
kind could never be recognised unless actual measure-

ments or careful drawings of the breeds in question
had been made long ago, which might serve for comparison. In some cases, however, unchanged, or but little
changed individuals of the same breed may be found
in less civilised districts, where the breed has been less
improved. There is reason to believe that King Charles's
spaniel has been unconsciously modified to a large extent
since the time of that monarch. Some highly competent
authorities are convinced that the setter is directly derived from the spaniel, and has probably been slowly
altered from it. It is known that the English pointer
has been greatly changed within the last century, and
in this case the change has, it is believed, been chiefly
effected by crosses with the fox-hound; but what concerns us is, that the change has been effected unconsciously and gradually, and yet so effectually, that, though
the old Spanish pointer certainly came from Spain, Mr.
Borrow has not seen, as I am informed by him, any
native dog in Spain like our pointer.

By a similar process of selection, and by careful training, the whole body of English racehorses have come to
surpass in fleetness and size the parent Arab stock, so
that the latter, by the regulations for the Goodwood Races,
are favoured in the weights they carry. Lord Spencer
and others have shown how the cattle of England have
increased in weight and in early maturity, compared with
the stock formerly kept in this country. By comparing
the accounts given in old pigeon treatises of carriers and
tumblers with these breeds as now existing in Britain,
India, and Persia, we can, I think, clearly trace the stages
through which they have insensibly passed, and come to
differ so greatly from the rock-pigeon.

Youatt gives an excellent illustration of the effects of
a course of selection, which may be considered as unconsciously followed, in so far that the breeders could

never have expected or even have wished to have produced the result which ensued—namely, the production
of two distinct strains. The two flocks of Leicester sheep
kept by Mr. Buckley and Mr. Burgess, as Mr. Youatt remarks, "have been purely bred from the original stock of
Mr. Bakewell for upwards of fifty years. There is not
a suspicion existing in the mind of any one at all acquainted with the subject that the owner of either of
them has deviated in any one instance from the pure
blood of Mr. Bakewell's flock, and yet the difference
between the sheep possessed by these two gentlemen is
so great that they have the appearance of being quite
different varieties."

If there exist savages so barbarous as never to think
of the inherited character of the offspring of their domestic animals, yet any one animal particularly useful
to them, for any special purpose, would be carefully
preserved during famines and other accidents, to which
savages are so liable, and such choice animals would thus
generally leave more offspring than the inferior ones;
so that in this case there would be a kind of unconscious selection going on. We see the value set on
animals even by the barbarians of Tierra del Fuego, by
their killing and devouring their old women, in times of
dearth, as of less value than their dogs.

In plants the same gradual process of improvement,
through the occasional preservation of the best individuals, whether or not sufficiently distinct to be ranked
at their first appearance as distinct varieties, and whether
or not two or more species or races have become
blended together by crossing, may plainly be recognised
in the increased size and beauty which we now see in the
varieties of the heartsease, rose, pelargonium, dahlia, and
other plants, when compared with the older varieties or
with their parent-stocks. No one would ever expect to

get a first-rate heartsease or dahlia from the seed of a
wild plant. No one would expect to raise a first-rate
melting pear from the seed of the wild pear, though
he might succeed from a poor seedling growing wild,
if it had come from a garden-stock. The pear, though
cultivated in classical times, appears, from Pliny's description, to have been a fruit of very inferior quality.
I have seen great surprise expressed in horticultural
works at the wonderful skill of gardeners, in having
produced such splendid results from such poor materials; but the art has been simple, and, as far as the
final result is concerned, has been followed almost unconsciously. It has consisted in always cultivating the
best known variety, sowing its seeds, and, when a slightly
better variety has chanced to appear, selecting it, and
so onwards. But the gardeners of the classical period,
who cultivated the best pear they could procure, never
thought what splendid fruit we should eat; though we
owe our excellent fruit, in some small degree, to their
having naturally chosen and preserved the best varieties
they could anywhere find.

A large amount of change in our cultivated plants,
thus slowly and unconsciously accumulated, explains, as
I believe, the well-known fact, that in a vast number of
cases we cannot recognise, and therefore do not know,
the wild parent-stocks of the plants which have been
longest cultivated in our flower and kitchen gardens.
If it has taken centuries or thousands of years to improve or modify most of our plants up to their present
standard of usefulness to man, we can understand how
it is that neither Australia, the Cape of Good Hope, nor
any other region inhabited by quite uncivilised man, has
afforded us a single plant worth culture. It is not that
these countries, so rich in species, do not by a strange
chance possess the aboriginal stocks of any useful plants,

but that the native plants have not been improved by
continued selection up to a standard of perfection comparable with that given to the plants in countries anciently civilised.

In regard to the domestic animals kept by uncivilised
man, it should not be overlooked that they almost
always have to struggle for their own food, at least
during certain seasons. And in two countries very differently circumstanced, individuals of the same species,
having slightly different constitutions or structure, would
often succeed better in the one country than in the
other; and thus by a process of "natural selection," as
will hereafter be more fully explained, two sub-breeds
might be formed. This, perhaps, partly explains what
has been remarked by some authors, namely, that the
varieties kept by savages have more of the character of
species than the varieties kept in civilised countries.

On the view here given of the all-important part which
selection by man has played, it becomes at once obvious,
how it is that our domestic races show adaptation in their
structure or in their habits to man's wants or fancies.
We can, I think, further understand the frequently
abnormal character of our domestic races, and likewise
their differences being so great in external characters
and relatively so slight in internal parts or organs.
Man can hardly select, or only with much difficulty, any
deviation of structure excepting such as is externally
visible; and indeed he rarely cares for what is internal.
He can never act by selection, excepting on variations
which are first given to him in some slight degree by
nature. No man would ever try to make a fantail, till
he saw a pigeon with a tail developed in some slight
degree in an unusual manner, or a pouter till he saw a
pigeon with a crop of somewhat unusual size; and the
more abnormal or unusual any character was when it first

appeared, the more likely it would be to catch his attention. But to use such an expression as trying to make
a fantail, is, I have no doubt, in most cases, utterly incorrect. The man who first selected a pigeon with a
slightly larger tail, never dreamed what the descendants
of that pigeon would become through long-continued,
partly unconscious and partly methodical selection.
Perhaps the parent-bird of all fantails had only fourteen
tail-feathers somewhat expanded, like the present Java
fantail, or like individuals of other and distinct breeds,
in which as many as seventeen tail-feathers have been
counted. Perhaps the first pouter-pigeon did not inflate
its crop much more than the turbit now does the upper
part of its œsophagus,—a habit which is disregarded by
all fanciers, as it is not one of the points of the breed.

Nor let it be thought that some great deviation of
structure would be necessary to catch the fancier's eye :
he perceives extremely small differences, and it is in
human nature to value any novelty, however slight, in
one's own possession. Nor must the value which would
formerly be set on any slight differences in the individuals
of the same species, be judged of by the value which
would now be set on them, after several breeds have
once fairly been established. Many slight differences
might, and indeed do now, arise amongst pigeons, which
are rejected as faults or deviations from the standard of
perfection of each breed. The common goose has not
given rise to any marked varieties; hence the Toulouse
and the common breed, which differ only in colour, that
most fleeting of characters, have lately been exhibited
as distinct at our poultry-shows.

I think these views further explain what has sometimes
been noticed—namely, that we know nothing about the
origin or history of any of our domestic breeds. But, in
fact, a breed, like a dialect of a language, can hardly be

said to have had a definite origin. A man preserves and breeds from an individual with some slight deviation of structure, or takes more care than usual in matching his best animals and thus improves them, and the improved individuals slowly spread in the immediate neighbourhood. But as yet they will hardly have a distinct name, and from being only slightly valued, their history will be disregarded. When further improved by the same slow and gradual process, they will spread more widely, and will get recognised as something distinct and valuable, and will then probably first receive a provincial name. In semi-civilised countries, with little free communication, the spreading and knowledge of any new sub-breed would be a slow process. As soon as the points of value of the new sub-breed are once fully acknowledged, the principle, as I have called it, of unconscious selection will always tend,—perhaps more at one period than at another, as the breed rises or falls in fashion,—perhaps more in one district than in another, according to the state of civilisation of the inhabitants,—slowly to add to the characteristic features of the breed, whatever they may be. But the chance will be infinitely small of any record having been preserved of such slow, varying, and insensible changes.

I must now say a few words on the circumstances, favourable, or the reverse, to man's power of selection. A high degree of variability is obviously favourable, as freely giving the materials for selection to work on ; not that mere individual differences are not amply sufficient, with extreme care, to allow of the accumulation of a large amount of modification in almost any desired direction. But as variations manifestly useful or pleasing to man appear only occasionally, the chance of their appearance will be much increased by a large number of individuals being kept ; and hence this comes to be of

the highest importance to success. On this principle
Marshall has remarked, with respect to the sheep of parts
of Yorkshire, that "as they generally belong to poor
people, and are mostly *in small lots*, they never can be
improved." On the other hand, nurserymen, from raising
large stocks of the same plants, are generally far more
successful than amateurs in getting new and valuable
varieties. The keeping of a large number of individuals
of a species in any country requires that the species
should be placed under favourable conditions of life, so
as to breed freely in that country. When the individuals
of any species are scanty, all the individuals, whatever
their quality may be, will generally be allowed to breed,
and this will effectually prevent selection. But probably
the most important point of all is that the animal or
plant should be so highly useful to man, or so much
valued by him, that the closest attention should be
paid to even the slightest deviation in the qualities or
structure of each individual. Unless such attention
be paid nothing can be effected. I have seen it gravely
remarked, that it was most fortunate that the strawberry began to vary just when gardeners began to attend
closely to this plant. No doubt the strawberry had
always varied since it was cultivated, but the slight
varieties had been neglected. As soon, however, as
gardeners picked out individual plants with slightly
larger, earlier, or better fruit, and raised seedlings from
them, and again picked out the best seedlings and bred
from them, then, there appeared (aided by some crossing
with distinct species) those many admirable varieties of
the strawberry which have been raised during the last
thirty or forty years.

In the case of animals with separate sexes, facility in
preventing crosses is an important element of success
in the formation of new races,—at least, in a country

which is already stocked with other races. In this respect enclosure of the land plays a part. Wandering savages or the inhabitants of open plains rarely possess more than one breed of the same species. Pigeons can be mated for life, and this is a great convenience to the fancier, for thus many races may be improved and kept true, though mingled in the same aviary; and this circumstance must have largely favoured the formation of new breeds. Pigeons, I may add, can be propagated in great numbers and at a very quick rate, and inferior birds may be freely rejected, as when killed they serve for food. On the other hand, cats, from their nocturnal rambling habits, cannot be easily matched, and, although so much valued by women and children, we hardly ever see a distinct breed kept up; such breeds as we do sometimes see are almost always imported from some other country. Although I do not doubt that some domestic animals vary less than others, yet the rarity or absence of distinct breeds of the cat, the donkey, peacock, goose, &c., may be attributed in main part to selection not having been brought into play: in cats, from the difficulty in pairing them: in donkeys, from only a few being kept by poor people, and little attention paid to their breeding; for recently in certain parts of Spain and of the United States this animal has been surprisingly modified and improved by careful selection: in peacocks, from not being very easily reared and a large stock not kept: in geese, from being valuable only for two purposes, food and feathers, and more especially from no pleasure having been felt in the display of distinct breeds; but the goose seems to have a singularly inflexible organisation.

To sum up on the origin of our domestic races of animals and plants. I believe that the conditions of life, from their action on the reproductive system, are

so far of the highest importance as causing variability.
It is not probable that variability is an inherent and
necessary contingency, under all circumstances, with all
organic beings, as some authors have thought. The
effects of variability are modified by various degrees of
inheritance and of reversion. Variability is governed
by many unknown laws, more especially by that of correlation of growth. Something may be attributed to
the direct action of the conditions of life. Something
must be attributed to use and disuse. The final result
is thus rendered infinitely complex. In some cases the
intercrossing of species, aboriginally distinct, has probably played an important part in the origin of our
domestic breeds. When in any country several domestic
breeds have once been established, their occasional intercrossing, with the aid of selection, has, no doubt,
largely aided in the formation of new sub-breeds; but
the importance of the crossing of varieties has been
greatly exaggerated, both in regard to animals and to
those plants which are propagated by seed. In plants
which are temporarily propagated by cuttings, buds, &c.,
the importance of the crossing both of distinct species
and of varieties is immense; for the cultivator here
quite disregards the extreme variability both of hybrids
and mongrels, and the frequent sterility of hybrids; but
the cases of plants not propagated by seed are of little
importance to us, for their endurance is only temporary.
Over all these causes of Change I am convinced that
the accumulative action of Selection, whether applied
methodically and more quickly, or unconsciously and
more slowly, but more efficiently, is by far the predomi-
nant Power.

CHAPTER II.

VARIATION UNDER NATURE.

Variability — Individual differences — Doubtful species — Wide ranging, much diffused, and common species vary most — Species of the larger genera in any country vary more than the species of the smaller genera — Many of the species of the larger genera resemble varieties in being very closely, but unequally, related to each other, and in having restricted ranges.

BEFORE applying the principles arrived at in the last chapter to organic beings in a state of nature, we must briefly discuss whether these latter are subject to any variation. To treat this subject at all properly, a long catalogue of dry facts should be given; but these I shall reserve for my future work. Nor shall I here discuss the various definitions which have been given of the term species. No one definition has as yet satisfied all naturalists; yet every naturalist knows vaguely what he means when he speaks of a species. Generally the term includes the unknown element of a distinct act of creation. The term "variety" is almost equally difficult to define; but here community of descent is almost universally implied, though it can rarely be proved. We have also what are called monstrosities; but they graduate into varieties. By a monstrosity I presume is meant some considerable deviation of structure in one part, generally injurious to or not useful to the species. Some authors use the term "variation" in a technical sense, as implying a modification directly due to the physical conditions of life; and "variations" in this sense are supposed not to be inherited: but who can say that the dwarfed condition of shells in the brackish waters of the Baltic, or dwarfed plants on Alpine summits, or the thicker fur of an animal from far northwards, would not in some cases be inherited for at least some few generations? and in this case I presume that the form would be called a variety.

It may perhaps be doubted whether monstrosities, or such sudden and great deviations of structure as we occasionally see in our domestic productions, more especially with plants, are ever permanently propagated in a state of nature. Monsters are very apt to be sterile; and almost every part of every organic being, at least with animals, is so beautifully related to its complex conditions of life that it seems as improbable that any part should have been suddenly produced perfect, as that a complex machine should have been invented by man in a perfect state. I have not, at least, been able to find good cases of species in a state of nature presenting modifications of structure resembling monstrosities observed in allied forms. If such have occurred, their perpetuation will have been due to their beneficial nature, so that natural selection will have come into play. Many cases are known of plants which regularly produce on different branches, or on the circumference and in the centre of umbels, &c., flowers of a widely different structure; and if the plant ceased to produce flowers of the one kind, a great change might perhaps suddenly be effected in the specific character; but then we do not at present know by what steps, or for what good, a plant produces two kinds of flowers. With cultivated plants, in the few cases known of a variety habitually bearing flowers or fruit slightly different from each other, the production of the variety has been sudden.

Again, we have many slight differences which may be called individual differences, such as are known frequently to appear in the offspring from the same parents, or which may be presumed to have thus arisen, from being frequently observed in the individuals of the same species inhabiting the same confined locality. No one supposes that all the individuals of the same species are cast in the very same mould. These individual differences are highly important for us, for they are often inherited, as must be familiar to every one; and thus they afford materials for natural selection to accumulate, in the same manner as man accumulates in any given direction individual differences in his domesticated productions. These individual differences generally affect what naturalists consider unimportant parts; but I could show by a long catalogue of facts, that parts which must be called important, whether viewed under a physiological or classificatory point of view, sometimes vary in the individuals of the same species. I am convinced that the most experienced naturalist would be surprised at the number of the cases of variability, even in important parts of structure, which he could collect on good authority, as I have collected, during a course of years. It should be remembered that systematists are far from pleased at finding variability in important characters, and that there are not many men who will laboriously examine internal and important organs, and compare them in many specimens of the same species. It would never have been expected that the branching of the main nerves close to the great central ganglion of an insect would have been variable in the same species; it might have been thought that changes of this nature could have been effected only by slow degrees; yet quite recently Mr. Lubbock has shown a degree of variability in these main nerves in Coccus, which may almost be compared to the irregular branching of the stem of a tree. This philosophical naturalist,

I may add, 'has also quite recently shown that the muscles in the larvæ of certain insects are very far from uniform. Authors sometimes argue in a circle when they state that important organs never vary; for these same authors practically rank that character as important (as some few naturalists have honestly confessed) which does not vary; and, under this point of view, no instance of an important part varying will ever be found : but under any other point of view many instances assuredly can be given.

There is one point connected with individual differences, which is extremely perplexing': I refer to those genera which have sometimes been called "protean" or "polymorphic," in which the species present an inordinate amount of variation ; and hardly two naturalists can agree which forms to rank as species and which as varieties. We may instance Rubus, Rosa, and Hieracium amongst plants, several genera of insects, and several genera of Brachiopod shells. In most polymorphic genera some of the species have fixed and definite characters. Genera which are polymorphic in one country seem to be, with some few exceptions, polymorphic in other countries, and likewise, judging from Brachiopod shells, at former periods of time. These facts are very perplexing, for they seem to show that this kind of variability is independent of the conditions of life. I am inclined to suspect that we see in these polymorphic genera variations in points of structure which are of no service or disservice to the species, and which consequently have not been seized on and rendered definite by natural selection, as hereafter will be explained.

Those forms which possess in some considerable degree the character of species, but which are so closely similar to some other forms, or are so closely linked to them by intermediate gradations, that naturalists do not like to rank them as distinct species, are in several respects the most important for us. We have every reason to believe that many of these doubtful and closely-allied forms have permanently retained their characters in their own country for a long time; for as long, as far as we know, as have good and true species. Practically, when a naturalist can unite two forms together by others having intermediate characters, he treats the one as a variety of the other, ranking the most common, but sometimes the one first described, as the species, and the other as the variety. But cases of great difficulty, which I will not here enumerate, sometimes occur in deciding whether or not to rank one form as a variety of another, even when they are closely connected by intermediate links; nor will the commonly-assumed hybrid nature of the intermediate links always remove the difficulty. In very many cases, however, one form is ranked as a variety of another, not because the intermediate links have actually been found, but because analogy leads the observer to suppose either that they do now somewhere exist, or may formerly have existed; and here a wide door for the entry of doubt and conjecture is opened.

Hence, in determining whether a form should be ranked as a species or a variety, the opinion of naturalists having sound judgment and wide experience seems the only guide to follow. We must, however, in many cases, decide by a majority of naturalists, for few well-marked and well-known varieties can be named which have not been ranked as species by at least some competent judges.

That varieties of this doubtful nature are far from uncommon cannot be disputed. Compare the several floras of Great Britain, of France, or of the United States, drawn up by different botanists, and see what a

D

surprising number of forms have been ranked by one
botanist as good species, and by another as mere
varieties. Mr. H. C. Watson, to whom I lie under
deep obligation for assistance of all kinds, has marked
for me 182 British plants, which are generally considered as varieties, but which have all been ranked
by botanists as species; and in making this list he
has omitted many trifling varieties, but which nevertheless have been ranked by some botanists as species,
and he has entirely omitted several highly polymorphic
genera. Under genera, including the most polymorphic
forms, Mr. Babington gives 251 species, whereas Mr.
Bentham gives only 112,—a difference of 139 doubtful
forms! Amongst animals which unite for each birth,
and which are highly locomotive, doubtful forms, ranked
by one zoologist as a species and by another as a variety,
can rarely be found within the same country, but are
common in separated areas. How many of those birds
and insects in North America and Europe, which differ
very slightly from each other, have been ranked by
one eminent naturalist as undoubted species, and by
another as varieties, or, as they are often called, as
geographical races! Many years ago, when comparing,
and seeing others compare, the birds from the separate islands of the Galapagos Archipelago, both one
with another, and with those from the American mainland, I was much struck how entirely vague and arbitrary is the distinction between species and varieties.
On the islets of the little Madeira group there are
many insects which are characterized as varieties in
Mr. Wollaston's admirable work, but which certainly
would be ranked as distinct species by many entomologists. Even Ireland has a few animals, now generally
regarded as varieties, but which have been ranked as
species by some zoologists. Several most experienced

ornithologists consider our British red grouse as only a strongly-marked race of a Norwegian species, whereas the greater number rank it as an undoubted species peculiar to Great Britain. A wide distance between the homes of two doubtful forms leads many naturalists to rank both as distinct species; but what distance, it has been well asked, will suffice? if that between America and Europe is ample, will that between the Contiuent and the Azores, or Madeira, or the Canaries, or Ireland, be sufficient? Some few naturalists maintain that animals never present varieties; but then these same naturalists rank the slightest differences as of specific value; and when even the same identical form is met with in two distant countries, or in two quite distinct geological formations, they go so far as to believe that two separate species are hidden under the same dress. Finally, it cannot be disputed that many forms, considered by highly-competent judges as varieties, have so perfectly the character of species that they are ranked by other highly-competent judges as good and true species. But to discuss whether such slightly different forms are rightly called species or varieties, before any definition of these terms has been generally accepted, is vainly to beat the air.

Many of the cases of strongly-marked varieties or doubtful species well deserve consideration; for several interesting lines of argument, from geographical distribution, analogical variation, hybridism, &c., have been brought to bear on the attempt to determine their rank. I will here give only a single instance,—the well-known one of the primrose and cowslip, or Primula vulgaris and veris. These plants differ considerably in appearance; they have a different flavour, and emit a different odour; they flower at slightly different periods; they grow in somewhat different stations; they ascend

mountains to different heights; they have different
geographical ranges; and lastly, according to very
numerous experiments made during several years by
that most careful observer Gärtner, they can be crossed
only with much difficulty. We could hardly wish for
better evidence of the two forms being specifically distinct. On the other hand, they are united by many
intermediate links, and it is very doubtful whether these
links are hybrids; and there is a large amount of ex-
perimental evidence, showing that they descend from
common parents, and consequently must be ranked as
varieties.

Close investigation, in most cases, will bring naturalists
to an agreement how to rank doubtful forms. Yet it
must be confessed that it is in the best-known countries
that we find the greatest number of forms of doubtful
value. I have been struck with the fact, that if any
animal or plant in a state of nature be highly useful
to man, or from any cause closely attract his attention,
varieties of it will almost universally be found recorded.
These varieties, moreover, will be often ranked by some
authors as species. Look at the common oak, how
closely it has been studied; yet a German author
makes more than a dozen species out of forms, which
are almost universally considered as varieties; and in
this country the highest botanical authorities and practical men can be quoted to show that the sessile and
pedunculated oaks are either good and distinct species
or mere varieties.

When a young naturalist commences the study of a
group of organisms quite unknown to him, he is at first
much perplexed to determine what differences to consider
as specific, and what as varieties; for he knows nothing
of the amount and kind of variation to which the group
is subject; and this shows, at least, how very generally

there is some variation. But if he confine his attention
to one class within one country, he will soon make up
his mind how to rank most of the doubtful forms. His
general tendency will be to make many species, for he
will become impressed, just like the pigeon or poultry
fancier before alluded to, with the amount of difference
in the forms which he is continually studying; and he
has little general knowledge of analogical variation in
other groups and in other countries, by which to correct
his first impressions. As he extends the range of his
observations, he will meet with more cases of difficulty;
for he will encounter a greater number of closely-allied
forms. But if his observations be widely extended, he
will in the end generally be enabled to make up his own
mind which to call varieties and which species; but he
will succeed in this at the expense of admitting much
variation,—and the truth of this admission will often be
disputed by other naturalists. When, moreover, he
comes to study allied forms brought from countries not
now continuous, in which case he can hardly hope to
find the intermediate links between his doubtful forms,
he will have to trust almost entirely to analogy, and his
difficulties rise to a climax.

Certainly no clear line of demarcation has as yet
been drawn between species and sub-species—that is,
the forms which in the opinion of some naturalists come
very near to, but do not quite arrive at the rank of
species; or, again, between sub-species and well-marked
varieties, or between lesser varieties and individual differences. These differences blend into each other in an
insensible series; and a series impresses the mind with
the idea of an actual passage.

Hence I look at individual differences, though of small
interest to the systematist, as of high importance for us,
as being the first step towards such slight varieties as

are barely thought worth recording in works on natural history. And I look at varieties which are in any degree more distinct and permanent, as steps leading to more strongly marked and more permanent varieties; and at these latter, as leading to sub-species, and to species. The passage from one stage of difference to another and higher stage may be, in some cases, due merely to the long-continued action of different physical conditions in two different regions; but I have not much faith in this view; and I attribute the passage of a variety, from a state in which it differs very slightly from its parent to one in which it differs more, to the action of natural selection in accumulating (as will hereafter be more fully explained) differences of structure in certain definite directions. Hence I believe a well-marked variety may be called an incipient species; but whether this belief be justifiable must be judged of by the general weight of the several facts and views given throughout this work.

It need not be supposed that all varieties or incipient species necessarily attain the rank of species. They may whilst in this incipient state become extinct, or they may endure as varieties for very long periods, as has been shown to be the case by Mr. Wollaston with the varieties of certain fossil land-shells in Madeira. If a variety were to flourish so as to exceed in numbers the parent species, it would then rank as the species, and the species as the variety; or it might come to supplant and exterminate the parent species; or both might co-exist, and both rank as independent species. But we shall hereafter have to return to this subject.

From these remarks it will be seen that I look at the term species, as one arbitrarily given for the sake of convenience to a set of individuals closely resembling each other, and that it does not essentially differ from the term variety, which is given to less distinct and more fluctuating forms. The term variety, again, in comparison with mere individual differences, is also applied arbitrarily, and for mere convenience' sake.

Guided by theoretical considerations, I thought that some interesting results might be obtained in regard to the nature and relations of the species which vary most, by tabulating all the varieties in several well-worked floras. At first this seemed a simple task; but Mr. H. C. Watson, to whom I am much indebted for valuable advice and assistance on this subject, soon convinced me that there were many difficulties, as did subsequently Dr. Hooker, even in stronger terms. I shall reserve for my future work the discussion of these difficulties, and the tables themselves of the proportional numbers of the varying species. Dr. Hooker permits me to add, that after having carefully read my manuscript, and examined the tables, he thinks that the following statements are fairly well established. The whole subject, however, treated as it necessarily here is with much brevity, is rather perplexing, and allusions cannot be avoided to the "struggle for existence," "divergence of character," and other questions, hereafter to be discussed.

Alph. de Candolle and others have shown that plants which have very wide ranges generally present varieties; and this might have been expected, as they become exposed to diverse physical conditions, and as they come into competition (which, as we shall hereafter see, is a more important circumstance) with different sets of organic beings. But my tables further show that, in any limited country, the species which are most common, that is abound most in individuals, and the species which are most widely diffused within their own country (and this is a different consideration from wide range, and to a certain extent from commonness), oftenest

give rise to varieties sufficiently well-marked to have
been recorded in botanical works. Hence it is the
most flourishing, or, as they may be called, the dominant species,—those which range widely, are the most
diffused in their own country, and are the most numerous in individuals,—which oftenest produce well-marked
varieties, or, as I consider them, incipient species. And
this, perhaps, might have been anticipated; for, as
varieties, in order to become in any degree permanent,
necessarily have to struggle with the other inhabitants
of the country, the species which are already dominant
will be the most likely to yield offspring, which, though
in some slight degree modified, still inherit those advantages that enabled their parents to become dominant
over their compatriots. In these remarks on predomi-
nance, it should be understood that reference is made
only to those forms which come into competition with
each other, and more especially to the members of the
same genus or class having nearly similar habits of life.
With respect to commonness or the number of individuals of any species, the comparison of course relates
only to the members of the same group. A plant may
be said to be dominant if it be more numerous in
individuals and more widely diffused than the other
plants of the same country, not living under widely different conditions of life. Such a plant is not the less
dominant in the sense here used, because some conferva
inhabiting the water or some parasitic fungus is infinitely more numerous in individuals and more widely
diffused; if one kind of conferva or parasitic fungus
exceeded its allies in the above respects, it would be a
dominant form within its own class.

If the plants inhabiting a country and described in
any Flora be divided into two equal masses, all those in
the larger genera being placed on one side, and all those

in the smaller genera on the other side, a somewhat larger number of the very common and much diffused or dominant species will be found on the side of the larger genera. This, again, might have been anticipated; for the mere fact of many species of the same genus inhabiting any country, shows that there is something in the organic or inorganic conditions of that country favourable to the genus; and, consequently, we might have expected to have found in the larger genera, or those including many species, a large proportional number of dominant species. But so many causes tend to obscure this result, that I am surprised that my tables show even a small majority on the side of the larger genera. I will here allude to only two causes of obscurity. Fresh-water and salt-loving plants have generally very wide ranges and are much diffused, but this seems to be connected with the nature of the stations inhabited by them, and has little or no relation to the size of the genera to which the species belong. Again, plants low in the scale of organisation are generally much more widely diffused than plants higher in the scale; and here again there is no close relation to the size of the genera. The cause of lowly-organised plants ranging widely will be discussed in our chapter on Geographical Distribution.

From looking at species as only strongly-marked and well-defined varieties, I was led to anticipate that the species of the larger genera in each country would oftener present varieties, than the species of the smaller genera; for wherever many closely related species (*i. e.* species of the same genus) have been formed, many varieties or incipient species ought, as a general rule, to be now forming. Where many large trees grow, we expect to find saplings. Where many species of a genus have been formed through variation, circumstances have been favourable for variation; and hence we might expect that the circumstances would generally be still favourable to variation. On the other hand, if we look at each species as a special act of creation, there is no apparent reason why more varieties should occur in a group having many species, than in one having few.

To test the truth of this anticipation I have arranged the plants of twelve countries, and the coleopterous insects of two districts, into two nearly equal masses, the species of the larger genera on one side, and those of the smaller genera on the other side, and it has invariably proved to be the case that a larger proportion of the species on the side of the larger genera present varieties, than on the side of the smaller genera. Moreover, the species of the large genera which present any varieties, invariably present a larger average number of varieties than do the species of the small genera. Both these results follow when another division is made, and when all the smallest genera, with from only one to four species, are absolutely excluded from the tables. These facts are of plain signification on the view that species are only strongly marked and permanent varieties; for wherever many species of the same genus have been formed, or where, if we may use the expression, the manufactory of species has been active, we ought generally to find the manufactory still in action, more especially as we have every reason to believe the process of manufacturing new species to be a slow one. And this certainly is the case, if varieties be looked at as incipient species; for my tables clearly show as a general rule that, wherever many species of a genus have been formed, the species of that genus present a number of varieties, that is of incipient species, beyond the average. It is not that all large genera are now varying much, and are thus increasing in the number of their species,

or that no small genera are now varying and increasing;
for if this had been so, it would have been fatal to my
theory; inasmuch as geology plainly tells us that small
genera have in the lapse of time often increased greatly
in size; and that large genera have often come to their
maxima, declined, and disappeared. All that we want
to show is, that where many species of a genus have
been formed, on an average many are still forming;
and this holds good.

There are other relations between the species of large
genera and their recorded varieties which deserve notice.
We have seen that there is no infallible criterion by
which to distinguish species and well-marked varieties;
and in those cases in which intermediate links have not
been found between doubtful forms, naturalists are compelled to come to a determination by the amount of
difference between them, judging by analogy whether
or not the amount suffices to raise one or both to the
rank of species. Hence the amount of difference is one
very important criterion in settling whether two forms
should be ranked as species or varieties. Now Fries
has remarked in regard to plants, and Westwood in
regard to insects, that in large genera the amount of
difference between the species is often exceedingly small.
I have endeavoured to test this numerically by averages,
and, as far as my imperfect results go, they confirm the
view. I have also consulted some sagacious and experienced observers, and, after deliberation, they concur
in this view. In this respect, therefore, the species of
the larger genera resemble varieties, more than do the
species of the smaller genera. Or the case may be
put in another way, and it may be said, that in the
larger genera, in which a number of varieties or incipient species greater than the average are now manufacturing, many of the species already manufactured

still to a certain extent resemble varieties, for they differ from each other by a less than the usual amount of difference.

Moreover, the species of the large genera are related to each other, in the same manner as the varieties of any one species are related to each other. No naturalist pretends that all the species of a genus are equally distinct from each other; they may generally be divided into sub-genera, or sections, or lesser groups. As Fries has well remarked, little groups of species are generally clustered like satellites around certain other species. And what are varieties but groups of forms, unequally related to each other, and clustered round certain forms —that is, round their parent-species? Undoubtedly there is one most important point of difference between varieties and species; namely, that the amount of difference between varieties, when compared with each other or with their parent-species, is much less than that between the species of the same genus. But when we come to discuss the principle, as I call it, of Divergence of Character, we shall see how this may be explained, and how the lesser differences between varieties tend to increase into the greater differences between species.

There is one other point which is worth notice. Varieties generally have much restricted ranges: this statement is indeed scarcely more than a truism, for if a variety were found to have a wider range than that of its supposed parent-species, their denominations ought to be reversed. But there is also reason to believe that those species which are very closely allied to other species, and in so far resemble varieties, often have much restricted ranges. For instance, Mr. H. C. Watson has marked for me in the well-sifted London Catalogue of plants (4th edition) 63 plants which are therein

ranked as species, but which he considers as so closely
allied to other species as to be of doubtful value : these
63 reputed species range on an average over 6·9 of the
provinces into which Mr. Watson has divided Great
Britain. Now, in this same catalogue, 53 acknowledged
varieties are recorded, and these range over 7·7 provinces; whereas, the species to which these varieties
belong range over 14·3 provinces. So that the acknowledged varieties have very nearly the same restricted
average range, as have those very closely allied forms,
marked for me by Mr. Watson as doubtful species, but
which are almost universally ranked by British botanists
as good and true species.

Finally, then, varieties cannot be distinguished from
species,—except, firstly, by the discovery of intermediate
forms linking them together, and the occurrence of such
links does not affect the character of the forms which
they connect; and except, secondly, by a certain amount
of difference, for two forms, if differing very little, are
generally ranked as varieties, notwithstanding that intermediate linking forms have not been discovered; but
the amount of difference considered necessary to give
to two forms the rank of species is quite indefinite. In
genera having more than the average number of species
in any country, the species of these genera have more
than the average number of varieties. In large genera
the species are apt to be closely, but unequally allied
together, forming little clusters round certain other
species. Species very closely allied to other species
apparently have restricted ranges. In all these several
respects the species of large genera present a strong
analogy with varieties. And we can clearly understand
these analogies, if species have once existed as varieties,
and have thus originated; whereas, these analogies are

utterly inexplicable if each species has been independently created.

We have, also, seen that it is the most flourishing
or dominant species of the larger genera within each
class which on an average vary most; and varieties, as
wo shall hereafter see, tend to become converted into
new and distinct species. The larger genera thus tend
to become larger; and throughout nature the forms of
life which are now dominant tend to become still more
dominant by leaving many modified and dominant
descendants. But by steps hereafter to be explained,
the larger genera also tend to break up into smaller
genera. And thus, the forms of life throughout tho
universe become divided into groups subordinate to
groups.

CHAPTER III.

STRUGGLE FOR EXISTENCE.

Its bearing on natural selection — The term used in a wide sense — Geometrical ratio of increase — Rapid increase of naturalised animals and plants — Nature of the checks to increase — Competition universal — Effects of climate — Protection from the number of individuals — Complex relations of all animals and plants throughout nature — Struggle for life most severe between individuals and varieties of the same species; often severe between species of the same genus — The relation of organism to organism the most important of all relations.

BEFORE entering on the subject of this chapter, I must make a few preliminary remarks, to show how the struggle for existence bears on Natural Selection. It has been seen in the last chapter that amongst organic beings in a state of nature there is some individual variability : indeed I am not aware that this has ever been disputed. It is immaterial for us whether a multitude of doubtful forms be called species or sub-species or varieties ; what rank, for instance, the two or three hundred doubtful forms of British plants are entitled to hold, if the existence of any well-marked varieties be admitted. But the mere existence of individual variability and of some few well-marked varieties, though necessary as the foundation for the work, helps us but little in understanding how species arise in nature. How have all those exquisite adaptations of one part of the organisation to another. part, and to the conditions of life, and of one organic being to another being, been perfected ? We see these beautiful co-adaptations most plainly in

the woodpecker and the mistletoe ; and only a little less
plainly in the humblest parasite which clings to the
hairs of a quadruped or feathers of a bird ; in the structure of the beetle which dives through the water ; in the
plumed seed which is wafted by the gentlest breeze ; in
short, we see beautiful adaptations everywhere and in
every part of the organic world.

Again, it may be asked, how is it that varieties, which
I have called incipient species, become ultimately converted into good and distinct species, which in most
cases obviously differ from each other far more than do
the varieties of the same species? How do those groups
of species, which constitute what are called distinct
genera, and which differ from each other more than do
the species of the same genus, arise? All these results,
as we shall more fully see in the next chapter, follow
from the struggle for life. Owing to this struggle for
life, any variation, however slight, and from whatever
cause proceeding, if it be in any degree profitable to an
individual of any species, in its infinitely complex relations to other organic beings and to its physical conditions of life, will tend to the preservation of that individual, and will generally be inherited by its offspring.
The offspring, also, will thus have a better chance of
surviving, for, of the many individuals of any species
which are periodically born, but a small number can
survive. I have called this principle, by which each
slight variation, if useful, is preserved, by the term
Natural Selection, in order to mark its relation to man's
power of selection. We have seen that man by selection can certainly produce great results, and can adapt
organic beings to his own uses, through the accumulation of slight but useful variations, given to him by the
hand of Nature. But Natural Selection, as we shall
hereafter see, is a power incessantly ready for action,

and is as immeasurably superior to man's feeble efforts, as the works of Nature are to those of Art.

We will now discuss in a little more detail the struggle for existence. In my future work this subject shall be treated, as it well deserves, at much greater length. The elder De Candolle and Lyell have largely and philosophically shown that all organic beings are exposed to severe competition. In regard to plants, no one has treated this subject with more spirit and ability than W. Herbert, Dean of Manchester, evidently the result of his great horticultural knowledge. Nothing is easier than to admit in words the truth of the universal struggle for life, or more difficult—at least I have found it so—than constantly to bear this conclusion in mind. Yet unless it be thoroughly engrained in the mind, the whole economy of nature, with every fact on distribution, rarity, abundance, extinction, and variation, will be dimly seen or quite misunderstood. We behold the face of nature bright with gladness, we often see superabundance of food; we do not see, or we forget that the birds which are idly singing round us mostly live on insects or seeds, and are thus constantly destroying life; or we forget how largely these songsters, or their eggs, or their nestlings, are destroyed by birds and beasts of prey; we do not always bear in mind, that though food may be now superabundant, it is not so at all seasons of each recurring year.

I should premise that I use the term Struggle for Existence in a large and metaphorical sense, including dependence of one being on another, and including (which is more important) not only the life of the individual, but success in leaving progeny. Two canine animals in a time of dearth, may be truly said to struggle with each other which shall get food and live. But a plant on the edge of a desert is said to struggle

for life against the drought, though more properly it
should be said to be dependent on the moisture. A
plant which annually produces a thousand seeds, of
which on an average only one comes to maturity, may
be more truly said to struggle with the plants of the
same and other kinds which already clothe the ground.
The mistletoe is dependent on the apple and a few other
trees, but can only in a far-fetched sense be said to
struggle with these trees, for if too many of these parasites grow on the same tree, it will languish and die.
But several seedling mistletoes, growing close together
on the same branch, may more truly be said to struggle
with each other. As the mistletoe is disseminated by
birds, its existence depends on birds; and it may metaphorically be said to struggle with other fruit-bearing
plants, in order to tempt birds to devour and thus dis-
seminate its seeds rather than those of other plants.
In these several senses, which pass into each other, I
use for convenience' sake the general term of struggle
for existence.

A struggle for existence inevitably follows from the
high rate at which all organic beings tend to increase.
Every being, which during its natural lifetime produces
several eggs or seeds, must suffer destruction during some
period of its life, and during some season or occasional
year, otherwise, on the principle of geometrical increase,
its numbers would quickly become so inordinately great
that no country could support the product. Hence, as
more individuals are produced than can possibly survive, there must in every case be a struggle for existence, either one individual with another of the same
species, or with the individuals of distinct species, or
with the physical conditions of life. It is the doctrine
of Malthus applied with manifold force to the whole
animal and vegetable kingdoms; for in this case there

can be no artificial increase of food, and no prudential restraint from marriage. Although some species may be now increasing, more or less rapidly, in numbers, all cannot do so, for the world would not hold them.

There is no exception to the rule that every organic being naturally increases at so high a rate, that if not destroyed, the earth would soon be covered by the progeny of a single pair. Even slow-breeding man has doubled in twenty-five years, and at this rate, in a few thousand years, there would literally not be standing-room for his progeny. Linnæus has calculated that if an annual plant produced only two seeds—and there is no plant nearly so unproductive as this—and their seedlings next year produced two, and so on, then in twenty years there would be a million plants. The elephant is reckoned the slowest breeder of all known animals, and I have taken some pains to estimate its probable minimum rate of natural increase: it will be under the mark to assume that it breeds when thirty years old, and goes on breeding till ninety years old, bringing forth three pair of young in this interval; if this be so, at the end of the fifth century there would be alive fifteen million elephants, descended from the first pair.

But we have better evidence on this subject than mere theoretical calculations, namely, the numerous recorded cases of the astonishingly rapid increase of various animals in a state of nature, when circumstances have been favourable to them during two or three following seasons. Still more striking is the evidence from our domestic animals of many kinds which have run wild in several parts of the world: if the statements of the rate of increase of slow-breeding cattle and horses in South America, and latterly in Australia, had not been well authenticated, they would have been incredible. So it is with plants: cases could be given of

introduced plants which have become common throughout whole islands in a period of less than ten years.
Several of the plants, such as the cardoon and a tall
thistle, now most numerous over the wide plains of La
Plata, clothing square leagues of surface almost to the
exclusion of all other plants, have been introduced from
Europe; and there are plants which now range in India,
as I hear from Dr. Falconer, from Cape Comorin to
the Himalaya, which have been imported from America
since its discovery. In such cases, and endless instances
could be given, no one supposes that the fertility of
these animals or plants has been suddenly and temporarily increased in any sensible degree. The obvious
explanation is that the conditions of life have been very
favourable, and that there has consequently been less
destruction of the old and young, and that nearly all the
young have been enabled to breed. In such cases the
geometrical ratio of increase, the result of which never
fails to be surprising, simply explains the extraordinarily
rapid increase and wide diffusion of naturalised productions in their new homes.

In a state of nature almost every plant produces seed,
and amongst animals there are very few which do not
annually pair. Hence we may confidently assert, that
all plants and animals are tending to increase at a geometrical ratio, that all would most rapidly stock every
station in which they could any how exist, and that the
geometrical tendency to increase must be checked by
destruction at some period of life. Our familiarity with
the larger domestic animals tends, I think, to mislead
us: we see no great destruction falling on them, and we
forget that thousands are annually slaughtered for food,
and that in a state of nature an equal number would
have somehow to be disposed of.

The only difference between organisms which annually

produce eggs or seeds by the thousand, and those which
produce extremely few, is, that the slow-breeders would
require a few more years to people, under favourable
conditions, a whole district, let it be ever so large. The
condor lays a couple of eggs and the ostrich a score, and
yet in the same country the condor may be the more
numerous of the two : the Fulmar petrel lays but one
egg, yet it is believed to be the most numerous bird in
the world. One fly deposits hundreds of eggs, and another, like the hippobosca, a single one ; but this difference does not determine how many individuals of the
two species can be supported in a district. A large
number of eggs is of some importance to those species
which depend on a rapidly fluctuating amount of food,
for it allows them rapidly to increase in number. But
the real importance of a large number of eggs or seeds
is to make up for much destruction at some period of
life ; and this period in the great majority of cases is an
early one. If an animal can in any way protect its own
eggs or young, a small number may be produced, and
yet the average stock be fully kept up ; but if many
eggs or young are destroyed, many must be produced,
or the species will become extinct. It would suffice to
keep up the full number of a tree, which lived on an
average for a thousand years, if a single seed were produced once in a thousand years, supposing that this seed
were never destroyed, and could be ensured to germinate in a fitting place. So that in all cases, the average
number of any animal or plant depends only indirectly
on the number of its eggs or seeds.

In looking at Nature, it is most necessary to keep
the foregoing considerations always in mind—never to
forget that every single organic being around us may
be said to be striving to the utmost to increase in numbers ; that each lives by a struggle at some period of

its life; that heavy destruction inevitably falls either
on the young or old, during each generation or at
recurrent intervals. Lighten any check, mitigate the
destruction ever so little, and the number of the
species will almost instantaneously increase to any
amount.

The causes which check the natural tendency of each
species to increase in number are most obscure. Look
at the most vigorous species; by as much as it swarms
in numbers, by so much will its tendency to increase be
still further increased. We know not exactly what the
checks are in even one single instance. Nor will this
surprise any one who reflects how ignorant we are on
this head, even in regard to mankind, so incomparably
better known than any other animal. This subject has
been ably treated by several authors, and I shall, in my
future work, discuss some of the checks at considerable
length, more especially in regard to the feral animals of
South America. Here I will make only a few remarks,
just to recall to the reader's mind some of the chief
points. Eggs or very young animals seem generally to
suffer most, but this is not invariably the case. With
plants there is a vast destruction of seeds, but, from
some observations which I have made, I believe that it
is the seedlings which suffer most from germinating
in ground already thickly stocked with other plants.
Seedlings, also, are destroyed in vast numbers by various
enemies; for instance, on a piece of ground three feet
long and two wide, dug and cleared, and where there
could be no choking from other plants, I marked all the
seedlings of our native weeds as they came up, and out
of the 357 no less than 295 were destroyed, chiefly by
slugs and insects. If turf which has long been mown,
and the case would be the same with turf closely browsed
by quadrupeds, be let to grow, the more vigorous plants

gradually kill the less vigorous, though fully grown plants: thus out of twenty species growing on a little plot of turf (three feet by four) nine species perished from the other species being allowed to grow up freely.

The amount of food for each species of course gives the extreme limit to which each can increase; but very frequently it is not the obtaining food, but the serving as prey to other animals, which determines the average numbers of a species. Thus, there seems to be little doubt that the stock of partridges, grouse, and hares on any large estate depends chiefly on the destruction of vermin. If not one head of game were shot during the next twenty years in England, and, at the same time, if no vermin were destroyed, there would, in all probability, be less game than at present, although hundreds of thousands of game animals are now annually killed. On the other hand, in some cases, as with the elephant, none are destroyed by beasts of prey; for even the tiger in India most rarely dares to attack a young elephant protected by its dam.

Climate plays an important part in determining the average numbers of a species, and periodical seasons of extreme cold or drought seem to be the most effective of all checks. I estimated (chiefly from the greatly reduced numbers of nests in the spring) that the winter of 1854-55 destroyed four-fifths of the birds in my own grounds; and this is a tremendous destruction, when we remember that ten per cent. is an extraordinarily severe mortality from epidemics with man. The action of climate seems at first sight to be quite independent of the struggle for existence; but in so far as climate chiefly acts in reducing food, it brings on the most severe struggle between the individuals, whether of the same or of distinct species, which subsist on the same kind of food. Even when climate, for instance extreme cold, acts directly, it will be the least vigorous, or those which have got least food through the advancing winter, which will suffer most. When we travel from south to north, or from a damp region to a dry, we invariably see some species gradually getting rarer and rarer, and finally disappearing; and the change of climate being conspicuous, we are tempted to attribute the whole effect to its direct action. But this is a false view: we forget that each species, even where it most abounds, is constantly suffering enormous destruction at some period of its life, from enemies or from competitors for the same place and food; and if these enemies or competitors be in the least degree favoured by any slight change of climate, they will increase in numbers, and, as each area is already fully stocked with inhabitants, the other species will decrease. When we travel southward and see a species decreasing in numbers, we may feel sure that the cause lies quite as much in other species being favoured, as in this one being hurt. So it is when we travel northward, but in a somewhat lesser degree, for the number of species of all kinds, and therefore of competitors, decreases northwards; hence in going northward, or in ascending a mountain, we far oftener meet with stunted forms, due to the *directly* injurious action of climate, than we do in proceeding southwards or in descending a mountain. When we reach the Arctic regions, or snow-capped summits, or absolute deserts, the struggle for life is almost exclusively with the elements.

That climate acts in main part indirectly by favouring other species, we may clearly see in the prodigious number of plants in our gardens which can perfectly well endure our climate, but which never become naturalised, for they cannot compete with our native plants nor resist destruction by our native animals.

When a species, owing to highly favourable circumstances, increases inordinately in numbers in a small tract, epidemics—at least, this seems generally to occur with our game animals—often ensue : and here we have a limiting check independent of the struggle for life. But even some of these so-called epidemics appear to be due to parasitic worms, which have from some cause, possibly in part through facility of diffusion amongst the crowded animals, been disproportionally favoured : and here comes in a sort of struggle between the parasite and its prey.

On the other hand, in many cases, a large stock of individuals of the same species, relatively to the numbers of its enemies, is absolutely necessary for its preservation. Thus we can easily raise plenty of corn and rape-seed, &c., in our fields, because the seeds are in great excess compared with the number of birds which feed on them ; nor can the birds, though having a superabundance of food at this one season, increase in number proportionally to the supply of seed, as their numbers are checked during winter : but any one who has tried, knows how troublesome it is to get seed from a few wheat or other such plants in a garden : I have in this case lost every single seed. This view of the necessity of a large stock of the same species for its preservation, explains, I believe, some singular facts in nature, such as that of very rare plants being sometimes extremely abundant in the few spots where they do occur ; and that of some social plants being social, that is, abounding in individuals, even on the extreme confines of their range. For in such cases, we may believe, that a plant could exist only where the conditions of its life were so favourable that many could exist together, and thus save the species from utter destruction. I should add that the good effects of frequent intercrossing, and

E

the ill effects of close interbreeding, probably come
into play in some of these cases ; but on this intricate
subject I will not here enlarge.

Many cases are on record showing how complex and
unexpected are the checks and relations between organic
beings, which have to struggle together in the same
country. I will give only a single instance, which,
though a simple one, has interested me. In Staffordshire, on the estate of a relation, where I had ample
means of investigation, there was a large and extremely
barren heath, which had never been touched by the
hand of man ; but several hundred acres of exactly the
same nature had been enclosed twenty-five years previously and planted with Scotch fir. The change in the
native vegetation of the planted part of the heath was
most remarkable, more than is generally seen in passing
from one quite different soil to another : not only the
proportional numbers of the heath-plants were wholly
changed, but twelve species of plants (not counting
grasses and carices) flourished in the plantations, which
could not be found on the heath. The effect on the insects must have been still greater, for six insectivorous
birds were very common in the plantations, which were
not to be seen on the heath ; and the heath was fre-
quented by two or three distinct insectivorous birds.
Here we see how potent has been the effect of the introduction of a single tree, nothing whatever else having
been done, with the exception that the land had been
enclosed, so that cattle could not enter. But how important an element enclosure is, I plainly saw near
Farnham, in Surrey. Here there are extensive heaths,
with a few clumps of old Scotch firs on the distant hill-
tops : within the last ten years large spaces have been
enclosed, and self-sown firs are now springing up in
multitudes, so close together that all cannot live.

When I ascertained that these young trees had not been sown or planted, I was so much surprised at their numbers that I went to several points of view, whence I could examine hundreds of acres of the unenclosed heath, and literally I could not see a single Scotch fir, except the old planted clumps. But on looking closely between the stems of the heath, I found a multitude of seedlings and little trees, which had been perpetually browsed down by the cattle. In one square yard, at a point some hundred yards distant from one of the old clumps, I counted thirty-two little trees; and one of them, with twenty-six rings of growth, had during many years tried to raise its head above the stems of the heath, and had failed. No wonder that, as soon as the land was enclosed, it became thickly clothed with vigorously growing young firs. Yet the heath was so extremely barren and so extensive that no one would ever have imagined that cattle would have so closely and effectually searched it for food.

Here we see that cattle absolutely determine the existence of the Scotch fir; but in several parts of the world insects determine the existence of cattle. Perhaps Paraguay offers the most curious instance of this; for here neither cattle nor horses nor dogs have ever run wild, though they swarm southward and northward in a feral state; and Azara and Rengger have shown that this is caused by the greater number in Paraguay of a certain fly, which lays its eggs in the navels of these animals when first born. The increase of these flies, numerous as they are, must be habitually checked by some means, probably by other parasitic insects. Hence, if certain insectivorous birds were to decrease in Paraguay, the parasitic insects would probably increase; and this would lessen the number of the navel-frequenting flies—then cattle and horses would become feral, and this

E 2

would certainly greatly alter (as indeed I have observed in parts of South America) the vegetation: this again would largely affect the insects; and this, as we have just seen in Staffordshire, the insectivorous birds, and so onwards in ever-increasing circles of complexity. We began this series by insectivorous birds, and we have ended with them. Not that in nature the relations can ever be as simple as this. Battle within battle must ever be recurring with varying success; and yet in the long-run the forces are so nicely balanced, that the face of nature remains uniform for long periods of time, though assuredly the merest trifle would often give the victory to one organic being over another. Nevertheless, so profound is our ignorance, and so high our presumption, that we marvel when we hear of the extinction of an organic being; and as we do not see the cause, we invoke cataclysms to desolate the world, or invent laws on the duration of the forms of life!

I am tempted to give one more instance showing how plants and animals, most remote in the scale of nature, are bound together by a web of complex relations. I shall hereafter have occasion to show that the exotic Lobelia fulgens, in this part of England, is never visited by insects, and consequently, from its peculiar structure, never can set a seed. Many of our orchidaceous plants absolutely require the visits of moths to remove their pollen-masses and thus to fertilise them. I find from experiments that humble-bees are almost indispensable to the fertilisation of the heartsease (Viola tricolor), for other bees do not visit this flower. I have also found that the visits of bees are necessary for the fertilisation of some kinds of clover: for instance, 20 heads of Dutch clover (Trifolium repens) yielded 2,290 seeds; but 20 other heads protected from bees produced not one. Again, 100 heads of red clover (T. pratense) pro-

duced 2,700 seeds, but the same number of protected
heads produced not a single seed. Humble-bees alone
visit red clover, as other bees cannot reach the nectar.
It has been suggested that moths may serve to fertilise
the clovers; but I doubt this in the case of the red
clover, from their weight being apparently not sufficient
to depress the wing-petals. Hence we may infer as
highly probable that if the whole genus of humble-bees
became extinct or very rare in England, the heartsease
and red clover would become very rare, or wholly disappear. The number of humble-bees in any district
depends in a great degree on the number of field-mice,
which destroy their combs and nests; and Mr. H. Newman, who has long attended to the habits of humble-
bees, believes that "more than two-thirds of them are
thus destroyed all over England." Now the number of
mice is largely dependent, as every one knows, on the
number of cats; and Mr. Newman says, " Near villages
and small towns I have found the nests of humble-bees
more numerous than elsewhere, which I attribute to the
number of cats that destroy the mice." Hence it is
quite credible that the presence of a feline animal in
large numbers in a district might determine, through
the intervention first of mice and then of bees, the frequency of certain flowers in that district!

In the case of every species, many different checks,
acting at different periods of life, and during different
seasons or years, probably come into play; some one
check or some few being generally the most potent, but
all concur in determining the average number or even
the existence of the species. In some cases it can be
shown that widely-different checks act on the same
species in different districts. When we look at the
plants and bushes clothing an entangled bank, we are
tempted to attribute their proportional numbers and

kinds to what we call chance. But how false a view
is this! Every one has heard that when an American
forest is cut down, a very different vegetation springs
up; but it has been observed that ancient Indian ruins
in the Southern United States, which must formerly
have been cleared of trees, now display the same beautiful diversity and proportion of kinds as in the surrounding virgin forests. What a struggle between the several
kinds of trees must here have gone on during long centuries, each annually scattering its seeds by the thousand; what war between insect and insect—between
insects, snails, and other animals with birds and beasts
of prey—all striving to increase, and all feeding on each
other or on the trees or their seeds and seedlings, or on
the other plants which first clothed the ground and thus
checked the growth of the trees! Throw up a handful
of feathers, and all must fall to the ground according to
definite laws; but how simple is the problem where
each shall fall compared to that of the action and reaction of the innumerable plants and animals which
have determined, in the course of centuries, the proportional numbers and kinds of trees now growing on the
old Indian ruins!

The dependency of one organic being on another, as
of a parasite on its prey, lies generally between beings
remote in the scale of nature. This is often the case
with those which may strictly be said to struggle with
each other for existence, as in the case of locusts and
grass-feeding quadrupeds. But the struggle almost invariably will be most severe between the individuals of
the same species, for they frequent the same districts,
require the same food, and are exposed to the same
dangers. In the case of varieties of the same species,
the struggle will generally be almost equally severe,
and we sometimes see the contest soon decided: for

instance, if several varieties of wheat be sown together, and the mixed seed be resown, some of tho varieties which best suit tho soil or climate, or are naturally tho most fertile, will beat the others and so yield more seed, and will consequently in a few years quite supplant the other varieties. To keep up a mixed stock of even such extremely close varieties as the variously coloured sweet-peas, they must be each year harvested separately, and the seed then mixed in due proportion, otherwise the weaker kinds will steadily decrease in numbers and disappear. So again with the varieties of sheep: it has been asserted that certain mountain-varieties will starve out other mountain-varieties, so that they cannot be kept together. The same result has followed from keeping together different varieties of the medicinal leech. It may even be doubted whether the varieties of any one of our domestic plants or animals have so exactly tho same strength, habits, and constitution, that the original proportions of a mixed stock could be kept up for half-a-dozen generations, if they were allowed to struggle together, like beings in a state of nature, and if the seed or young were not annually sorted.

As species of the same genus have usually, though by no means invariably, some similarity in habits and constitution, and always in structure, tho struggle will generally be more severe between species of the same genus, when they come into competition with each other, than between species of distinct genera. We see this in the recent extension over parts of tho United States of one species of swallow having caused the decrease of another species. The recent increase of tho missel-thrush in parts of Scotland has caused the decrease of the song-thrush. How frequently we hear of one species of rat taking tho place of another species under the

most different climates! In Russia the small Asiatic
cockroach has everywhere driven before it its great
congener. One species of charlock will supplant another,
and so in other cases. We can dimly see why the competition should be most severe between allied forms,
which fill nearly the same place in the economy of nature;
but probably in no one case could we precisely say why
one species has been victorious over another in the
great battle of life.

A corollary of the highest importance may be deduced
from the foregoing remarks, namely, that the structure
of every organic being is related, in the most essential
yet often hidden manner, to that of all other organic
beings, with which it comes into competition for food or
residence, or from which it has to escape, or on which it
preys. This is obvious in the structure of the teeth and
talons of the tiger; and in that of the legs and claws of
the parasite which clings to the hair on the tiger's body.
But in the beautifully plumed seed of the dandelion,
and in the flattened and fringed legs of the water-beetle,
the relation seems at first confined to the elements of
air and water. Yet the advantage of plumed seeds no
doubt stands in the closest relation to the land being
already thickly clothed by other plants; so that the
seeds may be widely distributed and fall on unoccupied ground. In the water-beetle, the structure of
its legs, so well adapted for diving, allows it to
compete with other aquatic insects, to hunt for its own
prey, and to escape serving as prey to other animals.

The store of nutriment laid up within the seeds of
many plants seems at first sight to have no sort of
relation to other plants. But from the strong growth
of young plants produced from such seeds (as peas
and beans), when sown in the midst of long grass,
it may be suspected that the chief use of the nutriment

in the seed is to favour the growth of the young seedling, whilst struggling with other plants growing vigorously all around.

Look at a plant in the midst of its range, why does it not double or quadruple its numbers? We know that it can perfectly well withstand a little more heat or cold, dampness or dryness, for elsewhere it ranges into slightly hotter or colder, damper or drier districts. In this case we can clearly see that if we wished in imagination to give the plant the power of increasing in number, we should have to give it some advantage over its competitors, or over the animals which preyed on it. On the confines of its geographical range, a change of constitution with respect to climate would clearly be an advantage to our plant; but we have reason to believe that only a few plants or animals range so far, that they are destroyed by the rigour of the climate alone. Not until we reach the extreme confines of life, in the Arctic regions or on the borders of an utter desert, will competition cease. The land may be extremely cold or dry, yet there will be competition between some few species, or between the individuals of the same species, for the warmest or dampest spots.

Hence, also, we can see that when a plant or animal is placed in a new country amongst new competitors, though the climate may be exactly the same as in its former home, yet the conditions of its life will generally be changed in an essential manner. If we wished to increase its average numbers in its new home, we should have to modify it in a different way to what we should have to do in its native country; for we should have to give it some advantage over a different set of competitors or enemies.

It is good thus to try in our imagination to give any form some advantage over another. Probably in no single instance should we know what to do, so as to succeed. It will convince us of our ignorance on the mutual relations of all organic beings; a conviction as necessary, as it seems to be difficult to acquire. All that we can do, is to keep steadily in mind that each organic being is striving to increase in a geometrical ratio; that each at some period of its life, during some season of the year, during each generation or at intervals, has to struggle for life, and to suffer great destruction. When we reflect on this struggle, we may console ourselves with the full belief, that the war of nature is not incessant, that no fear is felt, that death is generally prompt, and that the vigorous, the healthy, and the happy survive and multiply.

CHAPTER IV.

NATURAL SELECTION.

Natural Selection — its power compared with man's selection — its power on characters of trifling importance — its power at all ages and on both sexes — Sexual Selection — On the generality of intercrosses between individuals of the same species — Circumstances favourable and unfavourable to Natural Selection, namely, intercrossing, isolation, number of individuals — Slow action — Extinction caused by Natural Selection — Divergence of Character, related to the diversity of inhabitants of any small area, and to naturalisation — Action of Natural Selection, through Divergence of Character and Extinction, on the descendants from a common parent — Explains the Grouping of all organic beings — Advance in organisation — Low forms preserved — Objections considered — Indefinite multiplication of species — Summary.

How will the struggle for existence, discussed too briefly in the last chapter, act in regard to variation? Can the principle of selection, which we have seen is so potent in the hands of man, apply in nature? I think we shall see that it can act most effectually. Let it be borne in mind in what an endless number of strange peculiarities our domestic productions, and, in a lesser degree, those under nature, vary; and how strong the hereditary tendency is. Under domestication, it may be truly said that the whole organisation becomes in some degree plastic. But the variability, which we almost universally meet with in our domestic productions, is not directly produced, as Hooker and Asa Gray have well remarked, by man; he can neither originate varieties, nor prevent their occurrence; he can only preserve and accumulate such as do occur; unintentionally he exposes organic beings to new and changing conditions of life, and variability ensues; but similar changes of conditions might and do occur under nature. Let it also be borne in mind how infinitely complex and close-fitting are the mutual relations of all organic beings to each other and to their physical conditions of life; and consequently what infinitely varied diversities of structure may be of use to each being under changing conditions of life. Can it, then, be thought improbable, seeing that variations useful to man have undoubtedly occurred, that other variations useful in some way to each being in the great and complex battle of life, should sometimes occur in the course of thousands of generations? If such do occur, can we doubt (remembering that many more individuals are born than can possibly survive) that individuals having any advantage, however slight, over others, would have the best chance of surviving and of procreating their kind? On the other hand, we may feel sure that any variation in the least degree injurious would be rigidly destroyed. This preservation of favourable variations and the rejection of injurious variations, I call Natural Selection. Variations neither useful nor injurious would not be affected by natural selection, and would be left a fluctuating element, as perhaps we see in the species called polymorphic.

Several writers have misapprehended or objected to the term Natural Selection. Some have even imagined that natural selection induces variability, whereas it implies only the preservation of such variations as occur and are beneficial to the being under its conditions of life. No one objects to agriculturists speaking of the potent effects of man's selection; and in this case the individual differences given by nature, which man for some object selects, must of necessity first occur. Others have objected that the term selection implies conscious choice in the animals which become modified; and it has even been urged that as plants have no volition, natural selection is not applicable to them! In the literal sense of the word, no doubt, natural selection is a misnomer; but who ever objected to chemists speaking of the elective affinities of the various elements?—and yet an acid cannot strictly be said to elect the base with which it will in preference combine. It has been said that I speak of natural selection as an active power or Deity; but who objects to an author speaking of the attraction of gravity as ruling the movements of the planets? Every one knows what is meant and is implied by such metaphorical expressions; and they are almost necessary for brevity. So again it is difficult to avoid personifying the word Nature; but I mean by Nature, only the aggregate action and product of many natural laws, and by laws the sequence of events as ascertained by us. With a little familiarity such superficial objections will be forgotten.

We shall best understand the probable course of natural selection by taking the case of a country undergoing some physical change, for instance, of climate. The proportional numbers of its inhabitants would almost immediately undergo a change, and some species might become extinct. We may conclude, from what we have seen of the intimate and complex manner in which the inhabitants of each country are bound together, that any change in the numerical proportions of some of the inhabitants, independently of the change of climate itself, would seriously affect many of the others. If the country were open on its borders, new forms would certainly immigrate, and this also would seriously disturb the relations of some of the former inhabitants. Let it be remembered how powerful the

influence of a single introduced tree or mammal has
been shown to be. But in the case of an island, or of a
country partly surrounded by barriers, into which new
and better adapted forms could not freely enter, we
should then have places in the economy of nature which
would assuredly be better filled up, if some of the original inhabitants were in some manner modified; for,
had the area been open to immigration, these same
places would have been seized on by intruders. In such
case, every slight modification, which in the course of
ages chanced to arise, and which in any way favoured
the individuals of any of the species, by better adapting
them to their altered conditions, would tend to be preserved; and natural selection would thus have free scope
for the work of improvement.

We have reason to believe, as stated in the first
chapter, that a change in the conditions of life, by
specially acting on the reproductive system, causes or
increases variability; and in the foregoing case the
conditions of life are supposed to have undergone a
change, and this would manifestly be favourable to
natural selection, by giving a better chance of profitable
variations occurring; and unless profitable variations do
occur, natural selection can do nothing. Not that any
extreme amount of variability is necessary; as man can
certainly produce great results by adding up in any
given direction mere individual differences, so could
natural selection, but far more easily, from having incomparably longer time for action. Nor do I believe
that any great physical change, as of climate, or any
unusual degree of isolation to check immigration, is
actually necessary to produce new and unoccupied places
for natural selection to fill up by modifying and improving some of the varying inhabitants. For as all
the inhabitants of each country are struggling together

with nicely balanced forces, extremely slight modifications in the structure or habits of one inhabitant
would often give it an advantage over others; and still
further modifications of the same kind would often still
further increase the advantage, as long as the being
continued under the same conditions of life and profited
by similar means of subsistence and defence. No country
can be named in which all the native inhabitants are
now so perfectly adapted to each other and to the
physical conditions under which they live, that none of
them could anyhow be improved; for in all countries,
the natives have been so far conquered by naturalised
productions, that they have allowed foreigners to take
firm possession of the land. And as foreigners have
thus everywhere beaten some of the natives, we may
safely conclude that the natives might have been modified with advantage, so as to have better resisted such
intruders.

As man can produce and certainly has produced a
great result by his methodical and unconscious means
of selection, what may not natural selection effect?
Man can act only on external and visible characters:
Nature (if I may be allowed thus to personify the natural preservation of varying and favoured individuals
during the struggle for existence) cares nothing for
appearances, except in so far as they are useful to
any being. She can act on every internal organ, on
every shade of constitutional difference, on the whole
machinery of life. Man selects only for his own good;
Nature only for that of the being which she tends.
Every selected character is fully exercised by her; and
the being is placed under well-suited conditions of life.
Man keeps the natives of many climates in the same
country; he seldom exercises each selected character
in some peculiar and fitting manner; he feeds a long

and a short beaked pigeon on the same food; he does not exercise a long-backed or long-legged quadruped in any peculiar manner; he exposes sheep with long and short wool to the same climate. He does not allow the most vigorous males to struggle for the females. He does not rigidly destroy all inferior animals, but protects during each varying season, as far as lies in his power, all his productions. He often begins his selection by some half-monstrous form; or at least by some modification prominent enough to catch his eye, or to be plainly useful to him. Under nature, the slightest difference of structure or constitution may well turn the nicely-balanced scale in the struggle for life, and so be preserved. How fleeting are the wishes and efforts of man! how short his time! and consequently how poor will his products be, compared with those accumulated by Nature during whole geological periods! Can we wonder, then, that Nature's productions should be far "truer" in character than man's productions; that they should be infinitely better adapted to the most complex conditions of life, and should plainly bear the stamp of far higher workmanship?

It may metaphorically be said that natural selection is daily and hourly scrutinising, throughout the world, every variation, even the slightest; rejecting that which is bad, preserving and adding up all that is good; silently and insensibly working, whenever and wherever opportunity offers, at the improvement of each organic being in relation to its organic and inorganic conditions of life. We see nothing of these slow changes in progress, until the hand of time has marked the long lapse of ages, and then so imperfect is our view into long past geological ages, that we only see that the forms of life are now different from what they formerly were.

In order that any great amount of modification should thus in the course of time be produced, it is necessary to believe that when a variety has once arisen, it again varies, after perhaps a long interval of time; and that its varieties, if favourable, are again preserved, and so onwards. That varieties more or less different from the parent-stock occasionally arise, few will deny; but that the process of variation should be thus indefinitely prolonged is an assumption, the truth of which must be judged of by how far the hypothesis accords with and explains the general phenomena of nature. On the other hand, the ordinary belief that the amount of possible variation is a strictly limited quantity is likewise a simple assumption.

Although natural selection can act only through and for the good of each being, yet characters and structures, which we are apt to consider as of very trifling importance, may thus be acted on. When we see leaf-eating insects green, and bark-feeders mottled-grey; the alpine ptarmigan white in winter, the red-grouse the colour of heather, and the black-grouse that of peaty earth, we must believe that these tints are of service to these birds and insects in preserving them from danger. Grouse, if not destroyed at some period of their lives, would increase in countless numbers; they are known to suffer largely from birds of prey; and hawks are guided by eyesight to their prey—so much so, that on parts of the Continent persons are warned not to keep white pigeons, as being the most liable to destruction. Hence I can see no reason to doubt that natural selection might be most effective in giving the proper colour to each kind of grouse, and in keeping that colour, when once acquired, true and constant. Nor ought we to think that the occasional destruction of an animal of any particular colour would produce little effect : we should remember how essential it is in a flock of white sheep to destroy every lamb with the faintest trace of black. We have seen how in Florida the colour of the hogs, when feeding on the "paint root," determines whether they shall live or die. In plants the down on the fruit and the colour of the flesh are considered by botanists as characters of the most trifling importance: yet we hear from an excellent horticulturist, Downing, that in the United States smooth-skinned fruits suffer far more from a beetle, a curculio, than those with down; that purple plums suffer far more from a certain disease than yellow plums; whereas another disease attacks yellow-fleshed peaches far more than those with other coloured flesh. If, with all the aids of art, these slight differences make a great difference in cultivating the several varieties, assuredly, in a state of nature, where the trees would have to struggle with other trees and with a host of enemies, such differences would effectually settle which variety, whether a smooth or downy, a yellow or purple fleshed fruit, should succeed.

In looking at many small points of difference between species, which, as far as our ignorance permits us to judge, seem quite unimportant, we must not forget that climate, food, &c., probably produce some slight and direct effect. It is, however, far more necessary to bear in mind that there are many unknown laws of correlation of growth, which, when one part of the organisation is modified through variation and the modifications are accumulated by natural selection for the good of the being, cause other modifications, often of the most unexpected nature.

As we see that those variations which under domestication appear at any particular period of life, tend to reappear in the offspring at the same period;—for instance, in the shape, size, and flavour of the seeds of the many varieties of our culinary and agricultural plants; in the caterpillar and cocoon stages of the varieties of the silkworm; in the eggs of poultry, and in the colour of the down of their chickens; in the horns of our sheep and cattle when nearly adult;—so in a state of nature, natural selection will be enabled to act on and modify organic beings at any age, by the accumulation of variations profitable at that age, and by their inheritance at a corresponding age. If it profit a plant to have its seeds more and more widely disseminated by the wind, I can see no greater difficulty in this being effected through natural selection, than in the cotton-planter increasing and improving by selection the down in the pods on his cotton-trees. Natural selection may modify and adapt the larva of an insect to a score of contingencies, wholly different from those which concern the mature insect. These modifications will no doubt affect, through the laws of correlation, the structure of the adult; and probably in the case of those insects which live only for a few hours, and which never feed, a large part of their structure is merely the correlated result of successive changes in the structure of their larvæ. So, conversely, modifications in the adult will probably often affect the structure of the larva; but in all cases natural selection will ensure that modifications consequent on other modifications at a different period of life, shall not be in the least degree injurious: for if they became so, they would cause the extinction of the species.

Natural selection will modify the structure of the young in relation to the parent, and of the parent in relation to the young. In social animals it will adapt the structure of each individual for the benefit of the community; if each in consequence profits by the selected change. What natural selection cannot do, is

to modify the structure of one species, without giving it
any advantage, for the good of another species; and
though statements to this effect may be found in works
of natural history, I cannot find one case which will
bear investigation. A structure used only once in an
animal's whole life, if of high importance to it, might
be modified to any extent by natural selection; for instance, the great jaws possessed by certain insects, used
exclusively for opening the cocoon—or the hard tip to
the beak of nestling birds, used for breaking the egg.
It has been asserted, that of the best short-beaked
tumbler-pigeons more perish in the egg than are able
to get out of it; so that fanciers assist in the act of
hatching. Now, if nature had to make the beak of a
full-grown pigeon very short for the bird's own advantage, the process of modification would be very slow,
and there would be simultaneously the most rigorous
selection of the young birds within the egg, which had
the most powerful and hardest beaks, for all with weak
beaks would inevitably perish: or, more delicate and
more easily broken shells might be selected, the thickness of the shell being known to vary like every other
structure.

Sexual Selection.—Inasmuch as peculiarities often
appear under domestication in one sex and become here-
ditarily attached to that sex, the same fact probably
occurs under nature, and if so, natural selection will be
able to modify one sex in its functional relations to the
other sex, or in relation to wholly different habits of life
in the two sexes, as is sometimes the case with insects.
And this leads me to say a few words on what I call
Sexual Selection. This depends, not on a struggle for
existence, but on a struggle between the males for possession of the females; the result is not death to the

unsuccessful competitor, but few or no offspring. Sexual selection is, therefore, less rigorous than natural selection. Generally, the most vigorous males, those which are best fitted for their places in nature, will leave most progeny. But in many cases, victory depends not on general vigour, but on having special weapons, confined to the male sex. A hornless stag or spurless cock would have a poor chance of leaving offspring. Sexual selection by always allowing the victor to breed might surely give indomitable courage, length to the spur, and strength to the wing to strike in the spurred leg, as well as the brutal cockfighter, who knows well that he can improve his breed by careful selection of the best cocks. How low in the scale of nature the law of battle descends, I know not; male alligators have been described as fighting, bellowing, and whirling round, like Indians in a war-dance, for the possession of the females; male salmons have been seen fighting all day long; male stag-beetles sometimes bear wounds from the huge mandibles of other males. The war is, perhaps, severest between the males of polygamous animals, and these seem oftenest provided with special weapons. The males of carnivorous animals are already well armed; though to them and to others, special means of defence may be given through means of sexual selection, as the mane to the lion, the shoulder-pad to the boar, and the hooked jaw to the male salmon; for the shield may be as important for victory, as the sword or spear.

Amongst birds, the contest is often of a more peaceful character. All those who have attended to the subject, believe that there is the severest rivalry between the males of many species to attract by singing the females. The rock-thrush of Guiana, birds of Paradise, and some others, congregate; and successive males display their

gorgeous plumage and perform strange antics before tho
females, which, standing by as spectators, at last choose
the most attractive partner. Those who have closely
attended to birds in confinement well know that they
often take individual preferences and dislikes: thus
Sir R. Heron has described how one pied peacock was
eminently attractive to all his hen birds. It may
appear childish to attribute any effect to such apparently weak means: I cannot here enter on tho details
necessary to support this view; but if man can in a
short time give elegant carriage and beauty to his
bantams, according to his standard of beauty, I can see
no good reason to doubt that female birds, by selecting,
during thousands of generations, the most melodious
or beautiful males, according to their standard of beauty,
might produce a marked effect. I strongly suspect that
some well-known laws, with respect to the plumage of
male and female birds, in comparison with the plumage
of the young, can be explained on the view of plumage
having been chiefly modified by sexual selection, acting
when the birds have come to the breeding age or
during the breeding season; the modifications thus
produced being inherited at corresponding ages or
seasons, either by the males alone, or by the males
and females; but I have not space here to enter on
this subject.

Thus it is, as I believe, that when the males and
females of any animal have the same general habits
of life, but differ in structure, colour, or ornament, such
differences have been mainly caused by sexual selection; that is, individual males have had, in successive
generations, some slight advantage over other males,
in their weapons, means of defence, or charms; and
have transmitted these advantages to their male offspring. Yet, I would not wish to attribute all such

sexual differences to this agency: for we see peculiarities arising and becoming attached to the male sex
in our domestic animals (as the wattle in male carriers,
horn-like protuberances in the cocks of certain fowls,
&c.), which we cannot believe to be either useful to the
males in battle, or attractive to the females. We see
analogous cases under nature, for instance, the tuft of
hair on the breast of the turkey-cock, which can hardly
be either useful or ornamental to this bird;—indeed,
had the tuft appeared under domestication, it would
have been called a monstrosity.

Illustrations of the action of Natural Selection.—In
order to make it clear how, as I believe, natural selection acts, I must beg permission to give one or two
imaginary illustrations. Let us take the case of a wolf,
which preys on various animals, securing some by craft,
some by strength, and some by fleetness; and let us
suppose that the fleetest prey, a deer for instance, had
from any change in the country increased in numbers,
or that other prey had decreased in numbers, during
that season of the year when the wolf is hardest pressed
for food. Under such circumstances the swiftest and
slimmest wolves would have the best chance of surviving, and so be preserved or selected,—provided
always that they retained strength to master their prey
at this or at some other period of the year, when they
might be compelled to prey on other animals. I can
see no more reason to doubt this, than that man can
improve the fleetness of his greyhounds by careful and
methodical selection, or by that unconscious selection
which results from each man trying to keep the best
dogs without any thought of modifying the breed.
    Even without any change in the proportional numbers
of the animals on which our wolf preyed, a cub might

be born with an innate tendency to pursue certain kinds of prey. Nor can this be thought very improbable; for we often observe great differences in the natural tendencies of our domestic animals; one cat, for instance, taking to catch rats, another mice; one cat, according to Mr. St. John, bringing home winged game, another hares or rabbits, and another hunting on marshy ground and almost nightly catching woodcocks or snipes. The tendency to catch rats rather than mice is known to be inherited. Now, if any slight innate change of habit or of structure benefited an individual wolf, it would have the best chance of surviving and of leaving offspring. Some of its young would probably inherit the same habits or structure, and by the repetition of this process, a new variety might be formed which would either supplant or coexist with the parent form of wolf. Or, again, the wolves inhabiting a mountainous district, and those frequenting the lowlands, would naturally be forced to hunt different prey; and from the continued preservation of the individuals best fitted for the two sites, two varieties would slowly be formed. These varieties would cross and blend where they met; but to this subject of intercrossing we shall soon have to return. I may add, that, according to Mr. Pierce, there are two varieties of the wolf inhabiting the Catskill Mountains in the United States, one with a light greyhound-like form, which pursues deer, and the other more bulky, with shorter legs, which more frequently attacks the shepherd's flocks.

Let us now take a more complex case. Certain plants excrete a sweet juice, apparently for the sake of eliminating something injurious from their sap: this is effected by glands at the base of the stipules in some Leguminosæ, and at the back of the leaf of the common laurel. This juice, though small in quantity, is greedily sought by insects. Let us now suppose a little sweet juice or nectar to be excreted by the inner bases of the petals of a flower. In this case insects in seeking the nectar would get dusted with pollen, and would certainly often transport the pollen from one flower to the stigma of another flower. The flowers of two distinct individuals of the same species would thus get crossed ; and the act of crossing, we have good reason to believe (as will hereafter be more fully alluded to), would produce very vigorous seedlings, which consequently would have the best chance of flourishing and surviving. Some of these seedlings would almost certainly inherit the nectar-excreting power. Those individual flowers which had the largest glands or nectaries, and which excreted most nectar, would be oftenest visited by insects, and would be oftenest crossed ; and so in the long-run would gain the upper hand. Those flowers, also, which had their stamens and pistils placed, in relation to the size and habits of the particular insects which visited them, so as to favour in any degree the transportal of their pollen from flower to flower, would likewise be favoured or selected. We might have taken the case of insects visiting flowers for the sake of collecting pollen instead of nectar ; and as pollen is formed for the sole object of fertilisation, its destruction appears a simple loss to the plant ; yet if a little pollen were carried, at first occasionally and then habitually, by the pollen-devouring insects from flower to flower, and a cross thus effected, although nine-tenths of the pollen were destroyed, it might still be a great gain to the plant ; and those individuals which produced more and more pollen, and had larger and larger anthers, would be selected.

When our plant, by this process of the continued preservation or natural selection of more and more attractive flowers, had been rendered highly attractive to insects, they would, unintentionally on their part,

F

regularly carry pollen from flower to flower; and that they can most effectually do this, I could easily show by many striking instances. I will give only one—not as a very striking case, but as likewise illustrating one step in the separation of the sexes of plants, presently to be alluded to. Some holly-trees bear only male flowers, which have four stamens producing a rather small quantity of pollen, and a rudimentary pistil; other holly-trees bear only female flowers; these have a full-sized pistil, and four stamens with shrivelled anthers, in which not a grain of pollen can be detected. Having found a female tree exactly sixty yards from a male tree, I put the stigmas of twenty flowers, taken from different branches, under the microscope, and on all, without exception, there were pollen grains, and on some a profusion of pollen. As the wind had set for several days from the female to the male tree, the pollen could not thus have been carried. The weather had been cold and boisterous, and therefore not favourable to bees, nevertheless every female flower which I examined had been effectually fertilised by the bees, accidentally dusted with pollen, having flown from tree to tree in search of nectar. But to return to our imaginary case: as soon as the plant had been rendered so highly attractive to insects that pollen was regularly carried from flower to flower, another process might commence. No naturalist doubts the advantage of what has been called the "physiological division of labour;" hence we may believe that it would be advantageous to a plant to produce stamens alone in one flower or on one whole plant, and pistils alone in another flower or on another plant. In plants under culture and placed under new conditions of life, sometimes the male organs and sometimes the female organs become more or less impotent; now if we suppose this to

occur in ever so slight a degree under nature, then as
pollen is already carried regularly from flower to flower,
and as a more complete separation of the sexes of our
plant would be advantageous on the principle of the
division of labour, individuals with this tendency more
and more increased, would be continually favoured or
selected, until at last a complete separation of the sexes
would be effected.

Let us now turn to the nectar-feeding insects in our
imaginary case: we may suppose the plant of which
we have been slowly increasing the nectar by continued
selection, to be a common plant; and that certain insects depended in main part on its nectar for food. I
could give many facts, showing how anxious bees are
to save time; for instance, their habit of cutting holes
and sucking the nectar at the bases of certain flowers,
which they can, with a very little more trouble, enter
by the mouth. Bearing such facts in mind, I can see
no reason to doubt that an accidental deviation in the
size and form of the body, or in the curvature and
length of the proboscis, &c., far too slight to be appre-
ciated by us, might profit a bee or other insect, so that
an individual so characterised would be able to obtain
its food more quickly, and so have a better chance of
living and leaving descendants. Its descendants would
probably inherit a tendency to a similar slight deviation
of structure. The tubes of the corollas of the common
red and incarnate clovers (Trifolium pratense and in-
carnatum) do not on a hasty glance appear to differ in
length; yet the hive-bee can easily suck the nectar out
of the incarnate clover, but not out of the common red
clover, which is visited by humble-bees alone; so that
whole fields of the red clover offer in vain an abundant
supply of precious nectar to the hive-bee. The differ-

ence in the length of the corolla which determines the
visits of the hive-bee must be very trifling; for I have
been informed, that when the red clover has been mown,
the flowers of the second crop are somewhat smaller,
aud that these are abundantly visited by hive-bees.
Thus it might be a great advantage to the hive-bee to
have a slightly longer or differently constructed proboscis. On the other hand, the fertility of clover, as
previously stated, depends on bees moving the petals,
so as to push the pollen on to the stigmatic surface.
Hence, again, if humble-bees were to become rare in
any country, it might be a great advantage to the red
clover to have a shorter or more deeply divided tube to
its corolla, so that the hive-bee could visit its flowers.
Thus I can understand how a flower and a bee might
slowly become, either simultaneously or one after the
other, modified and adapted in the most perfect manner
to each other, by the continued preservation of individuals presenting mutual and slightly favourable deviations of structure.

I am well aware that this doctrine of natural selection,
exemplified in the above imaginary instances, is open to
the same objections which were at first urged against
Sir Charles Lyell's noble views on "the modern changes
of the earth, as illustrative of geology;" but we now
seldom hear the action, for instance, of the coast-waves,
called a trifling and insignificant cause, when applied
to the excavation of gigantic valleys or to the formation
of the longest lines of inland cliffs. Natural selection
can act only by the preservation and accumulation of
small inherited modifications, each profitable to the
preserved being; and as modern geology has almost
banished such views as the excavation of a great
valley by a single diluvial wave, so will natural selec-

tion, if it be a true principle, banish the belief of the continued creation of new organic beings, or of any great and sudden modification in their structure.

On the Intercrossing of Individuals.—I must here introduce a short digression. In the case of animals and plants with separated sexes, it is of course obvious that two individuals must always (with the exception of the curious and not well-understood cases of partheno-genesis) unite for each birth; but in the case of her-maphrodites this is far from obvious. Nevertheless I am strongly inclined to believe that with all hermaphrodites two individuals, either occasionally or habitually, concur for the reproduction of their kind. This view was first suggested by Andrew Knight. We shall presently see its importance; but I must here treat the subject with extreme brevity, though I have the materials prepared for an ample discussion. All vertebrate animals, all insects, and some other large groups of animals, pair for each birth. Modern research has much diminished the number of supposed hermaphrodites, and of real hermaphrodites a large number pair; that is, two indivi-duals regularly unite for reproduction, which is all that concerns us. But still there are many hermaphrodite animals which certainly do not habitually pair, and a vast majority of plants are hermaphrodites. What reason, it may be asked, is there for supposing in these cases that two individuals ever concur in reproduction? As it is impossible here to enter on details, I must trust to some general considerations alone.

In the first place, I have collected so large a body of facts, showing, in accordance with the almost universal belief of breeders, that with animals and plants a cross between different varieties, or between individuals of the same variety but of another strain, gives vigour and fertility to the offspring; and on the other hand, that *close* interbreeding diminishes vigour and fertility; that these facts alone incline me to believe that it is a general law of nature (utterly ignorant though we be of the meaning of the law) that no organic being fertilises itself for an eternity of generations; but that a cross with another individual is occasionally—perhaps at very long intervals—indispensable.

On the belief that this is a law of nature, we can, I think, understand several large classes of facts, such as the following, which on any other view are inexplicable. Every hybridizer knows how unfavourable exposure to wet is to the fertilisation of a flower, yet what a multitude of flowers have their anthers and stigmas fully exposed to the weather! but if an occasional cross be indispensable, the fullest freedom for the entrance of pollen from another individual will explain this state of exposure, more especially as the plant's own anthers and pistil generally stand so close together that self-fertilisation seems almost inevitable. Many flowers, on the other hand, have their organs of fructification closely enclosed, as in the great papilionaceous or pea-family; but in most of these flowers there is a very curious adaptation between their structure and the manner in which bees suck the nectar; for, in doing this, they either push the flower's own pollen on the stigma, or bring pollen from another flower. So necessary are the visits of bees to many papilionaceous flowers, that I have found, by experiments published elsewhere, that their fertility is greatly diminished if these visits be prevented. Now, it is scarcely possible that bees should fly from flower to flower, and not carry pollen from one to the other, to the great good, as I believe, of the plant. Bees will act like a camel-hair pencil, and it is quite sufficient just to touch the anthers of one flower and

then the stigma of another with the same brush to ensure
fertilisation; but it must not be supposed that bees
would thus produce a multitude of hybrids between distinct species; for if you bring on the same brush a
plant's own pollen and pollen from another species, the
former will have such a prepotent effect, that it will
invariably and completely destroy, as has been shown by
Gärtner, any influence from the foreign pollen.

When the stamens of a flower suddenly spring towards
the pistil, or slowly move one after the other towards it,
the contrivance seems adapted solely to ensure self-
fertilisation; and no doubt it is useful for this end:
but, the agency of insects is often required to cause
the stamens to spring forward, as Kölreuter has shown
to be the case with the barberry; and in this very
genus, which seems to have a special contrivance for
self-fertilisation, it is well known that if closely-allied
forms or varieties are planted near each other, it is
hardly possible to raise pure seedlings, so largely do
they naturally cross. In many other cases, far from
there being any aids for self-fertilisation, there are
special contrivances, as I could show from the writings
of C. C. Sprengel and from my own observations,
which effectually prevent the stigma receiving pollen
from its own flower: for instance, in Lobelia fulgens,
there is a really beautiful and elaborate contrivance
by which all the infinitely numerous pollen-granules
are swept out of the conjoined anthers of each flower,
before the stigma of that individual flower is ready to
receive them; and as this flower is never visited, at least
in my garden, by insects, it never sets a seed, though by
placing pollen from one flower on the stigma of another,
I raised plenty of seedlings; and whilst another species
of Lobelia growing close by, which is visited by bees,
seeds freely. In very many other cases, though there

be no special mechanical contrivance to prevent the
stigma of a flower receiving its own pollen, yet, as
C. C. Sprengel has shown, and as I can confirm, either the
anthers burst before the stigma is ready for fertilisation,
or the stigma is ready before the pollen of that flower
is ready, so that these plants have in fact separated
sexes, and must habitually be crossed. How strange
are these facts! How strange that the pollen and stigmatic surface of the same flower, though placed so close
together, as if for the very purpose of self-fertilisation,
should in so many cases be mutually useless to each
other! How simply are these facts explained on the
view of an occasional cross with a distinct individual
being advantageous or indispensable!

If several varieties of the cabbage, radish, onion, and
of some other plants, be allowed to seed near each other,
a large majority, as I have found, of the seedlings thus
raised will turn out mongrels: for instance, I raised 233
seedling cabbages from some plants of different varieties
growing near each other, and of these only 78 were true
to their kind, and some even of these were not perfectly
true. Yet the pistil of each cabbage-flower is surrounded not only by its own six stamens, but by those
of the many other flowers on the same plant; and the
pollen of each flower readily gets on its own stigma
without insect-agency; for I have found that a plant
carefully protected produced the full number of pods.
How, then, comes it that such a vast number of the
seedlings are mongrelized? I suspect that it must arise
from the pollen of a distinct *variety* having a prepotent
effect over a flower's own pollen; and that this is part
of the general law of good being derived from the intercrossing of distinct individuals of the same species.
When distinct *species* are crossed the case is directly
the reverse, for a plant's own pollen is almost always

prepotent over foreign pollen; but to this subject we shall return in a future chapter.

In the case of a gigantic tree covered with innumerable flowers, it may be objected that pollen could seldom be carried from tree to tree, and at most only from flower to flower on the same tree, and that flowers on the same tree can be considered as distinct individuals only in a limited sense. I believe this objection to be valid, but that nature has largely provided against it by giving to trees a strong tendency to bear flowers with separated sexes. When the sexes are separated, although the male and female flowers may be produced on the same tree, we can see that pollen must be regularly carried from flower to flower; and this will give a better chance of pollen being occasionally carried from tree to tree. That trees belonging to all Orders have their sexes more often separated than other plants, I find to be the case in this country; and at my request Dr. Hooker tabulated the trees of New Zealand, and Dr. Asa Gray those of the United States, and the result was as I anticipated. On the other hand, Dr. Hooker has recently informed me that he finds that the rule does not hold in Australia; and I have made these few remarks on the sexes of trees simply to call attention to the subject.

Turning for a very brief space to animals: on the land there are some hermaphrodites, as land-mollusca and earth-worms; but these all pair. As yet I have not found a single case of a terrestrial animal which fertilises itself. We can understand this remarkable fact, which offers so strong a contrast with terrestrial plants, on the view of an occasional cross being indispensable, by considering the medium in which terrestrial animals live, and the nature of the fertilising element; for we know of no means, analogous to the action of insects

F 3

and of the wind in the case of plants, by which an
occasional cross could be effected with terrestrial animals without the concurrence of two individuals. Of
aquatic animals, there are many self-fertilising hermaphrodites; but here currents in the water offer an
obvious means for an occasional cross. And, as in the
case of flowers, I have as yet failed, after consultation
with one of the highest authorities, namely, Professor
Huxley, to discover a single case of an hermaphrodite
animal with the organs of reproduction so perfectly
enclosed within the body, that access from without and
the occasional influence of a distinct individual can be
shown to be physically impossible. Cirripedes long
appeared to me to present a case of very great difficulty
under this point of view; but I have been enabled, by
a fortunate chance, elsewhere to prove that two individuals, though both are self-fertilising hermaphrodites,
do sometimes cross.

It must have struck most naturalists as a strange
anomaly that, in the case of both animals and plants,
species of the same family and even of the same genus,
though agreeing closely with each other in almost their
whole organisation, yet are not rarely, some of them
hermaphrodites, and some of them unisexual. But if,
in fact, all hermaphrodites do occasionally intercross
with other individuals, the difference between hermaphrodites and unisexual species, as far as function is
concerned, becomes very small.

From these several considerations and from the many
special facts which I have collected, but which I am
not here able to give, I am strongly inclined to suspect
that, both in the vegetable and animal kingdoms, an
occasional intercross with a distinct individual is a law
of nature. I am well aware that there are, on this view,
many cases of difficulty, some of which I am trying to

investigate. Finally then, we may conclude that in many organic beings, a cross between two individuals is an obvious necessity for each birth; in many others it occurs perhaps only at long intervals; but in none, as I suspect, can self-fertilisation go on for perpetuity.

Circumstances favourable to Natural Selection.—This is an extremely intricate subject. A large amount of variability will evidently be favourable for the work of natural selection, but mere individual differences probably suffice. A large number of individuals, by giving a better chance for the appearance within any given period of profitable variations, will compensate for a lesser amount of variability in each individual, and is, I believe, an extremely important element of success. Though Nature grants vast periods of time for the work of natural selection, she does not grant an indefinite period; for as all organic beings are striving to seize on each place in the economy of nature, if any one species does not become modified and improved in a corresponding degree with its competitors, it will soon be exterminated. Unless favourable variations be inherited by some at least of the offspring, nothing can be effected by natural selection. Non-inheritance of any new character is, in fact, the same thing as reversion to the character of the grandparents or more remote ancestors; and no doubt the tendency to reversion may often have checked or prevented the action of natural selection; but its importance has been greatly exaggerated by some writers. If the tendency to reversion has not prevented man from creating innumerable hereditary races in the animal and vegetable kingdoms, why should it have stopped the process of natural selection?

In man's methodical selection, a breeder selects for

some definite object, and free intercrossing will wholly stop his work. But when many men, without intending to alter the breed, have a nearly common standard of perfection, and all try to get and breed from the best animals, much improvement and modification surely but slowly follow from this unconscious process of selection, notwithstanding a large amount of crossing with inferior animals. Thus it will be in nature; for within a confined area, with some place in its polity not so perfectly occupied as might be, natural selection will always tend to preserve all the individuals varying in the right direction, though in different degrees, so as better to fill up the unoccupied place. But if the area be large, its several districts will almost certainly present different conditions of life ; and then if natural selection be modifying and improving a species in the several districts, there will be intercrossing with the other individuals of the same species on the confines of each. And in this case the effects of intercrossing can hardly be counterbalanced by natural selection always tending to modify all the individuals in each district in exactly the same manner to the conditions of each ; for in a continuous area, the physical conditions will generally graduate away insensibly from one district to another. The intercrossing will most affect those animals which unite for each birth, which wander much, and which do not breed at a very quick rate. Hence in animals of this nature, for instance in birds, varieties will generally be confined to separated countries; and this I find to be the case. In hermaphrodite organisms which cross only occasionally, and likewise in animals which unite for each birth, but which wander little and which can increase at a very rapid rate, a new and improved variety might be quickly formed on any one spot, and might there maintain itself in a body, so that whatever intercrossing took place would be chiefly between the individuals of the same new variety. A local variety when once thus formed might subsequently slowly spread to other districts. On the above principle, nurserymen always prefer getting seed from a large body of plants of the same variety, as the chance of intercrossing with other varieties is thus lessened.

Even in the case of slow-breeding animals, which unite for each birth, we must not overrate the effects of intercrosses in retarding natural selection; for I can bring a considerable catalogue of facts, showing that within the same area, varieties of the same animal can long remain distinct, from haunting different stations, from breeding at slightly different seasons, or from varieties of the same kind preferring to pair together.

Intercrossing plays a very important part in nature in keeping the individuals of the same species, or of the same variety, true and uniform in character. It will obviously thus act far more efficiently with those animals which unite for each birth; but I have already attempted to show that we have reason to believe that occasional intercrosses take place with all animals and with all plants. Even if these take place only at long intervals, the young thus produced will gain so much in vigour and fertility over the offspring from long-continued self-fertilisation, that they will have a better chance of surviving and propagating their kind; and thus, in the long run, the influence of intercrosses, even at rare intervals, will be great. If there exist organic beings which never intercross, uniformity of character can be retained amongst them, as long as their conditions of life remain the same, only through the principle of inheritance, and through natural selection destroying any which depart from the proper type; but if their conditions of life change and they undergo modification, uniformity of character can be given to their modified offspring, solely by natural selection preserving the same favourable variations.

Isolation, also, is an important element in the process of natural selection. In a confined or isolated area, if not very large, the organic and inorganic conditions of life will generally be in a great degree uniform; so that natural selection will tend to modify all the individuals of a varying species throughout the area in the same manner in relation to the same conditions. Intercrosses, also, with the individuals of the same species, which otherwise would have inhabited the surrounding and differently circumstanced districts, will be prevented. But isolation probably acts more efficiently in checking the immigration of better adapted organisms, after any physical change, such as of climate or elevation of the land, &c.; and thus new places in the natural economy of the country are left open for the old inhabitants to struggle for, and become adapted to, through modifications in their structure and constitution. Lastly, isolation, by checking immigration and consequently competition, will give time for any new variety to be slowly improved; and this may sometimes be of importance in the production of new species. If, however, an isolated area be very small, either from being surrounded by barriers, or from having very peculiar physical conditions, the total number of the individuals supported on it will necessarily be very small; and fewness of individuals will greatly retard the production of new species through natural selection, by decreasing the chance of the appearance of favourable variations.

The mere lapse of time by itself does nothing either for or against natural selection. I state this because it has been erroneously asserted that the element of time is assumed by me to play an all-important part in natural selection, as if all species were necessarily undergoing

slow modification from some innate law. Lapse of time
is only so far highly important, as it gives a better chance
of beneficial variations arising, being selected, accumulated, and fixed, in relation to the slowly changing organic and inorganic conditions of life. It likewise favours
the *direct* action of new or changed physical conditions
of life.

If we turn to nature to test the truth of these remarks, and look at any small isolated area, such as an
oceanic island, although the total number of the species
inhabiting it, will be found to be small, as we shall see
in our chapter on Geographical Distribution; yet of
these species a very large proportion are endemic,—
that is, have been produced there, and nowhere else.
Hence an oceanic island at first sight seems to have
been highly favourable for the production of new species.
But we may thus greatly deceive ourselves, for to ascertain whether a small isolated area, or a large open area
like a continent, has been most favourable for the production of new organic forms, we ought to make the
comparison within equal times; and this we are incapable of doing.

Although isolation is of considerable importance in
the production of new species, on the whole I am inclined to believe that largeness of area is of more importance, especially in the production of species, which
will prove capable of enduring for a long period, and
of spreading widely. Throughout a great and open
area, not only will there be a better chance of favourable variations arising from the large number of individuals of the same species there supported, but the
conditions of life are infinitely complex from the large
number of already existing species; and if some of
these many species become modified and improved,
others will have to be improved in a corresponding

degree or they will be exterminated. Each new form, also, as soon as it has been much improved, will be able to spread over the open and continuous area, and will thus come into competition with many others. Hence more new places will be formed, and the competition to fill them will be more severe, on a large than on a small and isolated area. Moreover, great areas, though now continuous, owing to oscillations of level, will often have recently existed in a broken condition, so that the good effects of isolation will generally, to a certain extent, have concurred. Finally, I conclude that, although small isolated areas probably have been in some respects highly favourable for the production of new species, yet that the course of modification will generally have been more rapid on large areas; and what is more important, that the new forms produced on large areas, which already have been victorious over many competitors, will be those that will spread most widely, will give rise to most new varieties and species, and will thus play the most important part in the changing history of the organic world.

We can, perhaps, on these views, understand some facts which will be again alluded to in our chapter on Geographical Distribution; for instance, that the productions of the smaller continent of Australia have formerly yielded, and apparently are now yielding, before those of the larger Europæo-Asiatic area. Thus, also, it is that continental productions have everywhere become so largely naturalised on islands. On a small island, the race for life will have been less severe, and there will have been less modification and less extermination. Hence, perhaps, it comes that the flora of Madeira, according to Oswald Heer, resembles the extinct tertiary flora of Europe. All fresh-water basins, taken together, make a small area compared with that of the sea or of the land; and, consequently, the competition between fresh-water productions will have been less severe than elsewhere; new forms will have been more slowly formed, and old forms more slowly exterminated. And it is in fresh water that we find seven genera of Ganoid fishes, remnants of a once preponderant order: and in fresh water we find some of the most anomalous forms now known in the world, as the Ornithorhynchus and Lepidosiren, which, like fossils, connect to a certain extent orders now widely separated in the natural scale. These anomalous forms may almost be called living fossils; they have endured to the present day, from having inhabited a confined area, and from having thus been exposed to less severe competition.

To sum up the circumstances favourable and unfavourable to natural selection, as far as the extreme intricacy of the subject permits. I conclude that for terrestrial productions a large continental area, which has undergone many oscillations of level, and which consequently has existed for long periods in a broken condition, has been the most favourable for the production of many new forms of life, fitted to endure long and to spread widely. For the area first existed as a continent, and the inhabitants, at this period numerous in individuals and kinds, will have been subjected to very severe competition. When converted by subsidence into large separate islands, there will still have existed many individuals of the same species on each island: intercrossing on the confines of the range of each species will thus have been checked: after physical changes of any kind, immigration will have been prevented, so that new places in the polity of each island will have had to be filled up by modifications of the old inhabitants; and time will have been allowed for the varieties in each to become well modified and perfected. When, by renewed elevation, the islands were re-converted into a continental area, there will again have been severe competition : the most favoured or improved varieties will have been enabled to spread : there will have been much extinction of the less improved forms, and the relative proportional numbers of the various inhabitants of the renewed continent will again have been changed; and again there will have been a fair field for natural selection to improve still further the inhabitants, and thus produce new species.

That natural selection will always act with extreme slowness I fully admit. Its action depends on there being places in the polity of nature, which can be better occupied by some of the inhabitants of the country undergoing modification of some kind. The existence of such places will often depend on physical changes, which are generally very slow, and on the immigration of better adapted forms having been checked. But the action of natural selection will probably still oftener depend on some few of the inhabitants becoming slowly modified; the mutual relations of many of the other inhabitants being thus disturbed. Nothing can be effected, unless favourable variations occur, and varia-tion itself is apparently always a slow process. The process will often be greatly retarded by free inter-crossing. Many will exclaim that these several causes are amply sufficient wholly to stop the action of natural selection. I do not believe so. On the other hand, I do believe that natural selection always acts very slowly, often only at long intervals of time, and generally on only a very few of the inhabitants of the same region at the same time. I further believe, that this slow, inter-

mittent action of natural selection accords perfectly
well with what geology tells us of the rate and
manner at which the inhabitants of the world have
changed.

Slow though the process of selection may be, if feeble
man can do much by his powers of artificial selection,
I can see no limit to the amount of change, to the
beauty and infinite complexity of the coadaptations
between all organic beings, one with another and with
their physical conditions of life, which may be effected
in the long course of time by nature's power of selection.

Extinction.—This subject will be more fully discussed
in our chapter on Geology; but it must be here alluded
to from being intimately connected with natural selection. Natural selection acts solely through the preservation of variations in some way advantageous, which
consequently endure. From the high geometrical ratio
of increase of all organic beings, each area is already
stocked with the full number of its existing inhabitants,
and as most areas are already stocked with a great
diversity of forms, it follows that as each selected and
favoured form increases in number, so generally will
the less favoured forms decrease and become rare.
Rarity, as geology tells us, is the precursor to extinction.
We can, also, see that any form represented by few
individuals will, during fluctuations in the seasons or in
the number of its enemies, run a good chance of utter
extinction. But we may go further than this; for as
new forms are continually and slowly being produced,
unless we believe that the number of specific forms goes
on perpetually and almost indefinitely increasing, numbers inevitably must become extinct. That the number
of specific forms has not indefinitely increased, geology
tells us plainly; and we shall presently attempt to

show why it is that the number of species throughout the world has not become immeasurably great.

We have seen that the species which are most numerous in individuals have the best chance of producing within any given period favourable variations. We have evidence of this, in the facts stated in the second chapter, showing that it is the common species which offer the greatest number of recorded varieties, or incipient species. Hence, rare species will be less quickly modified or improved within any given period, and they will consequently be beaten in the race for life by the modified descendants of the commoner species.

From these several considerations I think it inevitably follows, that as new species in the course of time are formed through natural selection, others will become rarer and rarer, and finally extinct. The forms which stand in closest competition with those undergoing modification and improvement, will naturally suffer most. And we have seen in the chapter on the Struggle for Existence that it is the most closely-allied forms,—varieties of the same species, and species of the same genus or of related genera,—which, from having nearly the same structure, constitution, and habits, generally come into the severest competition with each other. Consequently, each new variety or species, during the progress of its formation, will generally press hardest on its nearest kindred, and tend to exterminate them. We see the same process of extermination amongst our domesticated productions, through the selection of improved forms by man. Many curious instances could be given showing how quickly new breeds of cattle, sheep, and other animals, and varieties of flowers, take the place of older and inferior kinds. In Yorkshire, it is historically known that the ancient black cattle were displaced by the long-horns, and that these "were swept away by the short-horns" (I quote the words of au agricultural writer) "as if by some murderous pestilence."

Divergence of Character.—The principle, which I have designated by this term, is of high importance, and explains, as I believe, several important facts. In the first place, varieties, even strongly-marked ones, though having somewhat of the character of species—as is shown by the hopeless doubts in many cases how to rank them—yet certainly differ from each other far less than do good aud distinct species. Nevertheless, according to my view, varieties are species in the process of formation, or are, as I have called them, incipient species. How, then, does the lesser difference between varieties become augmented into the greater difference between species? That this does habitually happen, we must infer from most of the innumerable species throughout nature presenting well-marked differences; whereas varieties, the supposed prototypes and parents of future well-marked species, present slight and ill-defined differences. Mere chance, as we may call it, might cause one variety to differ in some character from its parents, and the offspring of this variety again to differ from its parent in the very same character and in a greater degree; but this alone would never account for so habitual and large an amount of difference as that between well-marked varieties of the same species and species of the same genus.

As has always been my practice, let us seek light on this head from our domestic productions. We shall here find something analogous. It will be admitted that the production of races so different as short-horn and Hereford cattle, race and cart horses, the several breeds of pigeons, &c., could never have been effected by the mere chance accumulation of variations of a

similar character during many successive generations.
In practice, a fancier is, for instance, struck by a pigeon
having a slightly shorter beak; another fancier is struck
by a pigeon having a rather longer beak; and on the
acknowledged principle that "fanciers do not and will not
admire a medium standard, but like extremes," they both
go on (as has actually occurred with the sub-breeds of the
tumbler-pigeon) choosing and breeding from birds with
longer and longer beaks, or with shorter and shorter
beaks. Again, we may suppose that at an early period
one man preferred swifter horses; another stronger and
more bulky horses. The early differences would be
very slight; in the course of time, from the continued
selection of swifter horses by some breeders, and of
stronger ones by others, the differences would become
greater, and would be noted as forming two sub-breeds;
finally, after the lapse of centuries, the sub-breeds
would become converted into two well-established and
distinct breeds. As the differences slowly become
greater, the inferior animals with intermediate characters, being neither very swift nor very strong, will
have been neglected, and will have tended to disappear.
Here, then, we see in man's productions the action of
what may be called the principle of divergence, causing
differences, at first barely appreciable, steadily to increase, and the breeds to diverge in character both
from each other and from their common parent.

But how, it may be asked, can any analogous principle apply in nature? I believe it can and does apply
most efficiently (though it was a long time before I
saw how), from the simple circumstance that the more
diversified the descendants from any one species become
in structure, constitution, and habits, by so much will
they be better enabled to seize on many and widely
diversified places in the polity of nature, and so be
enabled to increase in numbers.

We can clearly discern this in tho case of animals with simple habits. Take the case of a carnivorous quadruped, of which the number that can be supported in any country has long ago arrived at its full average. If its natural powers of increase be allowed to act, it can succeed in increasing (the country not undergoing any change in its conditions) only by its varying descendants seizing on places at present occupied by other animals: some of them, for instance, being enabled to feed on new kinds of prey, either dead or alive; some inhabiting new stations, climbing trees, frequenting water, and some perhaps becoming less carnivorous. The more diversified in habits and structure the descendants of our carnivorous animal became, the more places they would be enabled to occupy. What applies to one animal will apply throughout all time to all animals—that is, if they vary—for otherwise natural selection can do nothing. So it will be with plants. It has been experimentally proved, that if a plot of ground be sown with one species of grass, and a similar plot be sown with several distinct genera of grasses, a greater number of plants and a greater weight of dry herbage can thus be raised. The same has been found to hold good when one variety and several mixed varieties of wheat have been sown on equal spaces of ground. Hence, if any one species of grass were to go on varying, and those varieties were continually selected which differed from each other in at all the same manner as distinct species and genera of grasses differ from each other, a greater number of individual plants of this species of grass, including its modified descendants, would succeed in living on the same piece of ground. And we well know that each species and each variety of grass is annually sowing almost countless seeds; and thus, as it may be said, is striving its utmost to increase its num-

bers. Consequently, in the course of many thousands
of generations, the most distinct varieties of any one
species of grass would always have the best chance of
succeeding and of increasing in numbers, and thus of
supplanting the less distinct varieties; and varieties,
when rendered very distinct from each other, take the
rank of species.

The truth of the principle that the greatest amount
of life can be supported by great diversification of
structure, is seen under many natural circumstances.
In an extremely small area, especially if freely open to
immigration, and where the contest between individual
and individual must be severe, we always find great
diversity in its inhabitants. For instance, I found that
a piece of turf, three feet by four in size, which had been
exposed for many years to exactly the same conditions,
supported twenty species of plants, and these belonged
to eighteen genera and to eight orders, which shows how
much these plants differed from each other. So it is with
the plants and insects on small and uniform islets; and
so in small ponds of fresh water. Farmers find that they
can raise most food by a rotation of plants belonging to
the most different orders: nature follows what may be
called a simultaneous rotation. Most of the animals and
plants which live close round any small piece of ground,
could live on it (supposing it not to be in any way peculiar
in its nature), and may be said to be striving to the utmost
to live there; but, it is seen, that where they come into
the closest competition with each other, the advantages
of diversification of structure, with the accompanying
differences of habit and constitution, determine that the
inhabitants, which thus jostle each other most closely,
shall, as a general rule, belong to what we call different
genera and orders.

The same principle is seen in the naturalisation of

plants through man's agency iu foreign lands. It might have been expected that the plants which have succeeded in becoming naturalised in any land would generally have been closely allied to the indigenes; for these are commonly looked at as specially created and adapted for their own country. It might, also, perhaps have been expected that naturalised plants would have belonged to a few groups more especially adapted to certain stations in their new homes. But the case is very different; and Alph. de Candolle has well remarked in his great and admirable work, that floras gain by naturalisation, proportionally with the number of the native genera and species, far more in new genera than in new species. To give a single instance : in the last edition of Dr. Asa Gray's ' Manual of the Flora of the Northern United States,' 260 naturalised plants are enumerated, and these belong to 162 genera. We thus see that these naturalised plants are of a highly diversified nature. They differ, moreover, to a large extent from the indigenes, for out of the 162 naturalised genera, no less than 100 genera are not there indigenous, and thus a large proportional addition is made to the endemic genera of the United States.

By considering the nature of the plants or animals which have struggled successfully with the indigenes of any country, and have there become naturalised, we may gain some crude idea in what manner some of the natives would have to be modified, in order to gain an advantage over the other natives; and we may at least safely infer that diversification of structure, amounting to new generic differences, would be profitable to them.

The advantage of diversification in the inhabitants of the same region is, in fact, the same as that of the physiological division of labour in the organs of the same

G

individual body—a subject so well elucidated by Milne Edwards. No physiologist doubts that a stomach adapted to digest vegetable matter alone, or flesh alone, draws most nutriment from these substances. So in the general economy of any land, the more widely and perfectly the animals and plants are diversified for different habits of life, so will a greater number of individuals be capable of there supporting themselves. A set of animals, with their organisation but little diversified, could hardly compete with a set more perfectly diversified in structure. It may be doubted, for instance, whether the Australian marsupials, which are divided into groups differing but little from each other, and feebly representing, as Mr. Waterhouse and others have remarked, our carnivorous, ruminant, and rodent mammals, could successfully compete with these well-pronounced orders. In the Australian mammals, we see the process of diversification in an early and incomplete stage of development.

After the foregoing discussion, which ought to have been much amplified, we may assume that the modified descendants of any one species will succeed by so much the better as they become more diversified in structure, and are thus enabled to encroach on places occupied by other beings. Now let us see how this principle of benefit being derived from divergence of character, combined with the principles of natural selection and of extinction, will tend to act.

The accompanying diagram will aid us in understanding this rather perplexing subject. Let A to L represent the species of a genus large in its own country; these species are supposed to resemble each other in unequal degrees, as is so generally the case in nature, and as is represented in the diagram by the letters standing at unequal distances. I have said a large

XIV
XIII
XII
XI
X
IX
VIII
VII
VI
V
IV
III
II
I

a^{14} q^{16} p^{16} h^{16} u^{16}

a^{11} X

a^{9} IX

a^{8} f^{8} z^{8} VIII

a^{7} f^{7} VII

a^{6} f^{6} VI

a^{5} d^{7} V

a^{4} d^{4} IV

a^{3} III

a^{2} II

d I

A

genus, because we have seen in the second chapter, that on an average more of the species of large genera vary than of small genera; and the varying species of the large genera present a greater number of varieties. We have, also, seen that the species, which are the commonest and the most widely diffused, vary more than rare species with restricted ranges. Let (A) be a common, widely-diffused, and varying species, belonging to a genus large in its own country. The little fan of diverging dotted lines of unequal lengths proceeding from (A), may represent its varying offspring. The variations are supposed to be extremely slight, but of the most diversified nature; they are not supposed all to appear simultaneously, but often after long intervals of time; nor are they all supposed to endure for equal periods. Only those variations which are in some way profitable will be preserved or naturally selected. And here the importance of the principle of benefit being derived from divergence of character comes in; for this will generally lead to the most different or divergent variations (represented by the outer dotted lines) being preserved and accumulated by natural selection. When a dotted line reaches one of the horizontal lines, and is there marked by a small numbered letter, a sufficient amount of variation is supposed to have been accumulated to have formed a fairly well-marked variety, such as would be thought worthy of record in a systematic work.

The intervals between the horizontal lines in the diagram, may represent each a thousand generations; but it would have been better if each had represented ten thousand generations. After a thousand generations, species (A) is supposed to have produced two fairly well-marked varieties, namely a^1 and m^1. These two varieties will generally continue to be exposed to the same conditions which made their parents variable, and the tendency to variability is in itself hereditary, consequently they will tend to vary, and generally to vary in nearly the same manner as their parents varied. Moreover, these two varieties, being only slightly modified forms, will tend to inherit those advantages which made their parent (A) more numerous than most of the other inhabitants of the same country; they will likewise partake of those more general advantages which made the genus to which the parent-species belonged, a large genus in its own country. And these circumstances we know to be favourable to the production of new varieties.

If, then, these two varieties be variable, the most divergent of their variations will generally be preserved during the next thousand generations. And after this interval, variety a^1 is supposed in the diagram to have produced variety a^2, which will, owing to the principle of divergence, differ more from (A) than did variety a^1. Variety m^1 is supposed to have produced two varieties, namely m^2 and s^2, differing from each other, and more considerably from their common parent (A). We may continue the process by similar steps for any length of time; some of the varieties, after each thousand generations, producing only a single variety, but in a more and more modified condition, some producing two or three varieties, and some failing to produce any. Thus the varieties or modified descendants, proceeding from the common parent (A), will generally go on increasing in number and diverging in character. In the diagram the process is represented up to the ten-thousandth generation, and under a condensed and simplified form up to the fourteen-thousandth generation.

But I must here remark that I do not suppose that the process ever goes on so regularly as is represented in, the diagram, though in itself made somewhat irregular. I am far from thinking that the most divergent varieties invariably prevail and multiply: a medium form may often long endure, and may or may not produce more than one modified descendant ; for natural selection will always act according to the nature of the places which are either unoccupied or not perfectly occupied by other beings; and this will depend on infinitely complex relations. But as a general rule, the more diversified in structure the descendants from any one species can be rendered, the more places they will be enabled to seize on, and the more their modified progeny will increase. In our diagram the line of succession is broken at regular intervals by small numbered letters marking the successive forms which have become sufficiently distinct to be recorded as varieties. But these breaks are imaginary, and might have been inserted anywhere, after intervals long enough to have allowed the accumulation of a considerable amount of divergent variation.

As all the modified descendants from a common and widely-diffused species, belonging to a large genus, will tend to partake of the same advantages which made their parent successful in life, they will generally go on multiplying in number as well as diverging in character: this is represented in the diagram by the several divergent branches proceeding from (A). The modified offspring from the later and more highly improved branches in the lines of descent, will, it is probable, often take the place of, and so destroy, the earlier and less improved branches: this is represented in the diagram by some of the lower branches not reaching to the upper horizontal lines. In some cases I do not doubt that the process of modification will be confined to a single line of descent, and the number of the

descendants will not be increased; although the amount
of divergent modification may have been increased in
the successive generations. This case would be represented in the diagram, if all the lines proceeding from
(A) were removed, excepting that from a^1 to a^{10}. In
the same way, for instance, the English race-horse and
English pointer have apparently both gone on slowly
diverging in character from their original stocks, without either having given off any fresh branches or races.

After ten thousand generations, species (A) is sup-
posed to have produced three forms, a^{10}, f^{10}, and m^{10},
which, from having diverged in character during the
successive generations, will have come to differ largely,
but perhaps unequally, from each other and from their
common parent. If we suppose the amount of change
between each horizontal line in our diagram to be excessively small, these three forms may still be only
well-marked varieties; or they may have arrived at the
doubtful category of sub-species; but we have only to
suppose the steps in the process of modification to be
more numerous or greater in amount, to convert these
three forms into well-defined species: thus the diagram
illustrates the steps by which the small differences
distinguishing varieties are increased into the larger
differences distinguishing species. By continuing the
same process for a greater number of generations (as
shown in the diagram in a condensed and simplified
manner), we get eight species, marked by the letters
between a^{14} and m^{14}, all descended from (A). Thus, as
I believe, species are multiplied and genera are formed.

In a large genus it is probable that more than one
species would vary. In the diagram I have assumed
that a second species (I) has produced, by analogous
steps, after ten thousand generations, either two well-
marked varieties (w^{10} and z^{10}) or two species, according

to the amount of change supposed to be represented between the horizontal lines. After fourteen thousand
generations, six new species, marked by the letters n^{14}
to z^{14}, are supposed to have been produced. The species
of a genus which are already extremely different in
character, will generally tend to produce the greatest
number of modified descendants; for they will have
the best chance of filling new and widely different
places in the polity of nature: hence in the diagram I
have chosen the extreme species (A), and the nearly
extreme species (I), as those which have largely varied,
and have given rise to new varieties and species. The
other nine species (marked by capital letters) of our
original genus, may for a long period continue to
transmit unaltered descendants; and this is shown in
the diagram by the dotted lines not prolonged far upwards from want of space.

But during the process of modification, represented
in the diagram, another of our principles, namely that
of extinction, will have played an important part. As in
each fully stocked country natural selection necessarily
acts by the selected form having some advantage in the
struggle for life over other forms, there will be a constant tendency in the improved descendants of any one
species to supplant and exterminate in each stage of
descent their predecessors and their original parent.
For it should be remembered that the competition will
generally be most severe between those forms which
are most nearly related to each other in habits, constitution, and structure. Hence all the intermediate forms
between the earlier and later states, that is between the
less and more improved states of a species, as well as
the original parent-species itself, will generally tend to
become extinct. So it probably will be with many whole
collateral lines of descent, which will be conquered by

later and improved lines of descent. If, however, the modified offspring of a species get into some distinct country, or become quickly adapted to some quite new station, in which child and parent do not come into competition, both may continue to exist.

If, then, our diagram be assumed to represent a considerable amount of modification, species (A) and all the earlier varieties will have become extinct, having been replaced by eight new species (a^{14} to m^{14}) ; and (I) will have been replaced by six (n^{14} to z^{14}) new species.

But we may go further than this. The original species of our genus were supposed to resemble each other in unequal degrees, as is so generally the case in nature ; species (A) being more nearly related to B, C, and D, than to the other species ; and species (I) more to G, H, K, L, than to the others. These two species (A) and (I), were also supposed to be very common and widely diffused species, so that they must originally have had some advantage over most of the other species of the genus. Their modified descendants, fourteen in number at the fourteen-thousandth generation, will probably have inherited some of the same advantages: they have also been modified and improved in a diversified manner at each stage of descent, so as to have become adapted to many related places in the natural economy of their country. It seems, therefore, extremely probable that they will have taken the places of, and thus exterminated, not only their parents (A) and (I), but likewise some of the original species which were most nearly related to their parents. Hence very few of the original species will have transmitted offspring to the fourteen-thousandth generation. We may suppose that only one (F), of the two species which were least closely related to the other nine original species, has transmitted descendants to this late stage of descent.

The new species in our diagram descended from the original eleven species, will now be fifteen in number. Owing to the divergent tendency of natural selection, the extreme amount of difference in character between species a^{14} and z^{14} will be much greater than that between the most different of the original eleven species. The new species, moreover, will be allied to each other in a widely different manner. Of the eight descendants from (A) the three marked a^{14}, q^{14}, p^{14}, will be nearly related from having recently branched off from a^{10}; b^{14}, and f^{14}, from having diverged at an earlier period from a^5, will be in some degree distinct from the three first-named species; and lastly, o^{14}, e^{14}, and m^{14}, will be nearly related one to the other, but from having diverged at the first commencement of the process of modification, will be widely different from the other five species, and may constitute a sub-genus or even a distinct genus.

The six descendants from (I) will form two sub-genera or even genera. But as the original species (I) differed largely from (A), standing nearly at the extreme points of the original genus, the six descendants from (I) will, owing to inheritance alone, differ considerably from the eight descendants from (A); the two groups, moreover, are supposed to have gone on diverging in different directions. The intermediate species, also (and this is a very important consideration), which connected the original species (A) and (I), have all become, excepting (F), extinct, and have left no descendants. Hence the six new species descended from (I), and the eight descended from (A), will have to be ranked as very distinct genera, or even as distinct sub-families.

Thus it is, as I believe, that two or more genera

are produced by descent with modification, from two
or more species of the same genus. And the two or
more parent-species are supposed to have descended from
some one species of an earlier genus. In our diagram,
this is indicated by the broken lines, beneath the capital
letters, converging in sub-branches downwards towards
a single point; this point representing a single species,
the supposed single parent of our several new sub-genera
and genera.

It is worth while to reflect for a moment on the character of the new species F^{14}, which is supposed not to
have diverged much in character, but to have retained
the form of (F), either unaltered or altered only in a
slight degree. In this case, its affinities to the other
fourteen new species will be of a curious and circuitous
nature. Having descended from a form which stood
between the two parent species (A) and (I), now supposed to be extinct and unknown, it will be in some
degree intermediate in character between the two groups
descended from these species. But as these two groups
have gone on diverging in character from the type of
their parents, the new species (F^{14}) will not be directly
intermediate between them, but rather between types
of the two groups; and every naturalist will be able to
bring some such case before his mind.

In the diagram, each horizontal line has hitherto been
supposed to represent a thousand generations, but each
may represent a million or hundred million generations,
and likewise a section of the successive strata of the
earth's crust including extinct remains. We shall, when
we come to our chapter on Geology, have to refer again
to this subject, and I think we shall then see that the
diagram throws light on the affinities of extinct beings,
which, though generally belonging to the same orders,
or families, or genera, with those now living, yet are

often, in some degree, intermediate in character between existing groups; and we can understand this fact, for the extinct species lived at very ancient epochs when the branching lines of descent had diverged less.

I see no reason to limit the process of modification, as now explained, to the formation of genera alone. If, in our diagram, we suppose the amount of change represented by each successive group of diverging dotted lines to be very great, the forms marked a^{14} to p^{14}, those marked b^{14} and f^{14}, and those marked o^{14} to m^{14}, will form three very distinct genera. We shall also have two very distinct genera descended from (I); and as these latter two genera, both from continued divergence of character and from inheritance from a different parent, will differ widely from the three genera descended from (A), the two little groups of genera will form two distinct families, or even orders, according to the amount of divergent modification supposed to be represented in the diagram. And the two new families, or orders, will have descended from two species of the original genus; and these two species are supposed to have descended from one species of a still more ancient and unknown genus.

We have seen that in each country it is the species of the larger genera which oftenest present varieties or incipient species. This, indeed, might have been expected; for as natural selection acts through one form having some advantage over other forms in the struggle for existence, it will chiefly act on those which already have some advantage; and the largeness of any group shows that its species have inherited from a common ancestor some advantage in common. Hence, the struggle for the production of new and modified descendants will mainly lie between the larger groups which are all trying to increase in number. One large group will slowly conquer another large group, reduce its numbers, and thus lessen its chance of further variation and improvement. Within the same large group, the later and more highly perfected sub-groups, from branching out and seizing on many new places in the polity of Nature, will constantly tend to supplant and destroy the earlier and less improved sub-groups. Small and broken groups and sub-groups will finally disappear. Looking to the future, we can predict that the groups of organic beings which are now large and triumphant, and which are least broken up, that is, which as yet have suffered least extinction, will for a long period continue to increase. But which groups will ultimately prevail, no man can predict; for we well know that many groups, formerly most extensively developed, have now become extinct. Looking still more remotely to the future, we may predict that, owing to the continued and steady increase of the larger groups, a multitude of smaller groups will become utterly extinct, and leave no modified descendants; and consequently that of the species living at any one period, extremely few will transmit descendants to a remote futurity. I shall have to return to this subject in the chapter on Classification, but I may add that on this view of extremely few of the more ancient species having transmitted descendants, and on the view of all the descendants of the same species making a class, we can understand how it is that there exist but very few classes in each main division of the animal and vegetable kingdoms. Although extremely few of the most ancient species now have living and modified descendants, yet at the most remote geological period, the earth may have been as well peopled with many species of many genera, families, orders, and classes, as at the present day.

On the degree to which Organisation tends to advance.—
Natural selection acts, as we have seen, exclusively by
the preservation and accumulation of variations, which
are beneficial under the organic and inorganic conditions
of life to which each creature is at each successive
period exposed. The ultimate result will be that each
creature will tend to become more and more improved
in relation to its conditions of life. This improvement
will, I think, inevitably lead to the gradual advancement
of the organisation of the greater number of living
beings throughout the world. But here we enter on a
very intricate subject, for naturalists have not defined
to each other's satisfaction what is meant by an advance
in organisation. Amongst the vertebrata the degree of
intellect and an approach in structure to man clearly
come into play. It might be thought that the amount
of change which the various parts and organs undergo
in their development from the embryo to maturity
would suffice as a standard of comparison ; but there are
cases, as with certain parasitic crustaceans, in which
several parts of the structure become less perfect, so
that the mature animal cannot be called higher than
its larva. Von Baer's standard seems the most widely
applicable and the best, namely, the amount of differentiation of the different parts (in the adult state, as I
should be inclined to add) and their specialisation for
different functions ; or, as Milne Edwards would express
it, the completeness of the division of physiological
labour. But we shall see how obscure a subject this is
if we look, for instance, to fish, amongst which some
naturalists rank those as highest which, like the sharks,
approach nearest to reptiles ; whilst other naturalists
rank the common bony or teleostean fishes as the highest, inasmuch as they are most strictly fish-like, and
differ most from the other vertebrate classes. Still more

plainly we see the obscurity of the subject by turning
to plants, with which the standard of intellect is of
course quite excluded; and here some botanists rank
those plants as highest which have every organ, as
sepals, petals, stamens, and pistils, fully developed in
each flower; whereas other botanists, probably with
more truth, look at the plants which have their several
organs much modified and somewhat reduced in number
as being of the highest rank.

If we look at the differentiation and specialisation of
the several organs of each being when adult (and this
will include the advancement of the brain for intellectual purposes) as the best standard of highness of
organisation, natural selection clearly leads towards
highness; for all physiologists admit that the specialisation of organs, inasmuch as they perform in this state
their functions better, is an advantage to each being;
and hence the accumulation of variations tending
towards specialisation is within the scope of natural
selection.   On the other hand, we can see, bearing in
mind that all organic beings are striving to increase at
a high ratio and to seize on every ill-occupied place in
the economy of nature, that it is quite possible for
natural selection gradually to fit an organic being to a
situation in which several organs would be superfluous
and useless: in such cases there might be retrogression
in the scale of organisation. Whether organisation on
the whole has actually advanced from the remotest
geological periods to the present day will be more con-
veniently discussed in our chapter on Geological
Succession.

But it may be objected that if all organic beings thus
tend to rise in the scale, how is it that throughout the
world a multitude of the lowest forms still exist; and
how is it that in each great class some forms are far

more highly developed than others? Why have not the
more highly developed forms everywhere supplanted and
exterminated the lower? Lamarck, who believed in an
innate and inevitable tendency towards perfection in all
organic beings, seems to have felt this difficulty so strongly, that he was led to suppose that new and simple forms
were continually being produced by spontaneous generation. I need hardly say that Science in her present
state does not countenance the belief that living creatures are now ever produced from inorganic matter.
On my theory the present existence of lowly organised
productions offers no difficulty; for natural selection includes no necessary and universal law of advancement or
development—it only takes advantage of such variations
as arise and are beneficial to each creature under its
complex relations of life. And it may be asked what
advantage, as far as we can see, would it be to an infu-
sorian animalcule—to an intestinal worm—or even to
an earth-worm, to be highly organised? If it were no
advantage, these forms would be left by natural selection
unimproved or but little improved; and might remain
for indefinite ages in their present little advanced condition. And geology tells us that some of the lowest
forms, as the infusoria and rhizopods, have remained for
an enormous period in nearly their present state. But
to suppose that most of the many now existing low forms
have not in the least advanced since the first dawn of
life would be rash; for every naturalist who has dissected
some of the beings now ranked as very low in the scale,
must have been struck with their really wondrous and
beautiful organisation.

Nearly the same remarks are applicable if we look to
the great existing differences in the grades of organisation which occur within almost every great group;
for instance, to the co-existence of mammals and fish

in the vertebrata,—to the co-existence of man and the
ornithorhynchus amongst mammalia—or of the shark
and amphioxus, which latter fish in the extreme simplicity of its structure closely approaches the invertebrate
classes. But mammals and fish hardly come into competition with each other; the advancement of certain
mammals or of the whole class to the highest grade of
organisation would not lead to their taking the place of,
and thus exterminating, fishes. Physiologists believe
that the brain must be bathed by warm blood to be
highly active, and this requires aërial respiration; so
that warm-blooded mammals when inhabiting the water
live under some disadvantages compared with fishes.
In this latter class, members of the shark family would
not, it is probable, tend to supplant the amphioxus; the
struggle for existence in the case of the amphioxus
apparently will lie with members of the invertebrate
classes. The three lowest orders of mammals, namely,
marsupials, edentata, and rodents, co-exist in South
America in the same region with numerous monkeys,
and probably interfere little with each other. Although
organisation, on the whole, may have advanced and be
advancing throughout the world, yet the scale will still
present all degrees of perfection; for the high advancement of certain whole classes, or of certain members of
each class, does not at all necessarily lead to the extinction of those groups with which they do not enter into
close competition. In some cases, as we shall hereafter
see, lowly organised forms seem to have been preserved
to the present day from inhabiting peculiar or isolated
stations, where they have been subjected to less severe
competition, and where they have existed in scanty
numbers, which, as already explained, retards the
chance of favourable variations arising.

Finally, I believe that lowly organised forms now

exist in numbers throughout the world, and in nearly every class, from various causes. In some cases favourable variations may never have arisen for natural selection to act on and accumulate. In no case, probably, has time sufficed for the utmost possible amount of development. In some few cases there may have been what we must call retrogression of organisation. But the main cause lies in the circumstance that under very simple conditions of life a high organisation would be of no service,—possibly would be of actual disservice, as being of a more delicate nature, and more liable to be put out of order and thus injured.

A difficulty, diametrically opposite to this which we have just been considering, has been advanced, namely, looking to the dawn of life, when all organic beings, as we may imagine, presented the simplest structure, how could the first steps in advancement or in the differentiation and specialisation of parts have arisen? I can make no sufficient answer; and can only say that as we have no facts to guide us, all speculation on the subject would be baseless and useless. It is, however, an error to suppose that there would be no struggle for existence, and, consequently, no natural selection, until many forms had been produced: variations in a single species inhabiting an isolated station might be beneficial, and through their preservation either the whole mass of individuals might become modified, or two distinct forms might arise. But I must recur to what was stated towards the close of the Introduction, where I say that no one ought to feel surprise at much remaining as yet unexplained on the origin of species, if due allowance be made for our profound ignorance on the mutual relations of the inhabitants of the world during the many past epochs in its history.

I will here notice a few miscellaneous objections which have been advanced against my views, as some of the previous discussions may perhaps thus be made clearer. It has been argued that as none of the animals and plants of Egypt, of which we know anything, have changed during the last 3000 years, so probably none have been modified in other parts of the world. The many animals which have remained unchanged since the commencement of the glacial period would have been an incomparably stronger case, for these have been exposed to great changes of climate and have migrated over great distances; whereas, in Egypt, during the last 3000 years, the conditions of life, as far as we know, have remained absolutely uniform. The fact of little or no modification having been effected since the glacial period would be of some avail against those who believe in the existence of an innate and necessary law of development, but is powerless against the doctrine of natural selection, which only implies that variations occasionally occurring in single species are under favourable conditions preserved. As Mr. Fawcett has well asked, what would be thought of a man who argued that because he could show that Mont Blanc and the other Alpine peaks had exactly the same height 3000 years ago as at present, consequently that these mountains had never been slowly upraised, and that the height of other mountains in other parts of the world had not recently been increased by slow degrees?

It has been objected, if natural selection be so powerful, why has not this or that organ been recently modified and improved? Why has not the proboscis of the hive-bee been lengthened so as to reach the nectar in the flower of the red-clover? Why has not the ostrich acquired the power of flight? But granting that these organs have happened to vary in the right direction, granting that there has been time sufficient for the slow work of natural selection, checked as it will be by inter-crossing and the tendency to reversion, who will pretend that he knows the natural history of any one organic being sufficiently well to say whether any particular change would be to its advantage? Can we feel sure that a long proboscis would not be a disadvantage to the hive-bee in sucking the innumerable small flowers which it frequents? Can we feel sure that a long proboscis would not, by correlation of growth, almost necessarily give increased size to other parts of the mouth, perhaps interfering with the delicate cell-constructing work? In the case of the ostrich a moment's reflection will show that an enormous supply of food would be necessary in this bird of the desert, to supply force to move its huge body through the air. But such ill-considered objections are hardly worth notice.

The celebrated palæontologist, Professor Bronn, in his German translation of this work, has advanced various good objections to my views, and other remarks in its favour. Of the objections, some seem to me unimportant, some few are owing to misapprehension, and some are incidentally noticed in various parts of this volume. On the erroneous supposition that all the species of a region are believed by me to be changing at the same time, he justly asks how it is that all the forms of life do not present a fluctuating and inextricably confused body? but it is sufficient for us if some few forms at any one time are variable, and few will dispute that this is the case. He asks, how can it be on the principle of natural selection that a variety should live in abundance side by side with the parent species; for the variety during its formation is supposed to have supplanted the intermediate forms between itself and the parent species, and yet it has not supplanted the parent species itself, for both are supposed now to live side by side? If the variety and parent species have become fitted to slightly different habits of life, they might live together; though in the case of animals which freely cross and move about, varieties seem to be almost always confined to distinct localities. But is it the case that varieties of plants and of the lower animals are often found in abundance side by side with the parent forms? Laying aside the polymorphic species in which the innumerable variations that occur seem neither advantageous nor disadvantageous to the species, and have not been fixed; laying aside also temporary variations, such as albinism, &c., my impression is that varieties and the supposed parent species are generally found, inhabiting either distinct stations, high land or low land, dry or moist districts, or distinct regions.

Again, Professor Bronn truly remarks, that distinct species do not differ from each other in single characters alone, but in many; and he asks, how it comes that natural selection should always have simultaneously affected many parts of the organisation? Probably the whole amount of difference has not been simultaneously effected; and the unknown laws of correlation will certainly account for, but not strictly explain, much simultaneous modification. Anyhow, we see in our domestic varieties the very same fact: though our domestic races may differ much in some one organ from the other races of the same species, yet the other parts of the organisation will always be found in some degree different. Professor Bronn likewise asks with striking effect how, for instance in the mouse or hare genus, natural selection will account for the several species (descended, I may remark, from a parent of unknown character) having longer or shorter tails, longer or shorter ears, and fur of different colours; how will it account for one species of plant having pointed and another species obtuse leaves? I can give no definite answer to such questions; but I might ask in return, were these differences, on the doctrine of independent creation, formed for no purpose? If of use, or if due to correlation of growth, they could assuredly be formed through the natural preservation of such useful or correlated variations. I believe in the doctrine of descent with modification, notwithstanding that this or that particular change of structure cannot be accounted for, because this doctrine groups together and explains, as we shall see in the latter chapters, many general phenomena of nature.

A distinguished botanist, Mr. H. C. Watson, believes that I have overrated the importance of the principle of divergence of character (in which, however, he apparently believes), and that convergence of character, as it may be called, has likewise played a part. This is an intricate subject which need not be here discussed. I will only say that if two species of two closely allied genera produced a number of new and divergent species, I can believe that these new forms might sometimes approach each other so closely that they would for convenience sake be classed in the same new genus, and thus two genera would converge into one; but from the strength of the principle of inheritance, it seems hardly credible that the two groups of new species would not at least form two sections of the supposed new single genus.

Mr. Watson has also objected that the continued action of natural selection with divergence of character will tend to make an indefinite number of specific forms. As far as mere inorganic conditions are concerned, it seems probable that a sufficient number of species would soon become adapted to all considerable diversities of heat, moisture, &c.; but I fully admit that the mutual

relations of organic beings are more important; and as
the number of species in any country goes on increasing,
the organic conditions of life will become more and
more complex. Consequently there seems at first sight
to be no limit to the amount of profitable diversification of
structure, and therefore no limit to the number of species
which might be produced. We do not know that even
the most prolific area is fully stocked with specific forms:
at the Cape of Good Hope and in Australia, which support such an astonishing number of species, many
European plants have become naturalised. But geology
shows us, at least within the whole immense tertiary
period, that the number of species of shells, and, probably, of mammals, has not greatly or at all increased.
What then checks an indefinite increase in the number
of species? The amount of life (I do not mean the
number of specific forms) supported on any area must
have a limit, depending so largely as it does on physical
conditions: therefore, if an area be inhabited by very
many species, each or nearly each species will be represented by few individuals; and such species will be liable
to extermination from accidental fluctuations in the
nature of the seasons or in the number of their enemies.
The process of extermination in these cases will be rapid,
whereas the production of new species will always be
slow. Imagine the extreme case of as many species as
individuals in England, and the first severe winter or
very dry summer would exterminate thousands on thousands of species. Rare species, and each species will
become rare if the number of species become in any
country indefinitely increased, will, on the principle often
explained, present within a given period few favourable variations; consequently, the process of giving
birth to new specific forms will thus be retarded. When
any species becomes very rare, close interbreeding will

help in exterminating it; at least authors have thought that this comes into play iu accounting for the deterioration of Aurochs in Lithuania, of Red Deer in Scotland, and of Bears in Norway, &c. As far as animals are concerned, some species are closely adapted to prey on some one other being; but if this other being had been rare, it would not have been any advantage to the animal to have been produced in close relation to its prey: therefore, it would not have been produced by natural selection. Lastly, and this I am inclined to think is the most important element, a dominant species, which has already beaten many competitors in its own home, will tend to spread aud supplant many others. Alph. de Candolle has shown that those species which spread widely tend generally to spread *very* widely; and, consequently, they will tend to exterminate several species in several areas, and thus check the inordinate increase of specific forms throughout the world. Dr. Hooker has recently shown that in the S. E. corner of Australia, where, apparently, there are many invaders from different quarters of the world, the endemic Australian species have been greatly reduced in number. How much weight to attribute to these several considerations I do not pretend to assign; but conjointly they must limit in each country the tendency to an indefinite augmentation of specific forms.

Summary of Chapter.—If, during the long course of ages and under varying conditions of life, organic beings vary at all in the several parts of their organisation, and I think this cannot be disputed; if there be, owing to the high geometrical ratio of increase of each species, a severe struggle for life at some age, season, or year, and this certainly cannot be disputed; then, considering the infinite complexity of the relations of all organic beings to each other and to their conditions of existence, causing an infinite diversity in structure, constitution, and habits, to be advantageous to them, it would be a most extraordinary fact if no variation had ever occurred useful to each being's own welfare, in the same manner as so many variations have occurred useful to man. But if variations useful to any organic being do occur, assuredly individuals thus characterised will have the best chance of being preserved in the struggle for life; and from the strong principle of inheritance, they will tend to produce offspring similarly characterised. This principle of preservation I have called, for the sake of brevity, Natural Selection. It leads to the improvement of each creature in relation to its organic and inorganic conditions of life; and consequently, in most cases, to what must be regarded as an advance in organisation. Nevertheless low and simple forms would long endure if well fitted for their simple conditions of life.

Natural selection, on the principle of qualities being inherited at corresponding ages, can modify the egg, seed, or young, as easily as the adult. Amongst many animals, sexual selection will give its aid to ordinary selection, by assuring to the most vigorous and best adapted males the greatest number of offspring. Sexual selection will also give characters useful to the males alone, in their struggles with other males.

Whether natural selection has really thus acted in nature, in modifying and adapting the various forms of life to their several conditions and stations, must be judged of by the general tenor and balance of evidence given in the following chapters. But we already see how it entails extinction; and how largely extinction has acted in the world's history, geology plainly declares. Natural selection, also, leads to divergence of character; for the more organic beings diverge in structure, habits, and constitution, by so much a greater number can be supported on the same area,—of which we see proof by looking to the inhabitants of any small spot or to naturalised productions. Therefore, during the modification of the descendants of any one species, and during the incessant struggle of all species to increase in numbers, the more diversified these descendants become, the better will be their chance of succeeding in the battle for life. Thus the small differences distinguishing varieties of the same species, steadily tend to increase till they come to equal the greater differences between species of the same genus, or even of distinct genera.

We have seen that it is the common, the widely-diffused, and widely-ranging species, belonging to the larger genera within each class, which vary most; and these tend to transmit to their modified offspring that superiority which now makes them dominant in their own countries. Natural selection, as has just been remarked, leads to divergence of character and to much extinction of the less improved and intermediate forms of life. On these principles, the nature of the affinities, and the generally well-defined distinctions of the innumerable organic beings in each class throughout the world, may be explained. It is a truly wonderful fact —the wonder of which we are apt to overlook from familiarity—that all animals and all plants throughout all time and space should be related to each other in group subordinate to group, in the manner which we everywhere behold—namely, varieties of the same species most closely related together, species of the same genus less closely and unequally related together, forming sections and sub-genera, species of distinct genera much less closely related, and genera related in different degrees, forming sub-families, families, orders, sub-classes, and classes. The several subordinate groups in any class cannot be ranked in a single file, but seem rather to be clustered round points, and these round other points, and so on in almost endless cycles. On the view that each species has been independently created, I can see no explanation of this great fact in the classification of all organic beings; but, to the best of my judgment, it is explained through inheritance and the complex action of natural selection, entailing extinction and divergence of character, as we have seen illustrated in the diagram.

The affinities of all the beings of the same class have sometimes been represented by a great tree. I believe this simile largely speaks the truth. The green and budding twigs may represent existing species; and those produced during each former year may represent the long succession of extinct species. At each period of growth all the growing twigs have tried to branch out on all sides, and to overtop and kill the surrounding twigs and branches, in the same manner as species and groups of species have tried to overmaster other species in the great battle for life. The limbs divided into great branches, and these into lesser and lesser branches, were themselves once, when the tree was small, budding twigs; and this connexion of the former and present buds by ramifying branches may well represent the classification of all extinct and living species in groups subordinate to groups. Of the many twigs which flourished when the tree was a mere bush, only two or three, now grown into great branches, yet survive and bear all the other branches; so with the species which lived during long-past geological periods, very few now have living and modified descendants. From the first growth of the tree, many a limb and branch has decayed and dropped off; and these lost branches of various sizes may represent those whole orders, families, and genera which have now no living representatives, and which are known to us only from having been found in a fossil state. As we here and there see a thin straggling branch springing from a fork low down in a tree, and which by some chance has been favoured and is still alive on its summit, so we occasionally see an animal like the Ornithorhynchus or Lepidosiren, which in some small degree connects by its affinities two large branches of life, and which has apparently been saved from fatal competition by having inhabited a protected station. As buds give rise by growth to fresh buds, and these, if vigorous, branch out and overtop on all sides many a feebler branch, so by generation I believe it has been with the great Tree of Life, which fills with its dead and broken branches the crust of the earth, and covers the surface with its ever branching and beautiful ramifications.

CHAPTER V.

LAWS OF VARIATION.

Effects of external conditions — Use and disuse, combined with natural selection ; organs of flight and of vision — Acclimatisation — Correlation of growth -- Compensation and economy of growth — False correlations — Multiple, rudimentary, and lowly organised structures variable — Parts developed in an unusual manner are highly variable : specific characters more variable than generic : secondary sexual characters variable — Species of the same genus vary in an analogous manner — Reversions to long-lost characters — Summary.

I HAVE hitherto sometimes spoken as if the variations —so common and multiform in organic beings under domestication, and in a lesser degree in those in a state of nature—had been due to chance. This, of course, is a wholly incorrect expression, but it serves to acknowledge plainly our ignorance of the cause of each particular variation. Some authors believe it to be as much the function of the reproductive system to produce individual differences, or very slight deviations of structure, as to make the child like its parents. But the much greater variability, as well as the greater frequency of monstrosities, under domestication or cultivation, than under nature, leads to the belief that deviations of structure are in some way due to the nature of the conditions of life, to which the parents and their more remote ancestors have been exposed during several generations. I have remarked in the first chapter—but a long catalogue of facts which cannot be here given would be necessary to show the truth of the remark—that the reproductive system is eminently susceptible to changes in the conditions of life; and to this system being functionally disturbed in the parents, I chiefly attribute the varying or plastic condition of the offspring. The male and female sexual elements seem to be affected before that union takes place which is to form a new being. In the case of "sporting" plants, the bud, which in its earliest condition does not apparently differ essentially from an ovule, is alone affected. But why, because the reproductive system is disturbed, this or that part should vary more or less, we are profoundly ignorant. Nevertheless, we can here and there dimly catch a faint ray of light, and we may feel sure that there must be some cause for each deviation of structure, however slight.

How much direct effect difference of climate, food, &c., produces on any being is extremely doubtful. My impression is, that the effect is small in the case of animals, but more in that of plants. We may, at least, safely conclude that such influences cannot have produced the many striking and complex co-adaptations of structure between one organic being and another, which we see everywhere throughout nature. Some little influence may be attributed to climate, food, &c.: thus, E. Forbes speaks confidently that shells at their southern limit, and when living in shallow water, vary and become more brightly coloured than those of the same species further north or from greater depths. Gould believes that birds of the same species are more brightly coloured under a clear atmosphere, that when living on islands or near the coast. So with insects, Wollaston is convinced that residence near the sea affects their colours. Moquin-Tandon gives a list of plants which when growing near the sea-shore have their leaves in some degree fleshy, though not elsewhere fleshy. Several other such cases could be given.

The fact of varieties of one species, when they range into the zone of habitation of other species, often acquiring in a very slight degree some of the characters of such species, accords with our view that species of all kinds are only well-marked and permanent varieties. Thus the species of shells which are confined to tropical and shallow seas are generally brighter-coloured than those confined to cold and deeper seas. The birds which are confined to continents are, according to Mr. Gould, brighter-coloured than those of islands. The insect-species confined to sea-coasts, as every collector knows, are often brassy or lurid. Plants which live exclusively on the sea-side are very apt to have fleshy leaves. He who believes in the creation of each species, will have to say that this shell, for instance, was created with bright colours for a warm sea; but that this other shell became bright-coloured by variation when it ranged into warmer or shallower waters.

When a variation is of the slightest use to a being, we cannot tell how much of it to attribute to the accumulative action of natural selection, and how much to the conditions of life. Thus, it is well known to furriers that animals of the same species have thicker and better fur the more severe the climate is under which they have lived; but who can tell how much of this difference may be due to the warmest-clad individuals having been favoured and preserved during many generations, and how much to the direct action of the severe climate? for it would appear that climate has some direct action on the hair of our domestic quadrupeds.

Instances could be given of the same variety being produced under conditions of life as different as can well be conceived; and, on the other hand, of different varieties being produced from the same species under apparently the same conditions. Such facts show how indirectly the conditions of life act. Again, innumerable instances are known to every naturalist of species keeping true, or not varying at all, although living under the most opposite climates. Such considerations as these incline me to lay very little weight on the direct action of the conditions of life. Indirectly, as already remarked, they seem to play an important part in affecting the reproductive system, and in thus inducing variability; and natural selection will then accumulate all profitable variations, however slight, until they become plainly developed and appreciable by us.

Effects of Use and Disuse.—From the facts alluded to in the first chapter, I think there can be little doubt that use in our domestic animals strengthens and enlarges certain parts, and disuse diminishes them; and that such modifications are inherited. Under free nature, we can have no standard of comparison, by which to judge of the effects of long-continued use or disuse, for we know not the parent-forms; but many animals have structures which can be explained by the effects of disuse. As Professor Owen has remarked, there is no greater anomaly in nature than a bird that cannot fly; yet there are several in this state. The logger-headed duck of South America can only flap along the surface of the water, and has its wings in nearly the same condition as the domestic Aylesbury duck. As the larger ground-feeding birds seldom take flight except to escape danger, I believe that the nearly wingless condition of several birds, which now inhabit or have lately inhabited several oceanic islands, tenanted by no beast of prey, has been caused by disuse. The ostrich indeed inhabits continents and is exposed to danger from which it cannot escape by flight, but by kicking it can defend

itself from enemies, as well as any of the smaller
quadrupeds. We may believe that the progenitor of the
ostrich genus had habits like those of a bustard, and
that as natural selection increased in successive generations the size and weight of its body, its legs were used
more, and its wings less, until they became incapable
of flight.

Kirby has remarked (and I have observed the same
fact) that the anterior tarsi, or feet, of many male dung-
feeding beetles are very often broken off; he examined
seventeen specimens in his own collection, and not one
had even a relic left. In the Onites apelles the tarsi
are so habitually lost, that the insect has been described
as not having them. In some other genera they are present, but in a rudimentary condition. In the Ateuchus
or sacred beetle of the Egyptians, they are totally deficient. The evidence that accidental mutilations can be
inherited is at present very scanty; but the remarkable
case observed by Brown-Séquard of epilepsy produced
by injuring the spinal chord of guinea-pigs, being
inherited, should make us cautious. So that it will
perhaps be safest to look at the entire absence of the
anterior tarsi in Ateuchus, and their rudimentary condition in some other genera, as due to the long-continued
effects of disuse in their progenitors; for as the tarsi
are almost always lost in many dung-feeding beetles,
they must be lost early in life, and therefore cannot be
of much importance or be much used by these insects.

In some cases we might easily put down to disuse
modifications of structure which are wholly, or mainly,
due to natural selection. Mr. Wollaston has discovered
the remarkable fact that 200 beetles out of the 550
species inhabiting Madeira are so far deficient in wings
that they cannot fly; and that of the twenty-nine
endemic genera, no less than twenty-three genera have

all their species in this condition ! Several facts, namely,
that leetles in many parts of the world are frequently
blown to sea and perish ; that the beetles in Madeira,
as observed by Mr. Wollaston, lie much concealed,
until the wind lulls and the sun shines ; that the proportion of wingless beetles is larger on the exposed
Desertas than in Madeira itself; and especially the
extraordinary fact, so strongly insisted on by Mr. Wollaston, of the almost entire absence of certain large
groups of beetles, elsewhere excessively numerous, and
which groups have habits of life almost necessitating
frequent flight ;—these several considerations have made
me believe that the wingless condition of so many
Madeira beetles is mainly due to the action of natural
selection, but combined probably with disuse. For
during many successive generations each individual
beetle which flew least, either from its wings having
been ever so little less perfectly developed or from
indolent habit, will have had the best chance of surviving from not being blown out to sea ; and, on the
other hand, those beetles which most readily took to
flight would oftenest have been blown to sea and thus
have been destroyed.

The insects in Madeira which are not ground-feeders,
and which, as the flower-feeding coleoptera and lepidop-
tera, must habitually use their wings to gain their subsistence, have, as Mr. Wollaston suspects, their wings not
at all reduced, but even enlarged. This is quite compatible with the action of natural selection. For when
a new insect first arrived on the island, the tendency
of natural selection to enlarge or to reduce the wings,
would depend on whether a greater number of individuals were saved by successfully battling with the winds,
or by giving up the attempt and rarely or never flying.
As with mariners shipwrecked near a coast, it would

have been better for the good swimmers if they had
been able to swim still further, whereas it would have
been better for the bad swimmers if they had not been
able to swim at all and had stuck to the wreck.

The eyes of moles and of some burrowing rodents are
rudimentary in size, and in some cases are quite covered
up by skin and fur. This state of the eyes is probably
due to gradual reduction from disuse, but aided perhaps
by natural selection. In South America, a burrowing
rodent, the tuco-tuco, or Ctenomys, is even more subterranean in its habits than the mole; and I was assured
by a Spaniard, who had often caught them, that they
were frequently blind; one which I kept alive was certainly in this condition, the cause, as appeared on dis-
section, having been inflammation of the nictitating
membrane. As frequent inflammation of the eyes must
be injurious to any animal, and as eyes are certainly
not indispensable to animals with subterranean habits,
a reduction in their size with the adhesion of the eyelids and growth of fur over them, might in such case be
an advantage; and if so, natural selection would constantly aid the effects of disuse.

It is well known that several animals, belonging to the
most different classes, which inhabit the caves of Carniola and of Kentucky, are blind. In some of the crabs the
foot-stalk for the eye remains, though the eye is gone;
the stand for the telescope is there, though the telescope
with its glasses has been lost. As it is difficult to imagine that eyes, though useless, could be in any way
injurious to animals living in darkness, I attribute their
loss wholly to disuse. In one of the blind animals,
namely, the cave-rat (Neotoma), two of which were captured by Professor Silliman at above half a mile distance
from the mouth of the cave, and therefore not in the
profoundest depths, the eyes were lustrous and of large

size ; but these animals, as I am informed by Professor Silliman, having been exposed for about a month to a graduated light, acquired a dim perception of objects when brought towards their eyes, and blinked.

It is difficult to imagine conditions of life more similar than deep limestone caverns under a nearly similar climate ; so that on the common view of the blind animals having been separately created for the American and European caverns, very close similarity in their organisation and affinities might have been expected. This is certainly not the case if we look at the two whole faunas ; and with respect to the insects alone, Schiödte has remarked, " We are accordingly prevented from considering the entire phenomenon in any other light than something purely local, and the similarity which is exhibited in a few forms between the Mammoth cave (in Kentucky) and the caves in Carniola, otherwise than as a very plain expression of that analogy which subsists generally between the fauna of Europe and of North America." On my view we must suppose that American animals, having in most cases ordinary powers of vision, slowly migrated by successive generations from the outer world into the deeper and deeper recesses of the Kentucky caves, as did European animals into the caves of Europe. We have some evidence of this gradation of habit; for, as Schiödte remarks, " We accordingly look upon the subterranean faunas as small ramifications which have penetrated into the earth from the geographically limited faunas of the adjacent tracts, and which, as they extended themselves into darkness, have been accommodated to surrounding circumstances. Animals not far remote from ordinary forms, prepare the transition from light to darkness. Next follow those that are constructed for twilight ; and, last of all, those destined for total darkness, and

whose formation is quite peculiar." These remarks of
Schiödte's of course apply not to the same, but to distinct species. By the time that an animal had reached,
after numberless generations, the deepest recesses, disuse will on this view have more or less perfectly
obliterated its eyes, and natural selection will often
have effected other changes, such as an increase in
tho length of the antennæ or palpi, as a compensation
for blindness. Notwithstanding such modifications,
we might expect still to see in the cave-animals of
America, affinities to the other inhabitants of that continent, and in those of Europe, to the inhabitants of
tho European continent. And this is the case with
some of the American cave-animals, as I hear from
Professor Dana; and some of the European cave-
insects are very closely allied to those of the surrounding country. It would be most difficult to give any
rational explanation of the affinities of the blind cave-
animals to the other inhabitants of the two continents
on the ordinary view of their independent creation.
That several of tho inhabitants of tho caves of the
Old and New Worlds should be closely related, wo
might expect from the well-known relationship of most
of their other productions. As a blind species of
Bathyscia is found in abundance on shady rocks out of
the caves, tho loss of vision in the cave-species has
probably had no relation to its dark habitation; and it
is very natural that an insect already deprived of
vision should readily become adapted to dark caverns.
Another blind genus (Anophthalmus) offers this remarkable peculiarity: the several distinct species, as Mr.
Murray has remarked, inhabit several distinct European
caves and likewise those of Kentucky, and the genus is
found nowhere except in caves; but it is possible that
the progenitor or progenitors of these several species

may formerly have ranged widely over both contiuents, and have since (like the elephant on both continents) become extinct, excepting in their present secluded habitations. Far from feeling surprise that some of the cave-animals should be very anomalous, as Agassiz has remarked in regard to the blind fish, the Amblyopsis, and as is the case with the blind Proteus with reference to the reptiles of Europe, I am only surprised that more wrecks of ancient life have not been preserved, owing to the less severe competition to which the iuhabitauts of these dark abodes will have been exposed.

Acclimatisation.—Habit is hereditary with plants, as in the period of flowering, in the amount of rain requisite for seeds to germinate, in the *time* of sleep, &c., and this leads me to say a few words on acclimatisation. As it is extremely common for species of the same genus to inhabit very hot and very cold countries, and as I believe that all the species of the same genus have descended from a single parent, if this view be correct, acclimatisation must be readily effected during long-continued descent. It is notorious that each species is adapted to the climate of its own home: species from an arctic or even from a temperate region cannot endure a tropical climate, or conversely. So again, many succulent plants cannot endure a damp climate. But the degree of adaptation of species to the climates under which they live is often overrated. We may infer this from our frequent inability to predict whether or not an imported plant will endure our climate, and from the number of plants and animals brought from warmer countries which here enjoy good health. We have reason to believe that species in a state of nature are closely limited in their ranges by the competition of other organic beings quite as much as,

or more than, by adaptation to particular climates. But
whether or not the adaptation be generally very close,
we have evidence, in the case of some few plants, of their
becoming, to a certain extent, naturally habituated to
different temperatures, or becoming acclimatised: thus
the pines and rhododendrons, raised from seed collected
by Dr. Hooker from trees growing at different heights
on the Himalaya, were found in this country to possess
different constitutional powers of resisting cold. Mr.
Thwaites informs me that he has observed similar facts
in Ceylon, and analogous observations have been made
by Mr. H. C. Watson on European species of plants
brought from the Azores to England. In regard to
animals, several authentic cases could be given of
species within historical times having largely extended
their range from warmer to cooler latitudes, and conversely; but we do not positively know that these animals were strictly adapted to their native climate, yet
in all ordinary cases we assume such to be the case;
nor do we know that they have subsequently become
acclimatised to their new homes.

As we may infer that our domestic animals were originally chosen by uncivilised man because they were useful and bred readily under confinement, and not because
they were subsequently found capable of far-extended
transportation, the common and extraordinary capacity
in our domestic animals of not only withstanding the
most different climates, but of being perfectly fertile
(a far severer test) under them, may be used as an
argument that a large proportion of other animals now
in a state of nature could easily be brought to bear
widely different climates. We must not, however, push
the foregoing argument too far, on account of the probable origin of some of our domestic animals from
several wild stocks: the blood, for instance, of a tropical

and arctic wolf or wild dog may perhaps be mingled in our domestic breeds. The rat and mouse cannot be considered as domestic animals, but they have been transported by man to many parts of the world, and now have a far wider range than any other rodent, living free under the cold climate of Faroe in the north and of the Falklands in the south, and on many islands in the torrid zones. Hence I am inclined to look at adaptation to any special climate as a quality readily grafted on an innate wide flexibility of constitution, which is common to most animals. On this view, the capacity of enduring the most different climates by man himself and by his domestic animals, and such facts as that former species of the elephant and rhinoceros were capable of enduring a glacial climate, whereas the living species are now all tropical or sub-tropical in their habits, ought not to be looked at as anomalies, but merely as examples of a very common flexibility of constitution, brought, under peculiar circumstances, into play.

How much of the acclimatisation of species to any peculiar climate is due to mere habit, and how much to the natural selection of varieties having different innate constitutions, and how much to both means combined, is an obscure question. That habit or custom has some influence I must believe, both from analogy and from the incessant advice given in agricultural works, even in the ancient Encyclopædias of China, to be very cautious in transposing animals from one district to another; for it is not likely that man should have succeeded in selecting so many breeds and sub-breeds with constitutions specially fitted for their own districts: the result must, I think, be due to habit. On the other hand, I can see no reason to doubt that natural selection will continually tend to preserve those individuals which are born with constitutions best adapted to their native countries. In treatises on many kinds of cultivated plants, certain varieties are said to withstand certain climates better than others : this is very strikingly shown in works on fruit trees published in the United States, in which certain varieties are habitually recommended for the northern and others for the southern States ; and as most of these varieties are of recent origin, they cannot owe their constitutional differences to habit. The case of the Jerusalem artichoke, which is never propagated by seed, and of which consequently new varieties have not been produced, has even been advanced—for it is now as tender as ever it was—as proving that acclimatisation cannot be effected ! The case, also, of the kidney-bean has been often cited for a similar purpose, and with much greater weight ; but until some one will sow, during a score of generations, his kidney-beans so early that a very large proportion are destroyed by frost, and then collect seed from the few survivors, with care to prevent accidental crosses, and then again get seed from these seedlings, with the same precautions, the experiment cannot ·be said to have been even tried. Nor let it be supposed that no differences in the constitution of seedling kidney-beans ever appear, for an account has been published how much more hardy some seedlings appeared to be than others.

On the whole, I think we may conclude that habit, use, and disuse, have, in some cases, played a considerable part in the modification of the constitution, and of the structure of various organs ; but that the effects of use and disuse have often been largely combined with, and sometimes overmastered by the natural selection of innate variations.

Correlation of Growth.—I mean by this expression
that the whole organisation is so tied together during its
growth and development, that when slight variations in
any one part occur, and are accumulated through natural
selection, other parts become modified.   This is a very
important subject, most imperfectly understood.   The
most obvious case is, that modifications accumulated
solely for the good of the young or larva, will, it may
safely be concluded, affect the structure of the adult ;
in the same manner as any malconformation affecting
the early embryo, seriously affects the whole organisation of the adult.   The several parts of the body which
are homologous, and which, at an early embryonic period,
are alike, seem liable to vary in an allied manner : we
see this in the right and left sides of the body varying
in the same manner ; in the front and hind legs, and
even in the jaws and limbs, varying together, for the
lower jaw is believed to be homologous with the limbs.
These tendencies, I do not doubt, may be mastered
more or less completely by natural selection : thus a
family of stags once existed with an antler only on
one side ; and if this had been of any great use to the
breed, it might probably have been rendered permanent
by natural selection.

Homologous parts, as has been remarked by some
authors, tend to cohere ; this is often seen in monstrous
plants ; and nothing is more common than the union of
homologous parts in normal structures, as the union of
the petals of the corolla into a tube.   Hard parts seem
to affect the form of adjoining soft parts ; it is believed
by some authors that the diversity in the shape of the
pelvis in birds causes the remarkable diversity in the
shape of their kidneys.   Others believe that the shape
of the pelvis in the human mother influences by pressure the shape of the head of the child.   In snakes,

according to Schlegel, the shape of the body and the manner of swallowing determine the position of several of the most important viscera.

The nature of the bond of correlation is very frequently quite obscure. M. Is. Geoffroy St. Hilaire has forcibly remarked, that certain malconformations very frequeutly, and that others rarely coexist, without our being able to assign any reason. What can be more singular than the relation between blue eyes and deafness in cats, and the tortoise-shell colour with the female sex ; the feathered feet and skin between the outer toes in pigeons, and the presence of more or less down on the young birds when first hatched, with the future colour of their plumage; or, again, the relation between the hair and teeth in the naked Turkish dog, though here probably homology comes into play? With respect to this latter case of correlation, I think it can hardly be accidental, that if we pick out the two orders of mammalia which are most abnormal in their dermal covering, viz. Cetacea (whales) and Edentata (armadilloes, scaly anteaters, &c.), that these are likewise the most abnormal in their teeth.

I know of no case better adapted to show the importance of the laws of correlation in modifying important structures, independently of utility and, therefore, of natural selection, than that of the difference between the outer and inner flowers in some Compositous and Umbelliferous plants. Every one knows the difference in the ray and central florets of, for instance, the daisy, and this difference is often accompanied with the abortion of parts of the flower. But, in some Compositous plants, the seeds also differ in shape and sculpture; and even the ovary itself, with its accessory parts, differs, as has been described by Cassini. These differences have been attributed by some authors to pressure, and the shape of the seeds in the ray-florets in some Compositæ countenances this idea ; but, in the case of the corolla of the Umbelliferæ, it is by no means, as Dr. Hooker informs me, in species with the densest heads that the inner and outer flowers most frequently differ. It might have been thought that the development of the ray-petals by drawing nourishment from certain other parts of the flower had caused their abortion ; but in some Compositæ there is a difference in the seeds of the outer and inner florets without any difference in the corolla. Possibly these several differences may be connected with some difference in the flow of nutriment towards the central and external flowers : we know, at least, that in irregular flowers, those nearest to the axis are said to be oftenest subject to peloria, and become regular. I may add, as an instance of this, and of a striking case of correlation, that I have recently observed in some garden pelargoniums, that the central flower of the truss often loses the patches of darker colour in the two upper petals ; and that when this occurs, the adherent nectary is quite aborted ; when the colour is absent from only one of the two upper petals, the nectary is only much shortened.

With respect to the difference in the corolla of the central and exterior flowers of a head or umbel, I do not feel at all sure that C. C. Spreugel's idea that the ray-florets serve to attract insects, whose agency is highly advantageous in the fertilisation of plants of these two orders, is so far-fetched, as it may at first appear : and if it be advantageous, natural selection may have come into play. But in regard to the differences both in the internal and external structure of the seeds, which are not always correlated with any differences in the flowers, it seems impossible that they can be in any way advantageous to the plant : yet in the

Umbelliferæ these differences are of such apparent importance—the seeds being in some cases, according to Tausch, orthospermous in the exterior flowers and cœlospermous in the central flowers,—that the elder De Candolle founded his main divisions of the order on analogous differences. Hence we see that modifications of structure, viewed by systematists as of high value, may be wholly due to unknown laws of correlated growth, and without being, as far as we can see, of the slightest service to the species.

We may often falsely attribute to correlation of growth structures which are common to whole groups of species, and which in truth are simply due to inheritance; for an ancient progenitor may have acquired through natural selection some one modification in structure, and, after thousands of generations, some other and independent modification; and these two modifications, having been transmitted to a whole group of descendants with diverse habits, would naturally be thought to be correlated in some necessary manner. So, again, some correlations, occurring throughout whole orders, are apparently due to the manner alone in which natural selection can act. For instance, Alph. de Candolle has remarked that winged seeds are never found in fruits which do not open: I should explain the rule by the fact that seeds could not gradually become winged through natural selection, except in fruits which opened; so that the individual plants producing seeds which were a little better fitted to be wafted further, might get an advantage over those producing seed less fitted for dispersal; and this process could not possibly go on in fruit which did not open.

The elder Geoffroy and Goethe propounded, at about the same period, their law of compensation or balancement of growth; or, as Goethe expressed it, "in order to spend on one side, nature is forced to economise on the other side." I think this holds true to a certain extent with our domestic productions: if nourishment flows to one part or organ in excess, it rarely flows, at least in excess, to another part; thus it is difficult to get a cow to give much milk and to fatten readily. The same varieties of the cabbage do not yield abundant and nutritious foliage and a copious supply of oil-bearing seeds. When the seeds in our fruits become atrophied, the fruit itself gains largely in size and quality. In our poultry, a large tuft of feathers on the head is generally accompanied by a diminished comb, and a large beard by diminished wattles. With species in a state of nature it can hardly be maintained that the law is of universal application; but many good observers, more especially botanists, believe in its truth. I will not, however, here give any instances, for I see hardly any way of distinguishing between the effects, on the one hand, of a part being largely developed through natural selection and another and adjoining part being reduced by this same process or by disuse, and, on the other hand, the actual withdrawal of nutriment from one part owing to the excess of growth in another and adjoining part.

I suspect, also, that some of the cases of compensation which have been advanced, and likewise some other facts, may be merged under a more general principle, namely, that natural selection is continually trying to economise in every part of the organisation. If under changed conditions of life a structure before useful becomes less useful, any diminution, however slight, in its development, will be seized on by natural selection, for it will profit the individual not to have its nutriment wasted in building up an useless structure. I can thus only understand a fact with which I was much struck when examining cirripedes, and of which many other

instances could be given : namely, that when a cirripede
is parasitic within another and is thus protected, it loses
more or less completely its own shell or carapace. This
is the case with the male Ibla, and in a truly extraordinary manner with the Proteolepas: for the carapace in
all other cirripedes consists of the three highly-important
anterior segments of the head enormously developed,
and furnished with great nerves and muscles ; but in
the parasitic and protected Proteolepas, the whole anterior part of the head is reduced to the merest rudiment
attached to the bases of the prehensile antennæ. Now
the saving of a large and complex structure, when rendered superfluous by the parasitic habits of the Proteolepas, though effected by slow steps, would be a decided
advantage to each successive individual of the species ;
for in the struggle for life to which every animal is exposed, each individual Proteolepas would have a better
chance of supporting itself, by less nutriment being
wasted in developing a structure now become useless.

Thus, as I believe, natural selection will always succeed in the long run in reducing and saving every part
of the organisation, as soon as it is rendered superfluous,
without by any means causing some other part to be
largely developed in a corresponding degree. And, conversely, that natural selection may perfectly well succeed in largely developing any organ, without requiring
as a necessary compensation the reduction of some adjoining part.

It seems to be a rule, as remarked by Is. Geoffroy
St. Hilaire, both in varieties and in species, that when
any part or organ is repeated many times in the structure of the same individual (as the vertebræ in snakes,
and the stamens in polyandrous flowers) the number is
variable ; whereas the number of the same part or organ,
when it occurs in lesser numbers, is constant. The same

author and some botanists have further remarked that
multiple parts are also very liable to variation in structure. Inasmuch as this "vegetative repetition," to use
Prof. Owen's expression, seems to be a sign of low organisation, the foregoing remark seems connected with
the very general opinion of naturalists, that beings low
in the scale of nature are more variable than those which
are higher. I presume that lowness in this case means
that the several parts of the organisation have been but
little specialised for particular functions; and as long as
the same part has to perform diversified work, we can
perhaps see why it should remain variable, that is, why
natural selection should have preserved or rejected each
little deviation of form less carefully than when the part
has to serve for one special purpose alone. In the same
way that a knife which has to cut all sorts of things
may be of almost any shape; whilst a tool for some
particular object had better be of some particular shape.
Natural selection, it should never be forgotten, can act
on each part of each being, solely through and for its
advantage.

Rudimentary parts, it has been stated by some
authors, and I believe with truth, are apt to be highly
variable. We shall have to recur to the general subject
of rudimentary and aborted organs; and I will here only
add that their variability seems to be owing to their
uselessness, and therefore to natural selection having
no power to check deviations in their structure. Thus
rudimentary parts are left to the free play of the
various laws of growth, to the effects of long-continued
disuse, and to the tendency to reversion.

*A part developed in any species in an extraordinary
degree or manner, in comparison with the same part in
allied species, tends to be highly variable.*—Several years

ago I was much struck with a remark, nearly to the
above effect, published by Mr. Waterhouse. I infer
also from an observation made by Professor Owen, with
respect to the length of the arms of the ourang-outang,
that he has come to a nearly similar conclusion. It is
hopeless to attempt to convince any one of the truth of
this proposition without giving the long array of facts
which I have collected, and which cannot possibly be
here introduced. I can only state my conviction that it is
a rule of high generality. I am aware of several causes
of error, but I hope that I have made due allowance for
them. It should be understood that the rule by no
means applies to any part, however unusually developed,
unless it be unusually developed in comparison with
the same part in closely allied species. Thus, the bat's
wing is a most abnormal structure in the class mammalia; but the rule would not here apply, because there
is a whole group of bats having wings; it would apply
only if some one species of bat had its wings developed
in some remarkable manner in comparison with the
other species of the same genus. The rule applies very
strongly in the case of secondary sexual characters, when
displayed in any unusual manner. The term, secondary
sexual characters, used by Hunter, applies to characters
which are attached to one sex, but are not directly
connected with the act of reproduction. The rule applies to males and females; but as females more rarely
offer remarkable secondary sexual characters, it applies
more rarely to them. The rule being so plainly applicable in the case of secondary sexual characters, may be
due to the great variability of these characters, whether
or not displayed in any unusual manner—of which fact
I think there can be little doubt. But that our rule is
not confined to secondary sexual characters is clearly
shown in the case of hermaphrodite cirripedes; and

I may here add, that I particularly attended to Mr. Waterhouse's remark, whilst investigating this Order, and I am fully convinced that the rule almost invariably holds good with cirripedes. I shall, in my future work, give a list of the more remarkable cases; I will here only briefly give one, as it illustrates the rule in its largest application. The opercular valves of sessile cirripedes (rock barnacles) are, in every sense of the word, very important structures, and they differ extremely little even in different genera; but in the several species of one genus, Pyrgoma, these valves present a marvellous amount of diversification; the homologous valves in the different species being sometimes wholly unlike in shape; and the amount of variation in the individuals of several of the species is so great, that it is no exaggeration to state that the varieties differ more from each other in the characters of these important valves than do other species of distinct genera.

As birds within the same country vary in a remarkably small degree, I have particularly attended to them, and the rule seems to me certainly to hold good in this class. I cannot make out that it applies to plants, and this would seriously have shaken my belief in its truth, had not the great variability in plants made it particularly difficult to compare their relative degrees of variability.

When we see any part or organ developed in a remarkable degree or manner in any species, the fair presumption is that it is of high importance to that species; nevertheless the part in this case is eminently liable to variation. Why should this be so? On the view that each species has been independently created, with all its parts as we now see them, I can see no explanation. But on the view that groups of species

I

have descended from other species, and have been modified through natural selection, I think we can obtain
some light. In our domestic animals, if any part, or
the whole animal, be neglected and no selection be applied, that part (for instance, the comb in the Dorking
fowl) or the whole breed will cease to have a nearly
uniform character. The breed will then bo said to have
degenerated. In rudimentary organs, and in those
which have been but little specialised for any particular
purpose, and perhaps in polymorphic groups, we see a
nearly parallel natural case; for in such cases natural
selection either has not or cannot come into full play,
and thus the organization is left in a fluctuating condition. But what here more especially concerns us is,
that in our domestic animals those points, which at the
present time are undergoing rapid change by continued
selection, are also eminently liable to variation. Look
at the breeds of the pigeon; see what a prodigious
amount of difference there is in the beak of the different tumblers, in the beak and wattle of the different
carriers, in the carriage and tail of our fantails, &c.,
these being the points now mainly attended to by English fanciers. Even in the sub-breeds, as in the short-
faced tumblers, it is notoriously difficult to breed them
nearly to perfection, and frequently individuals are born
which depart widely from the standard. There may be
truly said to be a constant struggle going on between,
on the one hand, the tendency to reversion to a less
modified state, as well as an innate tendency to further
variability of all kinds, and, on the other hand, the
power of steady selection to keep the breed true. In
the long run selection gains the day, and we do not
expect to fail so far as to breed a bird as coarse as a
common tumbler from a good short-faced strain. But
as long as selection is rapidly going on, there may

always be expected to be much variability in the structure undergoing modification. It further deserves notice that these variable characters, produced by man's selection, sometimes become attached, from causes quite unknown to us, more to one sex than to the other, generally to the male sex, as with the wattle of carriers and the enlarged crop of pouters.

Now let us turn to nature. When a part has been developed in an extraordinary manner in any one species, compared with the other species of the same genus, we may conclude that this part has undergone an extraordinary amount of modification since the period when the species branched off from the common progenitor of the genus. This period will seldom be remote in any extreme degree, as species very rarely endure for more than one geological period. An extraordinary amount of modification implies an unusually large and long-continued amount of variability, which has continually been accumulated by natural selection for the benefit of the species. But as the variability of the extraordinarily developed part or organ has been so great and long-continued within a period not excessively remote, we might, as a general rule, expect still to find more variability in such parts than in other parts of the organisation which have remained for a much longer period nearly constant. And this, I am convinced, is the case. That the struggle between natural selection on the one hand, and the tendency to reversion and variability on the other hand, will in the course of time cease; and that the most abnormally developed organs may be made constant, I can see no reason to doubt. Hence when an organ, however abnormal it may be, has been transmitted in approximately the same condition to many modified descendants, as in the case of the wing of the bat, it must

have existed, according to my theory, for an immense
period in nearly the same state; and thus it comes to
be no more variable than any other structure. It is
only in those cases in which the modification has been
comparatively recent and extraordinarily great that we
ought to find the *generative variability*, as it may be
called, still present in a high degree. For in this case
the variability will seldom as yet have been fixed by
the continued selection of the individuals varying in
the required manner and degree, and by the continued
rejection of those tending to revert to a former and less
modified condition.

The principle included in these remarks may be extended. It is notorious that specific characters are more
variable than generic. To explain by a simple example
what is meant. If some species in a large genus of
plants had blue flowers and some had red, the colour
would be only a specific character, and no one would be
surprised at one of the blue species varying into red, or
conversely; but if all the species had blue flowers, the
colour would become a generic character, and its variation would be a more unusual circumstance. I have
chosen this example because an explanation is not in
this case applicable, which most naturalists would advance, namely, that specific characters are more variable
than generic, because they are taken from parts of less
physiological importance than those commonly used for
classing genera. I believe this explanation is partly,
yet only indirectly, true; I shall, however, have to return to this subject in the chapter on Classification. It
would be almost superfluous to adduce evidence in support of the above statement, that specific characters are
more variable than generic; but I have repeatedly
noticed in works on natural history, that when an author has remarked with surprise that some *important*

organ or part, which is generally very constant throughout large groups of species, has *differed* considerably in
closely-allied species, that it has, also, been *variable* in
the individuals of some of the species. And this fact
shows that a character, which is generally of generic
value, when it sinks in value and becomes only of specific value, often becomes variable, though its physiological importance may remain the same. Something of
the same kind applies to monstrosities: at least Is.
Geoffroy St. Hilaire seems to entertain no doubt, that
the more an organ normally differs in the different
species of the same group, the more subject it is to
individual anomalies.

On the ordinary view of each species having been
independently created, why should that part of the
structure, which differs from the same part in other
independently-created species of the same genus, be
more variable than those parts which are closely alike
in the several species? I do not see that any explanation can be given. But on the view of species being
only strongly marked and fixed varieties, we might
surely expect to find them still often continuing to vary
in those parts of their structure which have varied
within a moderately recent period, and which have thus
come to differ. Or to state the case in another manner:—the points in which all the species of a genus
resemble each other, and in which they differ from the
species of some other genus, are called generic characters; and these characters in common I attribute to inheritance from a common progenitor, for it can rarely
have happened that natural selection will have modified
several species, fitted to more or less widely-different
habits, in exactly the same manner: and as these so-
called generic characters have been inherited from a
remote period, since that period when the species first

branched off from their common progenitor, and subsequently have not varied or come to differ in any degree,
or only in a slight degree, it is not probable that they
should vary at the present day. On the other hand, the
points in which species differ from other species of the
same genus, are called specific characters; and as these
specific characters have varied and come to differ within
the period of the branching off of the species from a
common progenitor, it is probable that they should still
often be in some degree variable,—at least more variable than those parts of the organization which have
for a very long period remained constant.

In connexion with the present subject, I will make
only two other remarks. I think it will be admitted,
without my entering on details, that secondary sexual
characters are very variable; I think it also will be
admitted that species of the same group differ from
each other more widely in their secondary sexual characters, than in other parts of their organisation; compare, for instance, the amount of difference between the
males of gallinaceous birds, in which secondary sexual
characters are strongly displayed, with the amount of
difference between their females; and the truth of this
proposition will be granted. The cause of the original
variability of secondary sexual characters is not manifest; but we can see why these characters should not
have been rendered as constant and uniform as other
parts of the organisation; for secondary sexual characters have been accumulated by sexual selection, which
is less rigid in its action than ordinary selection, as it
does not entail death, but only gives fewer offspring to
the less favoured males. Whatever the cause may be
of the variability of secondary sexual characters, as
they are highly variable, sexual selection will have had
a wide scope for action, and may thus readily have suc-

ceeded in giving to the species of the same group a greater amount of difference in their sexual characters, than in other parts of their structure.

It is a remarkable fact, that the secondary sexual differences between the two sexes of the same species are generally displayed in the very same parts of the organisation in which the different species of the same genus differ from each other. Of this fact I will give in illustration two instances, the first which happen to stand on my list; and as the differences in these cases are of a very unusual nature, the relation can hardly be accidental. The same number of joints in the tarsi is a character generally common to very large groups of beetles, but in the Engidæ, as Westwood has remarked, the number varies greatly; and the number likewise differs in the two sexes of the same species: again in fossorial hymenoptera, the manner of neuration of the wings is a character of the highest importance, because common to large groups; but in certain genera the neuration differs in the different species, and likewise in the two sexes of the same species. Mr. Lubbock has recently remarked, that several minute crustaceans offer excellent illustrations of this law. " In Pontella, for instance, the sexual characters are afforded mainly by the anterior antennæ and by the fifth pair of legs: the specific differences also are principally given by these organs." This relation has a clear meaning on my view of the subject: I look at all the species of the same genus as having as certainly descended from the same progenitor, as have the two sexes of any one of the species. Consequently whatever part of the structure of the common progenitor, or of its early descendants, became variable; variations of this part would, it is highly probable, be taken advantage of by natural and sexual selection, in order to fit the several species to

their several places in the economy of nature, and likewise to fit the two sexes of the same species to each
other, or to fit the males and females to different habits
of life, or the males to struggle with other males for the
possession of the females.

Finally, then, I conclude that the greater variability
of specific characters, or those which distinguish species
from species, than of generic characters, or those which
the species possess in common;—that the frequent extreme variability of any part which is developed in a
species in an extraordinary manner in comparison with
the same part in its congeners; and the slight degree
of variability in a part, however extraordinarily it may
be developed, if it be common to a whole group of species;—that the great variability of secondary sexual
characters, and the great amount of difference in these
same characters between closely allied species;—that
secondary sexual and ordinary specific differences are
generally displayed in the same parts of the organisation,—are all principles closely connected together. All
being mainly due to the species of the same group having descended from a common progenitor, from whom
they have inherited much in common,—to parts which
have recently and largely varied being more likely still
to go on varying than parts which have long been inherited and have not varied,—to natural selection having
more or less completely, according to the lapse of time,
overmastered the tendency to reversion and to further
variability,—to sexual selection being less rigid than
ordinary selection,—and to variations in the same parts
having been accumulated by natural and sexual selection, and having been thus adapted for secondary sexual,
and for ordinary specific purposes.

Distinct species present analogous variations; and a

variety of one species often assumes some of the characters of an allied species, or reverts to some of the characters of an early progenitor.—These propositions will be most readily understood by looking to our domestic races. The most distinct breeds of pigeons, in countries most widely apart, present sub-varieties with reversed feathers on the head and feathers on the feet,—characters not possessed by the aboriginal rock-pigeon; these then are analogous variations in two or more distinct races. The frequent presence of fourteen or even sixteen tail-feathers in the pouter, may be considered as a variation representing the normal structure of another race, the fantail. I presume that no one will doubt that all such analogous variations are due to the several races of the pigeon having inherited from a common parent the same constitution and tendency to variation, when acted on by similar unknown influences. In the vegetable kingdom we have a case of analogous variation, in the enlarged stems, or roots as commonly called, of the Swedish turnip and Ruta baga, plants which several botanists rank as varieties produced by cultivation from a common parent: if this be not so, the case will then be one of analogous variation in two so-called distinct species; and to these a third may be added, namely, the common turnip. According to the ordinary view of each species having been independently created, we should have to attribute this similarity in the enlarged stems of these three plants, not to the *vera causa* of community of descent, and a consequent tendency to vary in a like manner, but to three separate yet closely related acts of creation.

With pigeons, however, we have another case, namely, the occasional appearance in all the breeds, of slaty-blue birds with two black bars on the wings, a white croup, a bar at the end of the tail, with the outer feathers externally edged near their bases with white. As all these marks are characteristic of the parent rock-pigeon, I presume that no one will doubt that this is a case of reversion, and not of a new yet analogous variation appearing in the several breeds. We may, I think, confidently come to this conclusion, because, as we have seen, these coloured marks are eminently liable to appear in the crossed offspring of two distinct and differently coloured breeds; and in this case there is nothing in the external conditions of life to cause the reappearance of the slaty-blue, with the several marks, beyond the influence of the mere act of crossing on the laws of inheritance.

No doubt it is a very surprising fact that characters should reappear after having been lost for many, perhaps for hundreds of generations. But when a breed has been crossed only once by some other breed, the offspring occasionally show a tendency to revert in character to the foreign breed for many generations—some say, for a dozen or even a score of generations. After twelve generations, the proportion of blood, to use a common expression, of any one ancestor, is only 1 in 2048; and yet, as we see, it is generally believed that a tendency to reversion is retained by this very small proportion of foreign blood. In a breed which has not been crossed, but in which *both* parents have lost some character which their progenitor possessed, the tendency, whether strong or weak, to reproduce the lost character might be, as was formerly remarked, for all that we can see to the contrary, transmitted for almost any number of generations. When a character which has been lost in a breed, reappears after a great number of generations, the most probable hypothesis is, not that the offspring suddenly takes after an ancestor some hundred generations distant, but that in each successive generation there has been a tendency to reproduce the character in question, which at last, under unknown favourable conditions, gains an ascendancy. For instance, it is probable that in each generation of the barb-pigeon, which produces most rarely a blue and black-barred bird, there has been a tendency in each generation in the plumage to assume this colour. This view is hypothetical, but could be supported by some facts; and I can see no more abstract improbability in a tendency to produce any character being inherited for an endless number of generations, than in quite useless or rudimentary organs being, as we all know them to be, thus inherited. Indeed, we may sometimes observe a mere tendency to produce a rudiment inherited: for instance, in the common snapdragon (Antirrhinum) a rudiment of a fifth stamen so often appears, that this plant must have an inherited tendency to produce it.

As all the species of the same genus are supposed, on my theory, to have descended from a common parent, it might be expected that they would occasionally vary in an analogous manner; so that a variety of one species would resemble in some of its characters another species; this other species being on my view only a well-marked and permanent variety. But characters thus gained would probably be of an unimportant nature, for the presence of all important characters will be governed by natural selection, in accordance with the diverse habits of the species, and will not be left to the mutual action of the conditions of life and of a similar inherited constitution. It might further be expected that the species of the same genus would occasionally exhibit reversions to lost ancestral characters. As, however, we never know the exact character of the common ancestor of a group, we could not distinguish these two cases: if, for instance, we did not know that the rock-

pigeon was not feather-footed or turn-crowned, we could
not have told, whether these characters in our domestic
breeds were reversions or only analogous variations; but
we might have inferred that the blueness was a case of
reversion, from the number of the markings, which are
correlated with the blue tint, and which it does not appear probable would all appear together from simple
variation. More especially we might have inferred this,
from the blue colour and marks so often appearing
when distinct breeds of diverse colours are crossed.
Hence, though under nature it must generally bo left
doubtful, what cases are reversions to an anciently existing character, and what are new but analogous variations, yet we ought, on my theory, sometimes to find
the varying offspring of a species assuming characters
(either from reversion or from analogous variation)
which already occur in some other members of the same
group. And this undoubtedly is the case in nature.

A considerable part of the difficulty in recognising a
variable species in our systematic works, is due to its
varieties mocking, as it were, some of the other species of the same genus. A considerable catalogue, also,
could be given of forms intermediate between two other
forms, which themselves must be doubtfully ranked as
either varieties or species; and this shows, unless all
these forms be considered as independently created
species, that the one in varying has assumed some of
the characters of the other, so as to produce the intermediate form. But the best evidence is afforded by
parts or organs of an important and uniform nature
occasionally varying so as to acquire, in some degree,
the character of the same part or organ in an allied
species. I have collected a long list of such cases; but
here, as before, I lie under a great disadvantage in
not being able to give them. I can only repeat that

such cases certainly do occur, and seem to me very remarkable.

I will, however, give one curious and complex case, not indeed as affecting any important character, but from occurring in several species of the same genus, partly under domestication and partly under nature. It is a case apparently of reversion. The ass not rarely has very distinct transverse bars on its legs, like those on the legs of the zebra: it has been asserted that these are plainest in the foal, and from inquiries which I have made, I believe this to be true. It has also been asserted that the stripe on each shoulder is sometimes double. The shoulder-stripe is certainly very variable in length and outline. A white ass, but *not* an albino, has been described without either spinal or shoulder stripe; and these stripes are sometimes very obscure, or actually quite lost, in dark-coloured asses. The koulan of Pallas is said to have been seen with a double shoulder-stripe. The hemionus has no shoulder-stripe; but traces of it, as stated by Mr. Blyth and others, occasionally appear: and I have been informed by Colonel Poole that the foals of this species are generally striped on the legs, and faintly on the shoulder. The quagga, though so plainly barred like a zebra over the body, is without bars on the legs; but Dr. Gray has figured one specimen with very distinct zebra-like bars on the hocks.

With respect to the horse, I have collected cases in England of the spinal stripe in horses of the most distinct breeds, and of *all* colours; transverse bars on the legs are not rare in duns, mouse-duns, and in one instance in a chestnut: a faint shoulder-stripe may sometimes be seen in duns, and I have seen a trace in a bay horse. My son made a careful examination and sketch for me of a dun Belgian cart-horse with a double

stripe on each shoulder and with leg-stripes; I have
myself seen a dun Devonshire pony, and a small dun
Welch pony has been carefully described to me, both
with *three* parallel stripes on each shoulder.

In the north-west part of India the Kattywar breed
of horses is so generally striped, that, as I hear from
Colonel Poole, who examined the breed for the Indian
Government, a horse without stripes is not considered as
purely-bred. The spine is always striped; the legs are
generally barred; and the shoulder-stripe, which is sometimes double and sometimes treble, is common; the side
of the face, moreover, is sometimes striped. The stripes
are often plainest in the foal; and sometimes quite
disappear in old horses. Colonel Poole has seen both gray
and bay Kattywar horses striped when first foaled. I
have, also, reason to suspect, from information given
me by Mr. W. W. Edwards, that with the English race-
horse the spinal stripe is much commoner in the foal
than in the full-grown animal. Without here entering
on further details, I may state that I have collected cases
of leg and shoulder stripes in horses of very different
breeds, in various countries from Britain to Eastern
China; and from Norway in the north to the Malay
Archipelago in the south. In all parts of the world
these stripes occur far oftenest in duns and mouse-duns;
by the term dun a large range of colour is included,
from one between brown and black to a close approach
to cream colour.

I am aware that Colonel Hamilton Smith, who has
written on this subject, believes that the several breeds
of the horse have descended from several aboriginal
species—one of which, the dun, was striped; and that
the above-described appearances are all due to ancient
crosses with the dun stock. But I am not at all satisfied
with this theory, and should be loth to apply it to breeds

so distinct as the heavy Belgian cart-horse, Welch ponies, cobs, the lanky Kattywar race, &c., inhabiting the most distant parts of the world.

Now let us turn to the effects of crossing the several species of the horse-genus. Rollin asserts, that the common mule from the ass and horse is particularly apt to have bars on its legs: according to Mr. Gosse, in certain parts of the United States about nine out of ten mules have striped legs. I once saw a mule with its legs so much striped that any one would at first have thought that it must have been the product of a zebra; and Mr. W. C. Martin, in his excellent treatise on the horse, has given a figure of a similar mule. In four coloured drawings, which I have seen, of hybrids between the ass and zebra, the legs wore much more plainly barred than the rest of the body; and in one of them there was a double shoulder-stripe. In Lord Morton's famous hybrid from a chestnut mare and male quagga, the hybrid, and even the pure offspring subsequently produced from the mare by a black Arabian sire, were much more plainly barred across the legs than is even the pure quagga. Lastly, and this is another most remarkable case, a hybrid has been figured by Dr. Gray (and he informs me that he knows of a second case) from the ass and the hemionus; and this hybrid, though the ass seldom has stripes on his legs and the hemionus has none and has not even a shoulder-stripe, nevertheless had all four legs barred, and had three short shoulder-stripes, like those on the dun Devonshire and Welch ponies, and even had some zebra-like stripes on the sides of its face. With respect to this last fact, I was so convinced that not even a stripe of colour appears from what would commonly be called an accident, that I was led solely from the occurrence of the face-stripes on this hybrid from the ass and hemionus to ask Colonel Poole whether such face-stripes ever occur in the eminently striped Kattywar breed of horses, and was, as we have seen, answered in the affirmative.

What now are we to say to these several facts? We see several very distinct species of the horse-genus becoming, by simple variation, striped on the legs like a zebra, or striped on the shoulders like an ass. In the horse we see this tendency strong whenever a dun tint appears—a tint which approaches to that of the general colouring of the other species of the genus. The appearance of the stripes is not accompanied by any change of form or by any other new character. We see this tendency to become striped most strongly displayed in hybrids from between several of the most distinct species. Now observe the case of the several breeds of pigeons: they are descended from a pigeon (including two or three sub-species or geographical races) of a bluish colour, with certain bars and other marks; and when any breed assumes by simple variation a bluish tint, these bars and other marks invariably reappear; but without any other change of form or character. When the oldest and truest breeds of various colours are crossed, we see a strong tendency for the blue tint and bars and marks to reappear in the mongrels. I have stated that the most probable hypothesis to account for the reappearance of very ancient characters, is—that there is a *tendency* in the young of each successive generation to produce the long-lost character, and that this tendency, from unknown causes, sometimes prevails. And we have just seen that in several species of the horse-genus the stripes are either plainer or appear more commonly in the young than in the old. Call the breeds of pigeons, some of which have bred true for centuries, species; and how exactly parallel is the case with that of the species of the horse-genus! For myself,

I venture confidently to look back thousands on thousands of generations, and I see an animal striped like a zebra, but perhaps otherwise very differently constructed, the common parent of our domestic horse, whether or not it be descended from one or more wild stocks, of the ass, the hemionus, quagga, and zebra.

He who believes that each equine species was independently created, will, I presume, assert that each species has been created with a tendency to vary, both under nature and under domestication, in this particular manner, so as often to become striped like other species of the genus; and that each has been created with a strong tendency, when crossed with species inhabiting distant quarters of the world, to produce hybrids resembling in their stripes, not their own parents, but other species of the genus. To admit this view is, as it seems to me, to reject a real for an unreal, or at least for an unknown, cause. It makes the works of God a mere mockery and deception; I would almost as soon believe with the old and ignorant cosmogonists, that fossil shells had never lived, but had been created in stone so as to mock the shells now living on the sea-shore.

Summary.—Our ignorance of the laws of variation is profound. Not in one case out of a hundred can we pretend to assign any reason why this or that part differs, more or less, from the same part in the parents. But whenever we have the means of instituting a comparison, the same laws appear to have acted in producing the lesser differences between varieties of the same species, and the greater differences between species of the same genus. The external conditions of life, as climate and food, &c., seem to have induced some slight modifications. Habit in producing constitutional differences, and use in strengthening and disuse in weakening and diminishing organs, seem to have been more potent in their effects. Homologous parts tend to vary in the same way, and homologous parts tend to cohere. Modifications in hard parts and in external parts sometimes affect softer and internal parts. When one part is largely developed, perhaps it tends to draw nourishment from the adjoining parts; and every part of the structure which can be saved without detriment to the individual, will be saved. Changes of structure at an early age will generally affect parts subsequently developed; and there are very many other correlations of growth, the nature of which we are utterly unable to understand. Multiple parts are variable in number and in structure, perhaps arising from such parts not having been closely specialised to any particular function, so that their modifications have not been closely checked by natural selection. It is probably from this same cause that organic beings low in the scale of nature are more variable than those which have their whole organisation more specialised, and are higher in the scale. Rudimentary organs, from being useless, will be disregarded by natural selection, and hence probably are variable. Specific characters—that is, the characters which have come to differ since the several species of the same genus branched off from a common parent—are more variable than generic characters, or those which have long been inherited, and have not differed within this same period. In these remarks we have referred to special parts or organs being still variable, because they have recently varied and thus come to differ; but we have also seen in the second Chapter that the same principle applies to the whole individual; for in a district where many species of any genus are found—that is, where there has been much former

variation and differentiation, or where the manufactory
of new specific forms has been actively at work—in that
district and amongst these species, we now find, on an
average, most varieties. Secondary sexual characters
are highly variable, and such characters differ much in
the species of the same group. Variability in the same
parts of the organisation has generally been taken advantage of in giving secondary sexual differences to the
sexes of the same species, and specific differences to the
several species of the same genus. Any part or organ
developed to an extraordinary size or in an extraordinary manner, in comparison with the same part or
organ in the allied species, must have gone through an
extraordinary amount of modification since the genus
arose; and thus we can understand why it should often
still be variable in a much higher degree than other
parts; for variation is a long-continued and slow process, and natural selection will in such cases not as
yet have had time to overcome the tendency to further
variability and to reversion to a less modified state. But
when a species with any extraordinarily-developed organ
has become the parent of many modified descendants
—which on my view must be a very slow process,
requiring a long lapse of time—in this case, natural
selection may readily have succeeded in giving a fixed
character to the organ, in however extraordinary a
manner it may be developed. Species inheriting nearly
the same constitution from a common parent and exposed to similar influences will naturally tend to present
analogous variations, and these same species may occasionally revert to some of the characters of their ancient
progenitors. Although new and important modifications may not arise from reversion and analogous variation, such modifications will add to the beautiful and
harmonious diversity of nature.

Whatever the cause may be of each slight difference in the offspring from their parents—and a cause for each must exist—it is the steady accumulation, through natural selection, of such differences, when beneficial to the individual, that gives rise to all the more important modifications of structure, by which the innumerable beings on the face of this earth are enabled to struggle with each other, and the best adapted to survive.

CHAPTER VI.

DIFFICULTIES ON THEORY.

Difficulties on the theory of descent with modification — Transitions — Absence or rarity of transitional varieties — Transitions in habits of life — Diversified habits in the same species — Species with habits widely different from those of their allies — Organs of extreme perfection — Means of transition — Cases of difficulty — Natura non facit saltum — Organs of small importance — Organs not in all cases absolutely perfect — The law of Unity of Type and of the Conditions of Existence embraced by the theory of Natural Selection.

LONG before having arrived at this part of my work, a crowd of difficulties will have occurred to the reader. Some of them are so serious that to this day I can never reflect on them without being staggered; but, to the best of my judgment, the greater number are only apparent, and those that are real are not, I think, fatal to my theory.

These difficulties and objections may be classed under the following heads :—Firstly, why, if species have descended from other species by insensibly fine gradations, do we not everywhere see innumerable transitional forms? Why is not all nature in confusion, instead of the species being, as we see them, well defined?

Secondly, is it possible that an animal having, for instance, the structure and habits of a bat, could have been formed by the modification of some animal with wholly different habits? Can we believe that natural selection could produce, on the one hand, organs of trifling importance, such as the tail of a giraffe, which serves as a fly-flapper, and, on the other hand, organs of

such wonderful structure, as the eye, of which we hardly
as yet fully understand the inimitable perfection?

Thirdly, can instincts be acquired and modified through
natural selection? What shall we say to so marvellous
an instinct as that which leads the bee to make cells,
which has practically anticipated the discoveries of profound mathematicians?

Fourthly, how can we account for species, when
crossed, being sterile and producing sterile offspring,
whereas, when varieties are crossed, their fertility is
unimpaired?

The two first heads shall be here discussed—Instinct
and Hybridism in separate chapters.

On the absence or rarity of transitional varieties.—
As natural selection acts solely by the preservation of
profitable modifications, each new form will tend in a
fully-stocked country to take the place of, and finally to
exterminate, its own less improved parent or other less-
favoured forms with which it comes into competition.
Thus extinction and natural selection will, as we have
seen, go hand in hand. Hence, if we look at each species
as descended from some other unknown form, both the
parent and all the transitional varieties will generally
have been exterminated by the very process of formation and perfection of the new form.

But, as by this theory innumerable transitional forms
must have existed, why do we not find them embedded
in countless numbers in the crust of the earth? It will
be much more convenient to discuss this question in the
chapter on the Imperfection of the Geological Record;
and I will here only state that I believe the answer
mainly lies in the record being incomparably less perfect
than is generally supposed; the imperfection of the
record being chiefly due to organic beings not inhabiting

profound depths of the sea, and to their remains being embedded and preserved to a future age only in masses of sediment sufficiently thick and extensive to withstand an enormous amount of future degradation ; and such fossiliferous masses can be accumulated only where much sediment is deposited on the shallow bed of the sea, whilst it slowly subsides. These contingencies will concur only rarely, and after enormously long intervals. Whilst the bed of the sea is stationary or is rising, or when very little sediment is being deposited, there will be blanks in our geological history. The crust of the earth is a vast museum ; but the natural collections have been made only at intervals of time immensely remote.

But it may be urged that when several closely-allied species inhabit the same territory, we surely ought to find at the present time many transitional forms. Let us take a simple case : in travelling from north to south over a continent, we generally meet at successive intervals with closely allied or representative species, evidently filling nearly the same place in the natural œconomy of the land. These representative species often meet and interlock ; and as the one becomes rarer and rarer, the other becomes more and more frequent, till the one replaces the other. But if we compare these species where they intermingle, they are generally as absolutely distinct from each other in every detail of structure as are specimens taken from the metropolis inhabited by each. By my theory these allied species have descended from a common parent ; and during the process of modification, each has become adapted to the conditions of life of its own region, and has supplanted and exterminated its original parent and all the transitional varieties between its past and present states. Hence we ought not to expect at the

present time to meet with numerous transitional varieties in each region, though they must have existed
there, and may be embedded there in a fossil condition.
But in the intermediate region, having intermediate
conditions of life, why do we not now find closely-linking
intermediate varieties? This difficulty for a long time
quite confounded me. But I think it can be in large
part explained.

In the first place we should be extremely cautious in
inferring, because an area is now continuous, that it
has been continuous during a long period. Geology
would lead us to believe that almost every continent
has been broken up into islands even during the later
tertiary periods; and in such islands distinct species
might have been separately formed without the possibility of intermediate varieties existing in the intermediate zones. By changes in the form of the land and
of climate, marine areas now continuous must often
have existed within recent times in a far less continuous
and uniform condition than at present. But I will pass
over this way of escaping from the difficulty; for I
believe that many perfectly defined species have been
formed on strictly continuous areas; though I do not
doubt that the formerly broken condition of areas now
continuous has played an important part in the formation of new species, more especially with freely-crossing
and wandering animals.

In looking at species as they are now distributed
over a wide area, we generally find them tolerably
numerous over a large territory, then becoming somewhat abruptly rarer and rarer on the confines, and
finally disappearing. Hence the neutral territory between two representative species is generally narrow in
comparison with the territory proper to each. We see
the same fact in ascending mountains, and sometimes

it is quite remarkable how abruptly, as Alph. de Candolle has observed, a common alpine species disappears. The same fact has been noticed by E. Forbes in sounding the depths of the sea with the dredge. To those who look at climate and the physical conditions of life as the all-important elements of distribution, these facts ought to cause surprise, as climate and height or depth graduate away insensibly. But when we bear in mind that almost every species, even in its metropolis, would increase immensely in numbers, were it not for other competing species; that nearly all either prey on or serve as prey for others; in short, that each organic being is either directly or indirectly related in the most important manner to other organic beings, we must see that the range of the inhabitants of any country by no means exclusively depends on insensibly changing physical conditions, but in large part on the presence of other species, on which it depends, or by which it is destroyed, or with which it comes into competition; and as these species are already defined objects (however they may have become so), not blending one into another by insensible gradations, the range of any one species, depending as it does on the range of others, will tend to be sharply defined. Moreover, each species on the confines of its range, where it exists in lessened numbers, will, during fluctuations in the number of its enemies or of its prey, or in the seasons, be extremely liable to utter extermination; and thus its geographical range will come to be still more sharply defined.

If I am right in believing that allied or representative species, when inhabiting a continuous area, are generally so distributed that each has a wide range, with a comparatively narrow neutral territory between them, in which they become rather suddenly rarer and rarer; then, as varieties do not essentially differ from species,

K

the same rule will probably apply to both; and if we in imagination adapt a varying species to a very large area, we shall have to adapt two varieties to two large areas, and a third variety to a narrow intermediate zone. The intermediate variety, consequently, will exist in lesser numbers from inhabiting a narrow and lesser area; and practically, as far as I can make out, this rule holds good with varieties in a state of nature. I have met with striking instances of the rule in the case of varieties intermediate between well-marked varieties in the genus Balanus. And it would appear from information given me by Mr. Watson, Dr. Asa Gray, and Mr. Wollaston, that generally when varieties intermediate between two other forms occur, they are much rarer numerically than the forms which they connect. Now, if we may trust these facts and inferences, and therefore conclude that varieties linking two other varieties together have generally existed in lesser numbers than the forms which they connect, then, I think, we can understand why intermediate varieties should not endure for very long periods;—why as a general rule they should be exterminated and disappear, sooner than the forms which they originally linked together.

For any form existing in lesser numbers would, as already remarked, run a greater chance of being exterminated than one existing in large numbers; and in this particular case the intermediate form would be eminently liable to the inroads of closely allied forms existing on both sides of it. But a far more important consideration, as I believe, is that, during the process of further modification, by which two varieties are supposed on my theory to be converted and perfected into two distinct species, the two which exist in larger numbers from inhabiting larger areas, will have a great advantage over the intermediate variety, which exists in smaller numbers in a narrow and intermediate zone. For forms existing in larger numbers will always have a better chance, within any given period, of presenting further favourable variations for natural selection to seize on, than will the rarer forms which exist in lesser numbers. Hence, the more common forms, in the race for life, will tend to beat and supplant the less common forms, for these will be more slowly modified and improved. It is the same principle which, as I believe, accounts for the common species in each country, as shown in the second chapter, presenting on an average a greater number of well-marked varieties than do the rarer species. I may illustrate what I mean by supposing three varieties of sheep to be kept, one adapted to an extensive mountainous region; a second to a comparatively narrow, hilly tract; and a third to wide plains at the base; and that the inhabitants are all trying with equal steadiness and skill to improve their stocks by selection; the chances in this case will be strongly in favour of the great holders on the mountains or on the plains improving their breeds more quickly than the small holders on the intermediate narrow, hilly tract; and consequently the improved mountain or plain breed will soon take the place of the less improved hill breed; and thus the two breeds, which originally existed in greater numbers, will come into close contact with each other, without the interposition of the supplanted, intermediate hill-variety.

To sum up, I believe that species come to be tolerably well-defined objects, and do not at any one period present an inextricable chaos of varying and intermediate links: firstly, because new varieties are very slowly formed, for variation is a very slow process, and natural selection can do nothing until favourable variations

chance to occur, and until a place in the natural
polity of the country can be better filled by some
modification of some one or more of its inhabitants.
And such new places will depend on slow changes of
climate, or on the occasional immigration of new inhabitants, and, probably, in a still more important degree,
on some of the old inhabitants becoming slowly modified, with the new forms thus produced and the old
ones acting and reacting on each other. So that, in
any one region and at any one time, we ought only to
see a few species presenting slight modifications of structure in some degree permanent; and this assuredly we
do see.

Secondly, areas now continuous must often have existed within the recent period in isolated portions, in
which many forms, more especially amongst the classes
which unite for each birth and wander much, may havo
separately been rendered sufficiently distinct to rank as
representative species. In this case, intermediate varieties between the several representative species and
their common parent, must formerly have existed in
each broken portion of the land, but these links will
have been supplanted and exterminated during the process of natural selection, so that they will no longer
exist in a living state.

Thirdly, when two or moro varieties have been formed
in different portions of a strictly continuous area, intermediate varieties will, it is probable, at first have been
formed in tho intermediate zones, but they will generally have had a short duration. For these intermediate varieties will, from reasons already assigned
(namely from what we know of the actual distribution
of closely allied or representative species, and likewise
of acknowledged varieties), exist in the intermediate
zones in lesser numbers than the varieties which they

tend to connect. From this cause alone the intermediate varieties will be liable to accidental extermination; and during the process of further modification through natural selection, they will almost certainly be beaten and supplanted by the forms which they connect; for these from existing in greater numbers will, in the aggregate, present more variation, and thus be further improved through natural selection and gain further advantages.

Lastly, looking not to any one time, but to all time, if my theory be true, numberless intermediate varieties, linking closely all the species of the same group together, must assuredly have existed; but the very process of natural selection constantly tends, as has been so often remarked, to exterminate the parent-forms and the intermediate links. Consequently evidence of their former existence could be found only amongst fossil remains, which are preserved, as we shall in a future chapter attempt to show, in an extremely imperfect and intermittent record.

On the origin and transitions of organic beings with peculiar habits and structure.—It has been asked by the opponents of such views as I hold, how, for instance, a land carnivorous animal could have been converted into one with aquatic habits; for how could the animal in its transitional state have subsisted? It would be easy to show that within the same group carnivorous animals exist having every intermediate grade between truly aquatic and strictly terrestrial habits; and as each exists by a struggle for life, it is clear that each is well adapted in its habits to its place in nature. Look at the Mustela vison of North America, which has webbed feet and which resembles an otter in its fur, short legs, and form of tail; during summer this animal dives

for and preys on fish, but during the long winter it
leaves the frozen waters, and preys like other pole-cats
on mice and land animals. If a different case had been
taken, and it had been asked how an insectivorous
quadruped could possibly have been converted into a
flying bat, the question would have been far more difficult, and I could have given no answer. Yet I think
such difficulties have little weight.

Here, as on other occasions, I lie under a heavy disadvantage, for out of the many striking cases which I
have collected, I can give only one or two instances of
transitional habits and structures in closely allied species
of the same genus; and of diversified habits, either
constant or occasional, in the same species. And it
seems to me that nothing less than a long list of such
cases is sufficient to lessen the difficulty in any particular case like that of the bat.

Look at the family of squirrels; here we have the
finest gradation from animals with their tails only
slightly flattened, and from others, as Sir J. Richardson
has remarked, with the posterior part of their bodies
rather wide and with the skin on their flanks rather full,
to the so-called flying squirrels; and flying squirrels
have their limbs and even the base of the tail united by
a broad expanse of skin, which serves as a parachute
and allows them to glide through the air to an astonishing distance from tree to tree. We cannot doubt
that each structure is of use to each kind of squirrel in
its own country, by enabling it to escape birds or beasts
of prey, or to collect food more quickly, or, as there
is reason to believe, by lessening the danger from occasional falls. But it does not follow from this fact that
the structure of each squirrel is the best that it is possible to conceive under all natural conditions. Let the
climate and vegetation change, let other competing

rodents or new beasts of prey immigrate, or old ones become modified, and all analogy would lead us to believe that some at least of the squirrels would decrease in numbers or become exterminated, unless they also became modified and improved in structure in a corresponding manner. Therefore, I can see no difficulty, more especially under changing conditions of life, in the continued preservation of individuals with fuller and fuller flank-membranes, each modification being useful, each being propagated, until by the accumulated effects of this process of natural selection, a perfect so-called flying squirrel was produced.

Now look at the Galeopithecus or flying lemur, which formerly was falsely ranked amongst bats. It has an extremely wide flank-membrane, stretching from the corners of the jaw to the tail, and including the limbs and the elongated fingers: the flank-membrane is, also, furnished with an extensor muscle. Although no graduated links of structure, fitted for gliding through the air, now connect the Galeopithecus with the other Lemuridæ, yet I see no difficulty in supposing that such links formerly existed, and that each had been formed by the same steps as in the case of the less perfectly gliding squirrels; and that each grade of structure was useful to its possessor. Nor can I see any insuperable difficulty in further believing it possible that the membrane-connected fingers and forearm of the Galeopithecus might be greatly lengthened by natural selection; and this, as far as the organs of flight are concerned, would convert it into a bat. In bats which have the wing-membrane extended from the top of the shoulder to the tail, including the hind-legs, we perhaps see traces of an apparatus originally constructed for gliding through the air rather than for flight.

If about a dozen genera of birds had become extinct or were unknown, who would have ventured to surmise that birds might have existed which used their wings solely as flappers, like the logger-headed duck (Micropterus of Eyton); as fins in the water and front legs on the land, like the penguin; as sails, like the ostrich; and functionally for no purpose, like the Apteryx? Yet the structure of each of these birds is good for it, under the conditions of life to which it is exposed, for each has to live by a struggle; but it is not necessarily the best possible under all possible conditions. It must not be inferred from these remarks that any of the grades of wing-structure here alluded to, which perhaps may all have resulted from disuse, indicate the natural steps by which birds have acquired their perfect power of flight; but they serve, at least, to show what diversified means of transition are possible.

Seeing that a few members of such water-breathing classes as the Crustacea and Mollusca are adapted to live on the land; and seeing that we have flying birds and mammals, flying insects of the most diversified types, and formerly had flying reptiles, it is conceivable that flying-fish, which now glide far through the air, slightly rising and turning by the aid of their fluttering fins, might have been modified into perfectly winged animals. If this had been effected, who would have ever imagined that in an early transitional state they had been inhabitants of the open ocean, and had used their incipient organs of flight exclusively, as far as we know, to escape being devoured by other fish?

When we see any structure highly perfected for any particular habit, as the wings of a bird for flight, we should bear in mind that animals displaying early tran-

sitional grades of the structure will seldom continue to
exist to the present day, for they will have been supplanted by the very process of perfection through natural
selection.   Furthermore, we may conclude that transitional grades between structures fitted for very different
habits of life will rarely have been developed at an early
period in great numbers and under many subordinate
forms.   Thus, to return to our imaginary illustration of
the flying-fish, it does not seem probable that fishes
capable of true flight would have been developed under
many subordinate forms, for taking prey of many kinds
in many ways, on the land and in the water, until their
organs of flight had come to a high stage of perfection,
so as to have given them a decided advantage over other
animals in the battle for life.   Hence the chance of discovering species with transitional grades of structure in
a fossil condition will always be less, from their having
existed in lesser numbers, than in the case of species
with fully developed structures.

I will now give two or three instances of diversified
and of changed habits in the individuals of the same
species.   When either case occurs, it would be easy for
natural selection to fit the animal, by some modification
of its structure, for its changed habits, or exclusively
for one of its several different habits.   But it is difficult
to tell, and immaterial for us, whether habits generally
change first and structure afterwards; or whether slight
modifications of structure lead to changed habits; both
probably often change almost simultaneously.   Of cases
of changed habits it will suffice merely to allude to that
of the many British insects which now feed on exotic
plants, or exclusively on artificial substances.   Of diversified habits innumerable instances could be given: I
have often watched a tyrant flycatcher (Saurophagus
sulphuratus) in South America, hovering over one spot

and then proceeding to another, like a kestrel, and at
other times standing stationary on the margin of water,
and then dashing like a kingfisher at a fish. In our
own country the larger titmouse (Parus major) may bo
seen climbing branches, almost like a creeper; it often,
like a shrike, kills small birds by blows on the head;
and I have many times seen and heard it hammering
the seeds of the yew on a branch, and thus breaking
them like a nuthatch. In North America the black
bear was seen by Hearne swimming for hours with
widely open mouth, thus catching, almost like a whale,
insects in the water.

As we sometimes see individuals of a species following
habits widely different from those of their own species
and of the other species of the same genus, we might
expect, on my theory, that such individuals would occasionally have given rise to new species, having anomalous habits, and with their structure either slightly or
considerably modified from that of their proper type.
And such instances do occur in nature. Can a more
striking instance of adaptation be given than that of a
woodpecker for climbing trees and for seizing insects in
the chinks of the bark? Yet in North America there
are woodpeckers which feed largely on fruit, and others
with elongated wings which chase insects on the wing.
On the plains of La Plata, where not a tree grows, there
is a woodpecker (*Colaptes campestris*) which has two toes
before and two behind, a long pointed tongue, stiff
pointed tail feathers, but not so stiff as in the typical
woodpeckers (yet I have seen it use its tail when alighting vertically on a post), and a straight strong beak. The
beak, however, is not so straight or strong as in the
typical woodpeckers; but it is strong enough to bore
into wood; and I may mention, as another illustration of
the varied habits of the tribe, that a Mexican Colaptes

has been described by De Saussure as boring holes into
hard wood in order to lay up a store of acorns for its
future consumption! Hence the Colaptes of La Plata
in all the essential parts of its organization is a woodpecker, and until recently was classed in the same
typical genus. Even such trifling characters as its
colouring, the harsh tone of its voice, and undulatory
flight, all told me plainly of its close blood-relationship
to our common species; yet it is a woodpecker, as I can
assert not only from my own observation, but from
that of the accurate Azara, which never climbs a tree!

Petrels are the most aërial and oceanic of birds, but
in the quiet Sounds of Tierra del Fuego, the Puffinuria
berardi, in its general habits, in its astonishing power
of diving, its manner of swimming, and of flying when
unwillingly it takes flight, would be mistaken by any
one for an auk or a grebe; nevertheless, it is essentially
a petrel, but with many parts of its organisation profoundly modified in relation to its new habits of life;
whereas the woodpecker of La Plata has its structure
only slightly modified. In the case of the water-
ouzel, the acutest observer by examining its dead body
would never have suspected its sub-aquatic habits; yet
this anomalous member of the terrestrial thrush family
wholly subsists by diving,—grasping the stones with its
feet, and using its wings under water.

He who believes that each being has been created as
we now see it, must occasionally have felt surprise when
he has met with an animal having habits and structure
not at all in agreement. What can be plainer than
that the webbed feet of ducks and geese are formed for
swimming? yet there are upland geese with webbed
feet which rarely or never go near the water; and no
one except Audubon has seen the frigate-bird, which has
all its four toes webbed, alight on the surface of the sea.

On the other hand, grebes and coots are eminently aquatic, although their toes are only bordered by membrane. What seems plainer than that the long toes of grallatores are formed for walking over swamps and floating plants?—yet the water-hen is nearly as aquatic as the coot; and the landrail nearly as terrestrial as the quail or partridge. In such cases, and many others could be given, habits have changed without a corresponding change of structure. The webbed feet of the upland goose may be said to have become rudimentary in function, though not in structure. In the frigate-bird, the deeply-scooped membrane between the toes shows that structure has begun to change.

He who believes in separate and innumerable acts of creation will say, that in these cases it has pleased the Creator to cause a being of one type to take the place of one of another type; but this seems to me only re-stating the fact in dignified language. He who believes in the struggle for existence and in the principle of natural selection, will acknowledge that every organic being is constantly endeavouring to increase in numbers; and that if any one being vary ever so little, either in habits or structure, and thus gain an advantage over some other inhabitant of the country, it will seize on the place of that inhabitant, however different it may be from its own place. Hence it will cause him no surprise that there should be geese and frigate-birds with webbed feet, living on the dry land or most rarely alighting on the water; that there should be long-toed corncrakes living in meadows instead of in swamps; that there should be woodpeckers where not a tree grows; that there should be diving thrushes, and petrels with the habits of auks.

Organs of extreme perfection and complication.—To

suppose that the eye, with all its inimitable contrivances for adjusting the focus to different distances, for admitting different amounts of light, and for the correction of spherical and chromatic aberration, could have been formed by natural selection, seems, I freely confess, absurd in the highest possible degree. When it was first said that the sun stood still and the world turned round, the common sense of mankind declared the doctrine false; but the old saying of *Vox populi, vox Dei*, as every philosopher knows, can never be trusted in science. Reason tells me, that if numerous gradations from a perfect and complex eye to one very imperfect and simple, each grade being useful to its possessor, can be shown to exist; if further, the eye does vary ever so slightly, and the variations be inherited, which is certainly the case; and if any variation or modification in the organ be ever useful to an animal under changing conditions of life, then the difficulty of believing that a perfect and complex eye could be formed by natural selection, though insuperable by our imagination, can hardly be considered real. How a nerve comes to be sensitive to light, hardly concerns us more than how life itself first originated; but I may remark that several facts make me suspect that nerves sensitive to touch may be rendered sensitive to light, and likewise to those coarser vibrations of the air which produce sound.

In looking for the gradations by which an organ in any species has been perfected, we ought to look exclusively to its lineal ancestors; but this is scarcely ever possible, and we are forced in each case to look to species of the same group, that is to the collateral descendants from the same original parent-form, in order to see what gradations are possible, and for the chance of some gradations having been transmitted from the earlier stages of descent, in an unaltered or little altered condition. Amongst existing Vertebrata, we find but a small amount of gradation in the structure of the eye (though in the fish Amphioxus, the eye is in an extremely simple condition without a lens), and from fossil species we can learn nothing on this head. In this great class we should probably have to descend far beneath the lowest known fossiliferous stratum to discover the earlier stages, by which the eye has been perfected.

In the great kingdom of the Articulata, we can start from an optic nerve, simply coated with pigment, which sometimes forms a sort of pupil, but is destitute of a lens or any other optical mechanism. From this rudimentary eye, which can distinguish light from darkness, but nothing else, there is an advance towards perfection along two lines of structure, which Müller thought were fundamentally different; namely,—firstly, stemmata, or the so-called "simple eyes," which have a lens and cornea; and secondly, "compound eyes," which seem to act mainly by excluding all the rays from each point of the object viewed, except the pencil that comes in a line perpendicular to the convex retina. In compound eyes, besides endless differences in the form, proportion, number, and position of the transparent cones coated by pigment, and which act by exclusion, we have additions of a more or less perfect concentrating apparatus: thus in the eye of the Meloe the facets of the cornea are "slightly convex both externally and internally—that is, lens-shaped." In many crustaceans there are two corneæ—the external smooth, and the internal divided into facets—within the substance of which, as Milne Edwards says, "renflemens lenticulaires paraissent s'être développés;" and sometimes these lenses can be detached in a layer distinct from the cornea. The transparent cones coated with pigment, which were supposed by Müller to act solely by ex-

cludiug divergent pencils of light, usually adhere to
the cornea, but not rarely they are separate from it,
and have their free ends convex; and in this case they
must act as converging lenses. Altogether so diversified is the structure of the compound eyes, that Müller
makes three main classes, with no less than seven sub-
divisions of structure; he makes a fourth main class,
namely, "aggregates" of stemmata, and he adds that
"this is the transition-form between the mosaic-like
compound eyes unprovided with a concentratiug apparatus, and organs of vision with such an apparatus."

With these facts, here too briefly and imperfectly
given, which show how much graduated diversity there
is in the eyes of our existing crustaceans, and bearing in
mind how small the number of living animals is in
proportion to those which have become extinct, I can
see no very great difficulty (not more than in the case
of many other structures) in believing that natural
selection has converted the simple apparatus of an optic
nerve merely coated with pigment and invested by
transparent membrane, into an optical instrument as
perfect as is possessed by any member of the great
Articulate class.

He who will go thus far, if he find on finishing this
treatise that large bodies of facts, otherwise inexplicable, can be explained by the theory of descent, ought
not to hesitate to go further, and to admit that a structure even as perfect as the eye of an eagle might be
formed by natural selection, although in this case he
does not kuow any of the transitional grades. His reason
ought to conquer his imagination; though I have felt
the difficulty far too keenly to be surprised at any degree
of hesitation in extending the principle of natural selection to such startling lengths.

It is scarcely possible to avoid comparing the eye to

a telescope. We know that this instrument has been perfected by the long-continued efforts of the highest human intellects; and we naturally infer that the eye has been formed by a somewhat analogous process. But may not this inference be presumptuous? Have we any right to assume that the Creator works by intellectual powers like those of man? If we must compare the eye to an optical instrument, we ought in imagination to take a thick layer of transparent tissue, with spaces filled with fluid, and with a nerve sensitive to light beneath, and then suppose every part of this layer to be continually changing slowly in density, so as to separate into layers of different densities and thicknesses, placed at different distances from each other, and with the surfaces of each layer slowly changing in form. Further we must suppose that there is a power (natural selection) always intently watching each slight accidental alteration in the transparent layers; and carefully selecting each alteration which, under varied circumstances, may in any way, or in any degree, tend to produce a distincter image. We must suppose each new state of the instrument to be multiplied by the million; and each to be preserved till a better be produced, and then the old ones to be destroyed. In living bodies, variation will cause the slight alterations, generation will multiply them almost infinitely, and natural selection will pick out with unerring skill each improvement. Let this process go on for millions on millions of years; and during each year on millions of individuals of many kinds; and may we not believe that a living optical instrument might thus be formed as superior to one of glass, as the works of the Creator are to those of man?

If it could be demonstrated that any complex organ existed, which could not possibly have been formed by numerous, successive, slight modifications, my theory would absolutely break down. But I can find out no such case. No doubt many organs exist of which we do not know the transitional grades, more especially if we look to much-isolated species, round which, according to my theory, there has been much extinction. Or again, if we look to an organ common to all the members of a large class, for in this latter case the organ must have been first formed at an extremely remote period, since which all the many members of the class have been developed; and in order to discover the early transitional grades through which the organ has passed, we should have to look to very ancient ancestral forms, long since become extinct.

We should be extremely cautious in concluding that an organ could not have been formed by transitional gradations of some kind. Numerous cases could be given amongst the lower animals of the same organ performing at the same time wholly distinct functions; thus the alimentary canal respires, digests, and excretes in the larva of the dragon-fly and in the fish Cobites. In the Hydra, the animal may be turned inside out, and the exterior surface will then digest and the stomach respire. In such cases natural selection might specialise, if any advantage were thus gained, a part or organ, which had performed two functions, for one function alone, and thus wholly change its nature by insensible steps. Certain plants, as some Leguminosæ, Violaceæ, &c., bear two kinds of flowers; one having the normal structure of the order, the other kind being degraded, though sometimes more fertile than the perfect kind: if the plant ceased to bear its perfect flowers, and this did occur during several years with an imported specimen of Aspicarpa in France, a great and sudden transition would apparently be effected in the nature of the plant. Two distinct organs sometimes perform simultaneously the same function in the same individual; to give one instance, there are fish with gills or branchiae that breathe the air dissolved in the water, at the same time that they breathe free air in their swimbladders, this latter organ having a ductus pneumaticus for its supply, and being divided by highly vascular partitions. In these cases one of the two organs might be modified and perfected so as to perform all the work by itself, being aided during the process of modification by the other organ; and then this other organ might be modified for some other and quite distinct purpose, or be quite obliterated.

The illustration of the swimbladder in fishes is a good one, because it shows us clearly the highly important fact that an organ originally constructed for one purpose, namely flotation, may be converted into one for a wholly different purpose, namely respiration. The swimbladder has, also, been worked in as an accessory to the auditory organs of certain fish, or, for I do not know which view is now generally held, a part of the auditory apparatus has been worked in as a complement to the swimbladder. All physiologists admit that the swimbladder is homologous, or " ideally similar" in position and structure with the lungs of the higher vertebrate animals : hence there seems to me to be no extreme difficulty in believing that natural selection has actually converted a swimbladder into a lung, or organ used exclusively for respiration.

On this view it may be inferred that all vertebrate animals having true lungs have descended by ordinary generation from an ancient prototype, of which we know nothing, furnished with a floating apparatus or swimbladder. We can thus, as I infer from Professor Owen's interesting description of these parts, understand the strange fact that every particle of food and drink which we swallow has to pass over the orifice of the trachea, with some risk of falling into the lungs, notwithstanding the beautiful contrivance by which the glottis is closed. In the higher Vertebrata the branchiæ have wholly disappeared—the slits on the sides of the neck and the loop-like course of the arteries still marking in the embryo their former position. But it is conceivable that the now utterly lost branchiæ might have been gradually worked in by natural selection for some quite distinct purpose : in the same manner as, on the view entertained by some naturalists that the branchiæ and dorsal scales of Annelids are homologous with the wings and wing-covers of insects, it is probable that organs which at a very ancient period served for respiration have been actually converted into organs of flight.

In considering transitions of organs, it is so important to bear in mind the probability of conversion from one function to another, that I will give one more instance. Pedunculated cirripedes have two minute folds of skin, called by me the ovigerous frena, which serve, through the means of a sticky secretion, to retain the eggs until they are hatched within the sack. These cirripedes have no branchiæ, the whole surface of the body and sack, including the small frena, serving for respiration. The Balanidæ or sessile cirripedes, on the other hand, have no ovigerous frena, the eggs lying loose at the bottom of the sack, in the well-enclosed shell ; but they have, in the same relative position, large, much folded membranes, which freely communicate with the circulatory lacunæ of the sack and body, and which have been considered to be branchiæ by Prof. Owen and all other naturalists who have treated on the subject. Now I think no one will dispute that the ovigerous frena in the one family are strictly homologous with the branchiæ

of the other family; indeed, they graduate into each
other. Therefore I do not doubt that the two little folds
of skin, which originially served as ovigerous frena, but
which, likewise, very slightly aided the act of respiration, have been gradually converted by natural selection
into branchiæ, simply through an increase in their size
and the obliteration of their adhesive glands. If all
pedunculated cirripedes had become extinct, and they
have already suffered far more extinction than have
sessile cirripedes, who would ever have imagined that
the branchiæ in this latter family had originally existed
as organs for preventing the ova from being washed out
of the sack?

Although we must be extremely cautious in concluding that any organ could not possibly have been produced by successive transitional gradations, yet, undoubtedly, grave cases of difficulty occur, some of which
will be discussed in my future work.

One of the gravest is that of neuter insects, which
are often very differently constructed from either the
males or fertile females; but this case will be treated
of in the next chapter. The electric organs of fishes
offer another case of special difficulty; it is impossible
to conceive by what steps these wondrous organs have
been produced; but, as Owen and others have remarked,
their intimate structure closely resembles that of common
muscle; and as it has lately been shown that Rays have
an organ closely analogous to the electric apparatus, and
yet do not, as Matteucci asserts, discharge any electri-
city, we must own that we are far too ignorant to argue
that no transition of any kind is possible.

The electric organs offer another and even more
serious difficulty; for they occur in only about a dozen
fishes, of which several are widely remote in their
affinities. Generally when the same organ appears in

several members of the same class, especially if in members having very different habits of life, we may attribute its presence to inheritance from a common ancestor; and its absence in some of the members to its loss through disuse or natural selection. But if the electric organs had been inherited from one ancient progenitor thus provided, we might have expected that all electric fishes would have been specially related to each other. Nor does geology at all lead to the belief that formerly most fishes had electric organs, which most of their modified descendants have lost. The presence of luminous organs in a few insects, belonging to different families and orders, offers a parallel case of difficulty. Other cases could be given; for instance in plants, the very curious contrivance of a mass of pollen-grains, borne on a foot-stalk with a sticky gland at the end, is the same in Orchis and Asclepias,— genera almost as remote as possible amongst flowering plants. In all these cases of two very distinct species furnished with apparently the same anomalous organ, it should be observed that, although the general appearance and function of the organ may be the same, yet some fundamental difference can generally be detected. I am inclined to believe that in nearly the same way as two men have sometimes independently hit on the very same invention, so natural selection, working for the good of each being and taking advantage of analogous variations, has sometimes modified in very nearly the same manner two parts in two organic beings, which beings owe but little of their structure in common to inheritance from the same ancestor.

Although in many cases it is most difficult to conjecture by what transitions organs could have arrived at their present state; yet, considering that the proportion of living and known forms to the extinct and unknown

is very small, I have been astonished how rarely an
organ can be named, towards which no transitional
grade is known to lead. It certainly is not true, that
new organs often appear suddenly in any class. as if
created for some special purpose; as indeed is
shown by that old, but somewhat exaggerated, canon
in natural history of "Natura non facit saltum." We
meet with this admission in the writings of almost
every experienced naturalist; as Milne Edwards has
well expressed it, Nature is prodigal in variety, but
niggard in innovation. Why, on the theory of Creation,
should this be so? Why should all the parts and organs
of many independent beings, each supposed to have been
separately created for its proper place in nature, be so
commonly linked together by graduated steps? Why
should not Nature have taken a leap from structure to
structure? On the theory of natural selection, we can
clearly understand why she should not; for natural
selection can act only by taking advantage of slight
successive variations; she can never take a leap, but
must advance by short and slow steps.

Organs of little apparent importance.—As natural
selection acts by life and death,—by the preservation of
individuals with any favourable variation, and by the
destruction of those with any unfavourable deviation of
structure,—I have sometimes felt much difficulty in
understanding the origin of simple parts, of which the
importance does not seem sufficient to cause the preservation of successively varying individuals. I have
sometimes felt as much difficulty, though of a very different kind, on this head, as in the case of an organ as
perfect and complex as the eye.

In the first place, we are much too ignorant in regard
to the whole economy of any one organic being, to say

what slight modifications would be of importance or
not. In a former chapter I have given instances of
most trifling characters, such as the down on fruit and
the colour of its flesh, the colour of the skin and hair
of quadrupeds, which, from being correlated with constitutional differences or from determining the attacks
of insects, might assuredly be acted on by natural
selection. The tail of the giraffe looks like an artificially constructed fly-flapper; and it seems at first
incredible that this could have been adapted for its
present purpose by successive slight modifications, each
better and better, for so trifling an object as driving
away flies; yet we should pause before being too positive even in this case, for we know that the distribution
and existence of cattle and other animals in South
America absolutely depend on their power of resisting
the attacks of insects: so that individuals which could
by any means defend themselves from these small
enemies, would be able to range into new pastures and
thus gain a great advantage. It is not that the larger
quadrupeds are actually destroyed (except in some rare
cases) by flies, but they are incessantly harassed and
their strength reduced, so that they are more subject to
disease, or not so well enabled in a coming dearth to
search for food, or to escape from beasts of prey.

Organs now of trifling importance have probably in
some cases been of high importance to an early progenitor, and, after having been slowly perfected at a
former period, have been transmitted in nearly the same
state, although now become of very slight use; and any
actually injurious deviations in their structure will always
have been checked by natural selection. Seeing how
important an organ of locomotion the tail is in most
aquatic animals, its general presence and use for many
purposes in so many land animals, which in their lungs

or modified swimbladders betray their aquatic origin,
may perhaps be thus accounted for. A well-developed
tail having been formed in an aquatic animal, it might
subsequently come to be worked in for all sorts of purposes, as a fly-flapper, an organ of prehension, or as an
aid in turning, as with the dog, though the aid must be
slight, for the hare, with hardly any tail, can double
quickly enough.

In the second place, we may sometimes attribute
importance to characters which are really of very little
importance, and which have originated from quite
secondary causes, independently of natural selection.
We should remember that climate, food, &c., probably
have some little direct influence on the organisation;
that characters reappear from the law of reversion; that
correlation of growth will have had a most important
influence in modifying various structures; and finally,
that sexual selection will often have largely modified
the external characters of animals having a will, to
give one male an advantage in fighting with another
or in charming the females. Moreover when a modification of structure has primarily arisen from the above
or other unknown causes, it may at first have been of
no advantage to the species, but may subsequently
have been taken advantage of by the descendants of
the species under new conditions of life and with newly
acquired habits.

To give a few instances to illustrate these latter remarks. If green woodpeckers alone had existed, and
we did not know that there were many black and pied
kinds, I dare say that we should have thought that the
green colour was a beautiful adaptation to hide this
tree-frequenting bird from its enemies; and consequently that it was a character of importance and might
have been acquired through natural selection; as it is,

we caunot doubt that the colour is due to some quite distinct cause, probably to sexual selection. A trailiug palm in the Malay Archipelago climbs the loftiest trees by the aid of exquisitely constructed hooks clustered around the ends of the branches, and this contrivance, no doubt, is of the highest service to the plaut; but as we see nearly similar hooks on many trees which are not climbers, the hooks on the palm may have arisen from unknown laws of growth, and have been subsequently taken advantage of by the plaut undergoing further modification and becoming a climber. The naked skiu on the head of a vulture is generally considered as a direct adaptation for wallowing in putridity; and so it may be, or it may possibly be due to the direct action of putrid matter; but we should be very cautious in drawing any such inference, when we see that the skin on the head of the clean-feeding male turkey is likewise naked. The sutures in the skulls of young mammals have been advanced as a beautiful adaptation for aiding parturition, and no doubt they facilitate, or may be indispensable for this act; but as sutures occur iu the skulls of young birds and reptiles, which have only to escape from a broken egg, we may infer that this structure has arisen from the laws of growth, aud has been taken advantage of in the parturition of the higher animals.

We are profoundly iguorant of the causes producing slight and unimportant variations; aud we are immediately made conscious of this by reflecting on the differences in the breeds of our domesticated animals iu different countries,—more especially in the less civilised countries where there has been but little artificial selection. Animals kept by savages in different countries often have to struggle for their own subsisteuce, and would be exposed to a certain exteut to natural selec-

L

tion, and individuals with slightly different constitutions would succeed best under different climates. A good observer states that in cattle susceptibility to the attacks of flies is correlated with colour, as is the liability to be poisoned by certain plants; so that even colour would be thus subjected to the action of natural selection. Other observers are convinced that a damp climate affects the growth of the hair, and that with the hair the horns are correlated. Mountain breeds always differ from lowland breeds; and a mountainous country would probably affect the hind limbs from exercising them more, and possibly even the form of the pelvis; and then by the law of homologous variation, the front limbs and the head would probably be affected. The shape, also, of the pelvis might affect by pressure the shape of the head of the young in the womb. The laborious breathing necessary in high regions would, we have some reason to believe, increase the size of the chest; and again correlation would come into play. The effects on the whole organisation of lessened exercise with abundant food is probably still more important; and this, as H. von Nathusius has lately shown in his excellent Treatise, is apparently one chief cause of the great modification which the breeds of swine have undergone. But we are far too ignorant to speculate on the relative importance of the several known and unknown laws of variation; and I have here alluded to them only to show that, if we are unable to account for the characteristic differences of our domestic breeds, which nevertheless we generally admit to have arisen through ordinary generation, we ought not to lay too much stress on our ignorance of the precise cause of the slight analogous differences between species. I might have adduced for this same purpose the differences between the races of man, which are so strongly marked; I may add that some little light can apparently be thrown on the origin of these differences, chiefly through sexual selection of a particular kind, but without here entoring on copious details my reasoning would appear frivolous.

The foregoing remarks lead me to say a few words on the protest lately made by some naturalists, against the utilitarian doctrine that every detail of structure has been produced for the good of its possessor. They believe that very many structures have been created for beauty in the eyes of man, or for mere variety. This doctrine, if true, would be absolutely fatal to my theory. Yet I fully admit that many structures are of no direct use to their possessors. Physical conditions probably have had some little effect on structure, quite independently of any good thus gained. Correlation of growth has no doubt played a most important part, and a useful modification of one part will often have entailed on other parts diversified changes of no direct use. So again characters which formerly were useful, or which formerly had arisen from correlation of growth, or from other unknown cause, may reappear from the law of reversion, though now of no direct use. The effects of sexual selection, when displayed in beauty to charm the females, can be called useful only in rather a forced sense. But by far the most important consideration is that the chief part of the organisation of every being is simply due to inheritance; and consequently, though each being assuredly is well fitted for its place in nature, many structures now have no direct relation to the habits of life of each species. Thus, we can hardly believe that the webbed feet of the upland goose or of the frigate-bird are of special use to these birds; we cannot believe that the similar bones in the

arm of the monkey, in the fore-leg of the horse, in the
wing of the bat, and in the flipper of the seal, are of
special use to these animals. We may safely attribute
these structures to inheritance. But to the progenitor
of the upland goose and of the frigate-bird, webbed feet
no doubt were as useful as they now are to the most
aquatic of existing birds. So we may believe that the
progenitor of the seal had not a flipper, but a foot with
five toes fitted for walking or grasping; and we may
further venture to believe that the several bones in the
limbs of the monkey, horse, and bat, which have been
inherited from a common progenitor, were formerly of
more special use to that progenitor, or its progenitors,
than they now are to these animals having such widely
diversified habits. Therefore we may infer that these
several bones might have been acquired through natural
selection, subjected formerly, as now, to the several laws
of inheritance, reversion, correlation of growth, &c.
Hence every detail of structure in every living creature
(making some little allowance for the direct action of
physical conditions) may be viewed, either as having
been of special use to some ancestral form, or as being
now of special use to the descendants of this form—
either directly, or indirectly through the complex laws
of growth.

Natural selection cannot possibly produce any modification in any one species exclusively for the good of
another species; though throughout nature one species
incessantly takes advantage of, and profits by, the structure of another. But natural selection can and does
often produce structures for the direct injury of other
species, as we see in the fang of the adder, and in the
ovipositor of the ichneumon, by which its eggs are deposited in the living bodies of other insects. If it could
be proved that any part of the structure of any one

species had been formed for the exclusive good of another species, it would annihilate my theory, for such could not have been produced through natural selection. Although many statements may be found in works on natural history to this effect, I cannot find even one which seems to me of any weight. It is admitted that the rattlesnake has a poison-fang for its own defence and for the destruction of its prey; but some authors suppose that at the same time this snake is furnished with a rattle for its own injury, namely, to warn its prey to escape. I would almost as soon believe that the cat curls the end of its tail when preparing to spring, in order to warn the doomed mouse. But I have not space here to enter on this and other such cases.

Natural selection will never produce in a being anything injurious to itself, for natural selection acts solely by and for the good of each. No organ will be formed, as Paley has remarked, for the purpose of causing pain or for doing an injury to its possessor. If a fair balance be struck between the good and evil caused by each part, each will be found on the whole advantageous. After the lapse of time, under changing conditions of life, if any part comes to be injurious, it will be modified; or if it be not so, the being will become extinct, as myriads have become extinct.

Natural selection tends only to make each organic being as perfect as, or slightly more perfect than, the other inhabitants of the same country with which it has to struggle for existence. And we see that this is the degree of perfection attained under nature. The endemic productions of New Zealand, for instance, are perfect one compared with another; but they are now rapidly yielding before the advancing legions of plants and animals introduced from Europe. Natural selection will not produce absolute perfection, nor do we always

meet, as far as we can judge, with this high standard
under nature. The correction for the aberration of
light is said, on high authority, not to be perfect even
in that most perfect organ, the eye. If our reason
leads us to admire with enthusiasm a multitude of
inimitable contrivances in nature, this same reason tells
us, though we may easily err on both sides, that some
other contrivances are less perfect. Can we consider
the sting of the wasp or of the bee as perfect, which,
when used against many attacking animals, cannot be
withdrawn, owing to the backward serratures, and so
inevitably causes the death of the insect by tearing out
its viscera?

If we look at the sting of the bee, as having originally existed in a remote progenitor as a boring and
serrated instrument, like that in so many members of
the same great order, and which has been modified but
not perfected for its present purpose, with the poison
originally adapted for some purpose, such as to produce
galls, subsequently intensified, we can perhaps understand how it is that the use of the sting should so often
cause the insect's own death: for if on the whole the
power of stinging be useful to the community, it will
fulfil all the requirements of natural selection, though it
may cause the death of some few members. If we admire the truly wonderful power of scent by which the
males of many insects find their females, can we admire
the production for this single purpose of thousands of
drones, which are utterly useless to the community for
any other end, and which are ultimately slaughtered by
their industrious and sterile sisters? It may be difficult, but we ought to admire the savage instinctive
hatred of the queen-bee, which urges her to destroy the
young queens her daughters as soon as born, or to
perish herself in the combat; for undoubtedly this is

for the good of the community; and maternal love or maternal hatred, though the latter fortunately is most rare, is all the same to the inexorable principle of natural selection. If we admire the several ingenious contrivances, by which the flowers of the orchis and of many other plants are fertilised through insect agency, can we consider as equally perfect the elaboration by our fir-trees of dense clouds of pollen, in order that a few granules may be wafted by a chance breeze on to the ovules ?

Summary of Chapter.—We have in this chapter discussed some of the difficulties and objections which may be urged against my theory. Many of them are very serious ; but I think that in the discussion light has been thrown on several facts, which on the theory of independent acts of creation are utterly obscure. We have seen that species at any one period are not indefinitely variable, and are not linked together by a multitude of intermediate gradations, partly because the process of natural selection will always be very slow, and will act, at any one time, only on a very few forms ; and partly because the very process of natural selection almost implies the continual supplanting and extinction of preceding and intermediate gradations. Closely allied species, now living on a continuous area, must often have been formed when the area was not continuous, and when the conditions of life did not insensibly graduate away from one part to another. When two varieties are formed in two districts of a continuous area, an intermediate variety will often be formed, fitted for an intermediate zone ; but from reasons assigned, the intermediate variety will usually exist in lesser numbers than the two forms which it connects ; consequently the two latter, during the course of further modification, from

existing in greater numbers, will have a great advantage
over the less numerous intermediate variety, and will
thus generally succeed in supplanting and exterminating it.

We have seen in this chapter how cautious we should
be in concluding that the most different habits of life
could not graduate into each other; that a bat, for
instance, could not have been formed by natural selection from an animal which at first could only glide
through the air.

We have seen that a species may under new conditions of life change its habits, or have diversified habits,
with some habits very unlike those of its nearest congeners. Hence we can understand, bearing in mind
that each organic being is trying to live wherever it
can live, how it has arisen that there are upland geese
with webbed feet, ground woodpeckers, diving thrushes,
and petrels with the habits of auks.

Although the belief that an organ so perfect as the
eye could have been formed by natural selection, is
more than enough to stagger any one; yet in the case
of any organ, if we know of a long series of gradations
in complexity, each good for its possessor, then, under
changing conditions of life, there is no logical impossibility in the acquirement of any conceivable degree of
perfection through natural selection. In the cases in
which we know of no intermediate or transitional states,
we should be very cautious in concluding that none
could have existed, for the homologies of many organs
and their intermediate states show that wonderful meta-
morphoses in function are at least possible. For instance,
a swim-bladder has apparently been converted into an
air-breathing lung. The same organ having performed
simultaneously very different functions, and then having
been specialised for one function; and two very distinct

organs having performed at the same time the same function, the one having been perfected whilst aided by the other, must often have largely facilitated transitions.

We are far too ignorant, in almost every case, to be enabled to assert that any part or organ is so unimportant for the welfare of a species, that modifications in its structure could not have been slowly accumulated by means of natural selection. But we may confidently believe that many modifications, wholly due to the laws of growth, and at first in no way advautageous to a species, have been subsequently taken advantage of by the still further modified descendants of this species. We may, also, believe that a part formerly of high importance has often been retained (as the tail of an aquatic animal by its terrestrial descendants), though it has become of such small importance that it could not, in its present state, have been acquired by natural selection,—a power which acts solely by the preservation of profitable variations in the struggle for life.

Natural selection will produce nothing in one species for the exclusive good or injury of another; though it may well produce parts, organs, and excretions highly useful or even indispensable, or highly injurious to another species, but in all cases at the same time useful to the owner. Natural selection in each well-stocked country must act chiefly through the competition of the inhabitants one with another, and consequently will produce perfection, or strength in the battle for life, only according to the standard of that country. Hence the inhabitants of one country, generally the smaller one, will often yield, as we see they do yield, to the inhabitants of another and generally larger country. For in the larger country there will have existed more individuals, and more diversified forms, and the competition will have been severer, and thus the standard of perfection will have been rendered higher. Natural selection will not necessarily produce absolute perfection; nor, as far as we can judge by our limited faculties, can absolute perfection be everywhere found.

On the theory of natural selection we can clearly understand the full meaning of that old canon in natural history, "Natura non facit saltum." This canon, if we look only to the present inhabitants of the world, is not strictly correct, but if we include all those of past times, it must by my theory be strictly true.

It is generally acknowledged that all organic beings have been formed on two great laws—Unity of Type, and the Conditions of Existence. By unity of type is meant that fundamental agreement in structure which we see in organic beings of the same class, and which is quite independent of their habits of life. On my theory, unity of type is explained by unity of descent. The expression of conditions of existence, so often insisted on by the illustrious Cuvier, is fully embraced by the principle of natural selection. For natural selection acts by either now adapting the varying parts of each being to its organic and inorganic conditions of life; or by having adapted them during the long-past periods of time: the adaptations being aided in some cases by use and disuse, being slightly affected by the direct action of the external conditions of life, and being in all cases subjected to the several laws of growth. Hence, in fact, the law of the Conditions of Existence is the higher law; as it includes, through the inheritance of former adaptations, that of Unity of Type.

CHAPTER VII.

INSTINCT.

Instincts comparable with habits, but different in their origin —
Instincts graduated — Aphides and ants — Instincts variable —
Domestic instincts, their origin — Natural instincts of the cuckoo,
ostrich, and parasitic bees — Slave-making ants — Hive-bee, its
cell-making instinct — Changes of instinct and structure not
necessarily simultaneous — Difficulties on the theory of the
Natural Selection of instincts — Neuter or sterile insects —
Summary.

THE subject of instinct might have been worked into the
previous chapters; but I have thought that it would be
more convenient to treat the subject separately, especially as so wonderful an instinct as that of the hive-
bee making its cells will probably have occurred to
many readers, as a difficulty sufficient to overthrow my
whole theory. I must premise, that I have nothing to
do with the origin of the primary mental powers, any
more than I have with that of life itself. We are
concerned only with the diversities of instinct and of
the other mental qualities of animals within the same
class.

I will not attempt any definition of instinct. It
would be easy to show that several distinct mental
actions are commonly embraced by this term; but every
one understands what is meant, when it is said that instinct impels the cuckoo to migrate and to lay her eggs
in other birds' nests. An action, which we ourselves
should require experience to enable us to perform, when
performed by an animal, more especially by a very

young one, without any experience, and when performed
by many individuals in the same way, without their
knowing for what purpose it is performed, is usually
said to be instinctive. But I could show that none of
these characters of instinct are universal. A little dose,
as Pierre Huber expresses it, of judgment or reason,
often comes into play, even in animals very low in the
scale of nature.

Frederick Cuvier and several of the older metaphy-
sicians have compared instinct with habit. This comparison gives, I think, a remarkably accurate notion of
the frame of mind under which an instinctive action is
performed, but not of its origin. How unconsciously
many habitual actions are performed, indeed not rarely
in direct opposition to our conscious will! yet they may
be modified by the will or reason. Habits easily become
associated with other habits, and with certain periods of
time and states of the body. When once acquired, they
often remain constant throughout life. Several other
points of resemblance between instincts and habits
could be pointed out. As in repeating a well-known
song, so in instincts, one action follows another by a sort
of rhythm ; if a person be interrupted in a song, or in
repeating anything by rote, he is generally forced to go
back to recover the habitual train of thought : so P.
Huber found it was with a caterpillar, which makes a
very complicated hammock ; for if he took a caterpillar
which had completed its hammock up to, say, the sixth
stage of construction, and put it into a hammock completed up only to the third stage, the caterpillar simply
re-performed the fourth, fifth, and sixth stages of construction. If, however, a caterpillar were taken out of
a hammock made up, for instance, to the third stage,
and were put into one finished up to the sixth stage, so
that much of its work was already done for it, far from

feeling the benefit of this, it was much embarrassed, and, in order to complete its hammock, seemed forced to start from the third stage, where it had left off, and thus tried to complete the already finished work.

If we suppose any habitual action to become inherited —and I think it can be shown that this does sometimes happen—then the resemblance between what originally was a habit and an instinct becomes so close as not to be distinguished. If Mozart, instead of playing the pianoforte at three years old with wonderfully little practice, had played a tune with no practice at all, he might truly be said to have done so instinctively. But it would be the most serious error to suppose that the greater number of instincts have been acquired by habit in one generation, and then transmitted by inheritance to succeeding generations. It can be clearly shown that the most wonderful instincts with which we are acquainted, namely, those of the hive-bee and of many ants, could not possibly have been thus acquired.

It will be universally admitted that instincts are as important as corporeal structure for the welfare of each species, under its present conditions of life. Under changed conditions of life, it is at least possible that slight modifications of instinct might be profitable to a species; and if it can be shown that instincts do vary ever so little, then I can see no difficulty in natural selection preserving and continually accumulating variations of instinct to any extent that was profitable. It is thus, as I believe, that all the most complex and wonderful instincts have originated. As modifications of corporeal structure arise from, and are increased by, use or habit, and are diminished or lost by disuse, so I do not doubt it has been with instincts. But I believe that the effects of habit are of quite subordinate importance to the effects of the natural selection of what

may be called accidental variations of instincts;—
that is of variations produced by the same unknown
causes which produce slight deviations of bodily structure.

No complex instinct can possibly be produced through
natural selection, except by the slow and gradual accumulation of numerous, slight, yet profitable, variations.
Hence, as in the case of corporeal structures, we ought
to find in nature, not the actual transitional gradations
by which each complex instinct has been acquired—for
these could be found only in the lineal ancestors of each
species—but we ought to find in the collateral lines of
descent some evidence of such gradations; or we ought
at least to be able to show that gradations of some kind
are possible; and this we certainly can do. I have been
surprised to find, making allowance for the instincts of
animals having been but little observed except in Europe
and North America, and for no instinct being known
amongst extinct species, how very generally gradations,
leading to the most complex instincts, can be discovered. Changes of instinct may sometimes be facilitated by the same species having different instincts at
different periods of life, or at different seasons of the
year, or when placed under different circumstances, &c.;
in which case either one or the other instinct might be
preserved by natural selection. And such instances of
diversity of instinct in the same species can be shown
to occur in nature.

Again as in the case of corporeal structure, and conformably with my theory, the instinct of each species is
good for itself, but has never, as far as we can judge,
been produced for the exclusive good of others. One
of the strongest instances of an animal apparently performing an action for the sole good of another, with
which I am acquainted, is that of aphides voluntarily

yielding, as was first observed by Huber, their sweet
excretion to ants: that they do so voluntarily, the following facts show.   I removed all the ants from a group
of about a dozen aphides on a dock-plant, and prevented
their attendance during several hours.   After this interval, I felt sure that the aphides would want to excrete.
I watched them for some time through a lens, but not
one excreted; I then tickled and stroked them with a
hair in the same manner, as well as I could, as the ants
do with their antennæ; but not one excreted.   Afterwards I allowed an ant to visit them, and it immediately
seemed, by its eager way of running about, to be well
aware what a rich flock it had discovered; it then began
to play with its antennæ on the abdomen first of one
aphis and then of another; and each, as soon as it felt
the antennæ, immediately lifted up its abdomen and
excreted a limpid drop of sweet juice, which was eagerly
devoured by the ant.   Even the quite young aphides
behaved in this manner, showing that the action was
instinctive, and not the result of experience.   It is certain, from the observations of Huber, that the aphides
show no dislike to the ants: if the latter be not
present, they are at last compelled to eject their excretion.   But as the excretion is extremely viscid, it is
probably a convenience to the aphides to have it removed; and therefore probably they do not instinctively
excrete for the sole good of the ants.   Although there
is no evidence that any animal performs an action for
the exclusive good of another of a distinct species, yet
each species tries to take advantage of the instincts of
others, as each takes advantage of the weaker bodily
structure of others.   So again, in some few cases, certain
instincts cannot be considered as absolutely perfect;
but as details on this and other such points are not indispensable, they may be here passed over.

As some degree of variation in instincts under a state of nature, and the inheritance of such variations, are indispensable for the action of natural selection, as many instances as possible ought to be here given; but want of space prevents me. I can only assert, that instincts certainly do vary—for instance, the migratory instinct, both in extent and direction, and in its total loss. So it is with the nests of birds, which vary partly in dependence on the situations chosen, and on the nature and temperature of the country inhabited, but often from causes wholly unknown to us: Audubon has given several remarkable cases of differences in the nests of the same species in the northern and southern United States. Why, it has been asked, if instinct be variable, has it not given to the bee "the ability to use some other material when wax was deficient?" But what other material could bees use? They will work with and use, as I have seen, wax hardened with vermilion and softened with lard. Andrew Knight observed that his bees, instead of laboriously collecting propolis, used a cement of wax and turpentine, with which he had covered decorticated trees. It has lately been shown that bees, instead of searching flowers for their pollen, will gladly use a very different substance, namely, oatmeal. Fear of any particular enemy is certainly an instinctive quality, as may be seen in nestling birds, though it is strengthened by experience, and by the sight of fear of the same enemy in other animals. But fear of man is slowly acquired, as I have elsewhere shown, by various animals inhabiting desert islands; and we may see an instance of this, even in England, in the greater wildness of all our large birds than of our small birds; for the large birds have been most persecuted by man. We may safely attribute the greater wildness of our large birds to this cause; for in

uninhabited islands large birds are not more fearful than
small; and the magpie, so wary in England, is tame in
Norway, as is the hooded crow in Egypt.

That the general disposition of individuals of the
same species, born in a state of nature, is extremely
diversified, can be shown by a multitude of facts.
Several cases, also, could be given of occasional and
strange habits in certain species, which might, if advantageous to the species, give rise, through natural selection, to quite new instincts. But I am well aware that
these general statements, without facts given in detail,
can produce but a feeble effect on the reader's mind.
I can only repeat my assurance, that I do not speak
without good evidence.

The possibility, or even probability, of inherited
variations of instinct in a state of nature will be strengthened by briefly considering a few cases under domestication. We shall thus also be enabled to see the
respective parts which habit and the selection of so-called
accidental variations have played in modifying the
mental qualities of our domestic animals. A number
of curious and authentic instances could be given of the
inheritance of all shades of disposition and tastes, and
likewise of the oddest tricks, associated with certain
frames of mind or periods of time. But let us look to
the familiar case of the several breeds of dogs: it cannot be doubted that young pointers (I have myself seen
a striking instance) will sometimes point and even back
other dogs the very first time that they are taken out;
retrieving is certainly in some degree inherited by re-
trievers; and a tendency to run round, instead of at, a
flock of sheep, by shepherd-dogs. I cannot see that
these actions, performed without experience by the
young, and in nearly the same manner by each individual, performed with eager delight by each breed, and

without the end being known—for the young pointer
can no more know that he points to aid his master, than
the white butterfly knows why she lays her eggs on the
leaf of the cabbage—I cannot see that these actions
differ essentially from true instincts. If we were to behold
one kind of wolf, when young and without any training,
as soon as it scented its prey, stand motionless like a
statue, and then slowly crawl forward with a peculiar
gait ; and another kind of wolf rushing round, instead of
at, a herd of deer, and driving them to a distant point, we
should assuredly call these actions instinctive. Domestic
instincts, as they may be called, are certainly far less
fixed than natural instincts ; but they have been acted
on by far less rigorous selection, and have been transmitted for an incomparably shorter period, under less
fixed conditions of life.

How strongly these domestic instincts, habits, and dis-
positions are inherited, and how curiously they become
mingled, is well shown when different breeds of dogs are
crossed. Thus it is known that a cross with a bull-dog
has affected for many generations the courage and ob-
stinacy of greyhounds ; and a cross with a greyhound
has given to a whole family of shepherd-dogs a tendency to hunt hares. These domestic instincts, when
thus tested by crossing, resemble natural instincts,
which in a like manner become curiously blended together, and for a long period exhibit traces of the
instincts of either parent: for example, Le Roy describes a dog, whose great-grandfather was a wolf, and
this dog showed a trace of its wild parentage only in one
way, by not coming in a straight line to his master
when called.

Domestic instincts are sometimes spoken of as actions
which have become inherited solely from long-continued
and compulsory habit, but this is not true. No one

would ever have thought of teaching, or probably could have taught, the tumbler-pigeon to tumble,—an action which, as I have witnessed, is performed by young birds, that have never seen a pigeon tumble. We may believe that some one pigeon showed a slight tendency to this strange habit, and that the long-continued selection of the best individuals in successive generations made tumblers what they now are; and near Glasgow there are house-tumblers, as I hear from Mr. Brent, which cannot fly eighteen inches high without going head over heels. It may be doubted whether any one would have thought of training a dog to point, had not some one dog naturally shown a tendency in this line; and this is known occasionally to happen, as I once saw, in a pure terrier: the act of pointing is probably, as many have thought, only the exaggerated pause of an animal preparing to spring on its prey. When the first tendency to point was once displayed, methodical selection and the inherited effects of compulsory training in each successive generation would soon complete the work; and unconscious selection is still in progress, as each man tries to procure, without intending to improve the breed, dogs which will stand and hunt best. On the other hand, habit alone in some cases has sufficed; hardly any animal is more difficult, in most cases, to tame than the young of the wild rabbit; scarcely any animal is tamer than the young of the tame rabbit; but I do not suppose that domestic rabbits have ever been selected for tameness; and I presume that we must attribute the whole of the inherited change from extreme wildness to extreme tameness, simply to habit and long-continued close confinement.

Natural instincts are lost under domestication: a remarkable instance of this is seen in those breeds of fowls which very rarely or never become "broody,"

that is, never wish to sit on their eggs. Familiarity
alone prevents our seeing how universally and largely
the minds of our domestic animals have been modified
by domestication. It is scarcely possible to doubt that
the love of man has become instinctive in the dog. All
wolves, foxes, jackals, and species of the cat genus, when
kept tame, are most eager to attack poultry, sheep, and
pigs; and this tendency has been found incurable in
dogs which have been brought home as puppies from
countries, such as Tierra del Fuego and Australia, where
the savages do not keep these domestic animals. How
rarely, on the other hand, do our civilised dogs, even
when quite young, require to be taught not to attack
poultry, sheep, and pigs! No doubt they occasionally
do make an attack, and are then beaten; and if not
cured, they are destroyed; so that habit, with some
degree of selection, has probably concurred in civilising
by inheritance our dogs. On the other hand, young
chickens have lost, wholly by habit, that fear of the dog
and cat which no doubt was originally instinctive in
them, in the same way as it is so plainly instinctive in
young pheasants, though reared under a hen. It is not
that chickens have lost all fear, but fear only of dogs
and cats, for if the hen gives the danger-chuckle, they
will run (more especially young turkeys) from under
her, and conceal themselves in the surrounding grass or
thickets; and this is evidently done for the instinctive
purpose of allowing, as we see in wild ground-birds, their
mother to fly away. But this instinct retained by our
chickens has become useless under domestication, for
the mother-hen has almost lost by disuse the power
of flight.

Hence, we may conclude, that under domestication
instincts have been acquired and natural instincts have
been lost partly by habit, and partly by man selecting

and accumulating during successive generations, peculiar mental habits and actions, which at first appeared from what we must in our ignorance call an accident. In some cases compulsory habit alone has sufficed to produce such inherited mental changes; in other cases compulsory habit has done nothing, and all has been the result of selection, pursued both methodically and unconsciously; but in most cases, probably, habit and selection have acted together.

We shall, perhaps, best understand how instincts in a state of nature have become modified by selection, by considering a few cases. I will select only three, out of the several which I shall have to discuss in my future work,—namely, the instinct which leads the cuckoo to lay her eggs in other birds' nests; the slave-making instinct of certain ants; and the comb-making power of the hive-bee: these two latter instincts have generally, and most justly, been ranked by naturalists as the most wonderful of all known instincts.

It is now commonly admitted that the more immediate and final cause of the cuckoo's instinct is, that she lays her eggs, not daily, but at intervals of two or three days; so that, if she were to make her own nest and sit on her own eggs, those first laid would have to be left for some time unincubated, or there would be eggs and young birds of different ages in the same nest. If this were the case, the process of laying and hatching might be inconveniently long, more especially as she has to migrate at a very early period; and the first hatched young would probably have to be fed by the male alone. But the American cuckoo is in this predicament; for she makes her own nest and has eggs and young successively hatched, all at the same time. It has been asserted that the American cuckoo occasionally lays her eggs in other birds' nests; but I hear on the high authority of Dr. Brewer, that this is a mistake. Nevertheless, I could give several instances of various birds which have been known occasionally to lay their eggs in other birds' nests. Now let us suppose that the ancient progenitor of our European cuckoo had the habits of the American cuckoo; but that occasionally she laid an egg in another bird's nest. If the old bird profited by this occasional habit, or if the young were made more vigorous by advantage having been taken of the mistaken maternal instinct of another bird, than by their own mother's care, encumbered as she can hardly fail to be by having eggs and young of different ages at the same time, and by having to migrate at a very early period; then the old birds or the fostered young would gain an advantage. And analogy would lead me to believe, that the young thus reared would be apt to follow by inheritance the occasional and aberrant habit of their mother, and in their turn would be apt to lay their eggs in other birds' nests, and thus be successful in rearing their young. By a continued process of this nature, I believe that the strange instinct of our cuckoo could be, and has been, generated. I may add that, according to Dr. Gray and to some other observers, the European cuckoo has not utterly lost all maternal love and care for her own offspring.

The occasional habit of birds laying their eggs in other birds' nests, either of the same or of a distinct species, is not very uncommon with the Gallinaceæ; and this perhaps explains the origin of a singular instinct in the allied group of ostriches. For several hen ostriches, at least in the case of the American species, unite and lay first a few eggs in one nest and then in another; and these are hatched by the males. This instinct may probably be accounted for by the fact

of the hens laying a large number of eggs; but, as in
the case of the cuckoo, at intervals of two or three days.
This instinct, however, of the American ostrich has not
as yet been perfected; for a surprising number of eggs
lie strewed over the plains, so that in one day's hunting
I picked up no less than twenty lost and wasted eggs.

Many bees are parasitic, and always lay their eggs in
the nests of bees of other kinds. This case is more remarkable than that of the cuckoo; for these bees have
not only their instincts but their structure modified in
accordance with their parasitic habits; for they do not
possess the pollen-collecting apparatus which would be
necessary if they had to store food for their own young.
Some species, likewise, of Sphegidæ (wasp-like insects)
are parasitic on other species; and M. Fabre has lately
shown good reason for believing that although the
Tachytes nigra generally makes its own burrow and
stores it with paralysed prey for its own larvæ to feed
on, yet that when this insect finds a burrow already
made and stored by another sphex, it takes advantage
of the prize, and becomes for the occasion parasitic. In
this case, as with the supposed case of the cuckoo, I can
see no difficulty in natural selection making an occasional habit permanent, if of advantage to the species,
and if the insect whose nest and stored food are thus
feloniously appropriated, be not thus exterminated.

Slave-making instinct.—This remarkable instinct was
first discovered in the Formica (Polyerges) rufescens
by Pierre Huber, a better observer even than his celebrated father. This ant is absolutely dependent on its
slaves; without their aid, the species would certainly
become extinct in a single year. The males and fertile
females do no work of any kind, and the workers or sterile females, though most energetic and courageous in

capturing slaves, do no other work. They are incapable of making their own nests, or of feeding their own larvæ. When the old nest is found inconvenient, and they have to migrate, it is the slaves which determine the migration, and actually carry their masters in their jaws. So utterly helpless are the masters, that when Huber shut up thirty of them without a slave, but with plenty of the food which they like best, and with their larvæ and pupæ to stimulate them to work, they did nothing; they could not even feed themselves, and many perished of hunger. Huber then introduced a single slave (F. fusca), and she instantly set to work, fed and saved the survivors; made some cells and tended the larvæ, and put all to rights. What can be more extraordinary than these well-ascertained facts? If we had not known of any other slave-making ant, it would have been hopeless to have speculated how so wonderful an instinct could have been perfected.

Another species, Formica sanguinea, was likewise first discovered by P. Huber to be a slave-making ant. This species is found in the southern parts of England, and its habits have been attended to by Mr. F. Smith, of the British Museum, to whom I am much indebted for information on this and other subjects. Although fully trusting to the statements of Huber and Mr. Smith, I tried to approach the subject in a sceptical frame of mind, as any one may well be excused for doubting the truth of so extraordinary and odious an instinct as that of making slaves. Hence I will give the observations which I made, in some little detail. I opened fourteen nests of F. sanguinea, and found a few slaves in all. Males and fertile females of the slave-species (F. fusca) are found only in their own proper communities, and have never been observed in the nests of F. sanguinea. The slaves are black and not above half the size of their

red masters, so that the contrast in their appearauce is
very great. When the nest is slightly disturbed, the
slaves occasionally come out, and like their masters aro
much agitated and defend the nest : when the nest is
much disturbed and the larvæ and pupæ are exposed,
the slaves work energetically with their masters in carrying them away to a place of safety. Hence, it is clear,
that the slaves feel quite at home. During the months
of June and July, on three successive years, I have
watched for many hours several nests in Surrey and
Sussex, and never saw a slave either leave or enter
a nest. As, during these months, the slaves are very
few in number, I thought that they might behave differently when more numerous ; but Mr. Smith informs me
that he has watched the nests at various hours during
May, June, and August, both in Surrey and Hampshire,
and has never seen the slaves, though present in large
numbers in August, either leave or enter the nest.
Hence he considers them as strictly household slaves.
The masters, on the other hand, may be constantly
seen bringing in materials for the nest, and food of all
kinds. During the year 1860, however, in the month
of July, I came across a community with an unusnally
large stock of slaves, and I observed a few slaves mingled
with their masters leaving the nest, and marching along
the same road to a tall Scotch-fir-tree, twenty-five yards
distant, which they ascended together, probably in search
of aphides or cocci. According to Huber, who had ample
opportunities for observation, in Switzerland the slaves
habitually work with their masters in making the nest,
and they alone open and close the doors in the morning
and evening ; and, as Huber expressly states, their
principal office is to search for aphides. This difference in the usual habits of the masters and slaves
in the two countries, probably depends merely on the

M

slaves being captured in greater numbers in Switzerland than in England.

One day I fortunately witnessed a migration of F. sanguinea from one nest to another, and it was a most interesting spectacle to behold the masters carefully carrying their slaves in their jaws instead of being carried by them, as in the case of F. rufescens. Another day my attention was struck by about a score of the slave-makers haunting the same spot, and evidently not in search of food; they approached and were vigorously repulsed by an independent community of the slave-species (F. fusca); sometimes as many as three of these ants clinging to the legs of the slave-making F. sanguinea. The latter ruthlessly killed their small opponents, and carried their dead bodies as food to their nest, twenty-nine yards distant; but they were prevented from getting any pupæ to rear as slaves. I then dug up a small parcel of the pupæ of F. fusca from another nest, and put them down on a bare spot near the place of combat; they were eagerly seized, and carried off by the tyrants, who perhaps fancied that, after all, they had been victorious in their late combat.

At the same time I laid on the same place a small parcel of the pupæ of another species, F. flava, with a few of these little yellow ants still clinging to the fragments of the nest. This species is sometimes, though rarely, made into slaves, as has been described by Mr. Smith. Although so small a species, it is very courageous, and I have seen it ferociously attack other ants. In one instance I found to my surprise an independent community of F. flava under a stone beneath a nest of the slave-making F. sanguinea; and when I had accidentally disturbed both nests, the little ants attacked their big neighbours with surprising courage. Now I was curious to ascertain whether F. sanguinea

could distinguish the pupæ of F. fusca, which they
habitually make into slaves, from those of the little and
furious F. flava, which they rarely capture, and it was
evident that they did at once distinguish them : for we
have seen that they eagerly and instantly seized the
pupæ of F. fusca, whereas they were much terrified when
they came across the pupæ, or even the earth from the
nest of F. flava, and quickly ran away; but in about a
quarter of an hour, shortly after all the little yellow
ants had crawled away, they took heart and carried off
the pupæ.

One evening I visited another community of F. sanguinea, and found a number of these ants returning home
and entering their nests, carrying the dead bodies of F.
fusca (showing that it was not a migration) and numerous pupæ. I traced a long file of ants burthened with
booty, for about forty yards, to a very thick clump of
heath, whence I saw the last individual of F. sanguinea
emerge, carrying a pupa ; but I was not able to find the
desolated nest in the thick heath. The nest, however,
must have been close at hand, for two or three individuals of F. fusca were rushing about in the greatest
agitation, and one was perched motionless with its own
pupa in its mouth on the top of a spray of heath, an
image of despair, over its ravaged home.

Such are the facts, though they did not need confirmation by me, in regard to the wonderful instinct of
making slaves. Let it be observed what a contrast the
instinctive habits of F. sanguinea present with those of
the continental F. rufescens. The latter does not build
its own nest, does not determine its own migrations,
does not collect food for itself or its young, and cannot
even feed itself: it is absolutely dependent on its numerous slaves. Formica sanguinea, on the other hand,
possesses much fewer slaves, and in the early part of

the summer extremely few: the masters determine
when and where a new nest shall be formed, and when
they migrate, the masters carry the slaves. Both in
Switzerland and England the slaves seem to have the
exclusive care of the larvæ, and the masters alone go
on slave-making expeditions. In Switzerland the slaves
and masters work together, making and bringing materials for the nest: both, but chiefly the slaves, tend, and
milk as it may be called, their aphides; and thus both
collect food for the community. In England the masters
alone usually leave the nest to collect building materials
and food for themselves, their slaves and larvæ. So that
the masters in this country receive much less service
from their slaves than they do in Switzerland.

By what steps the instinct of F. sanguinea originated
I will not pretend to conjecture. But as ants, which are
not slave-makers, will, as I have seen, carry off pupæ of
other species, if scattered near their nests, it is possible
that such pupæ originally stored as food might become
developed; and the foreign ants thus unintentionally
reared would then follow their proper instincts, and do
what work they could. If their presence proved useful
to the species which had seized them—if it were more
advantageous to this species to capture workers than to
procreate them—the habit of collecting pupæ originally
for food might by natural selection be strengthened and
rendered permanent for the very different purpose of
raising slaves. When the instinct was once acquired,
if carried out to a much less extent even than in our
British F. sanguinea, which, as we have seen, is less
aided by its slaves than the same species in Switzerland, natural selection might increase and modify the
instinct—always supposing each modification to be of
use to the species—until an ant was formed as abjectly
dependent on its slaves as is the Formica rufescens.

Cell-making instinct of the Hive-Bee.—I will not here enter on minute details on this subject, but will merely give an outline of the conclusions at which I have arrived. He must be a dull man who can examine the exquisite structure of a comb, so beautifully adapted to its end, without enthusiastic admiration. We hear from mathematicians that bees have practically solved a recondite problem, and have made their cells of the proper shape to hold the greatest possible amount of honey, with the least possible consumption of precious wax in their construction. It has been remarked that a skilful workman, with fitting tools and measures, would find it very difficult to make cells of wax of the true form, though this is perfectly effected by a crowd of bees working in a dark hive. Grant whatever instincts you please, and it seems at first quite inconceivable how they can make all the necessary angles and planes, or even perceive when they are correctly made. But the difficulty is not nearly so great as it at first appears: all this beautiful work can be shown, I think, to follow from a few simple instincts.

I was led to investigate this subject by Mr. Waterhouse, who has shown that the form of the cell stands in close relation to the presence of adjoining cells; and the following view may, perhaps, be considered only as a modification of his theory. Let us look to the great principle of gradation, and see whether Nature does not reveal to us her method of work. At one end of a short series we have humble-bees, which use their old cocoons to hold honey, sometimes adding to them short tubes of wax, and likewise making separate and very irregular rounded cells of wax. At the other end of the series we have the cells of the hive-bee, placed in a double layer: each cell, as is well known, is an hexagonal prism, with the basal edges of its six sides bevelled so as to fit on to a pyramid, formed of three rhombs. These rhombs have certain angles, and the three which form the pyramidal base of a single cell on one side of the comb, enter into the composition of the bases of three adjoining cells on the opposite side. In the series between the extreme perfection of the cells of the hive-bee and the simplicity of those of the humble-bee, we have the cells of the Mexican Melipona domestica, carefully described and figured by Pierre Huber. The Melipona itself is intermediate in structure between the hive and humble bee, but more nearly related to the latter: it forms a nearly regular waxen comb of cylindrical cells, in which the young are hatched, and, in addition, some large cells of wax for holding honey. These latter cells are nearly spherical and of nearly equal sizes, and are aggregated into an irregular mass. But the important point to notice, is that these cells are always made at that degree of nearness to each other, that they would have intersected or broken into each other, if the spheres had been completed; but this is never permitted, the bees building perfectly flat walls of wax between the spheres which thus tend to intersect. Hence each cell consists of an outer spherical portion and of two, three, or more perfectly flat surfaces, according as the cell adjoins two, three, or more other cells. When one cell comes into contact with three other cells, which, from the spheres being nearly of the same size, is very frequently and necessarily the case, the three flat surfaces are united into a pyramid; and this pyramid, as Huber has remarked, is manifestly a gross imitation of the three-sided pyramidal bases of the cell of the hive-bee. As in the cells of the hive-bee, so here, the three plane surfaces in any one cell necessarily enter into the construction of three adjoining cells. It is obvious that the Melipona saves wax by this manner of building; for the flat walls between the adjoining cells are not double, but are of the same thickness as the outer spherical portions, and yet each flat portion forms a part of two cells.

Reflecting on this case, it occurred to me that if the Melipona had made its spheres at some given distance from each other, and had made them of equal sizes and had arranged them symmetrically in a double layer, the resulting structure would probably have been as perfect as the comb of the hive-bee. Accordingly I wrote to Professor Miller, of Cambridge, and this geometer has kindly read over the following statement, drawn up from his information, and tells me that it is strictly correct :—

If a number of equal spheres be described with their centres placed in two parallel layers; with the centre of each sphere at the distance of radius $\times \sqrt{2}$, or radius $\times$ 1·41421 (or at some lesser distance), from the centres of the six surrounding spheres in the same layer; and at the same distance from the centres of the adjoining spheres in the other and parallel layer; then, if planes of intersection between the several spheres in both layers be formed, there will result a double layer of hexagonal prisms united together by pyramidal bases formed of three rhombs; and the rhombs and the sides of the hexagonal prisms will have every angle identically the same with the best measurements which have been made of the cells of the hive-bee.

Hence we may safely conclude that if we could slightly modify the instincts already possessed by the Melipona, and in themselves not very wonderful, this bee would make a structure as wonderfully perfect as that of the hive-bee. We must suppose the Melipona to make her cells truly spherical, and of equal sizes; and this would not be very surprising, seeing that she

already does so to a certain extent, and seeing what
perfectly cylindrical burrows in wood many insects can
make, apparently by turning round on a fixed point.
We must suppose the Melipona to arrange her cells in
level layers, as she already does her cylindrical cells;
and we must further suppose, and this is the greatest
difficulty, that she can somehow judge accurately at
what distance to stand from her fellow-labourers when
several are making their spheres; but she is already
so far enabled to judge of distance, that she always
describes her spheres so as to intersect largely; and
then she unites the points of intersection by perfectly
flat surfaces. We have further to suppose, but this is
no difficulty, that after hexagonal prisms have been
formed by the intersection of adjoining spheres in the
same layer, she can prolong the hexagon to any length
requisite to hold the stock of honey; in the same way
as the rude humble-bee adds cylinders of wax to the
circular mouths of her old cocoons. By such modifications of instincts in themselves not very wonderful,—
hardly more wonderful than those which guide a bird
to make its nest,—I believe that the hive-bee has acquired, through natural selection, her inimitable architectural powers.

But this theory can be tested by experiment. Following the example of Mr. Tegetmeier, I separated two
combs, and put between them a long, thick, rectangular
strip of wax : the bees instantly began to excavate minute
circular pits in it; and as they deepened these little
pits, they made them wider and wider until they were
converted into shallow basins, appearing to the eye
perfectly true or parts of a sphere, and of about the
diameter of a cell. It was most interesting to me to observe that wherever several bees had begun to excavate
these basins near together, they had begun their work

at such a distance from each other, that by the time the
basins had acquired the above stated width (*i. e.* about
the width of an ordinary cell), and were in depth about
one sixth of the diameter of the sphere of which they
formed a part, the rims of the basins intersected or
broke into each other. As soon as this occurred, the
bees ceased to excavate, and began to build up flat walls
of wax on the lines of intersection between the basins,
so that each hexagonal prism was built upon the scal-
loped edge of a smooth basin, instead of on the straight
edges of a three-sided pyramid as in the case of ordinary cells.

I then put into the hive, instead of a thick, rectangular
piece of wax, a thin and narrow, knife-edged ridge,
coloured with vermilion. The bees instantly began on
both sides to excavate little basins near to each other,
in the same way as before; but the ridge of wax was so
thin, that the bottoms of the basins, if they had been
excavated to the same depth as in the former experiment, would have broken into each other from the
opposite sides. The bees, however, did not suffer this
to happen, and they stopped their excavations in due
time; so that the basins, as soon as they had been a
little deepened, came to have flat bottoms; and these
flat bottoms, formed by thin little plates of the vermilion
wax having been left ungnawed, were situated, as far
as the eye could judge, exactly along the planes of
imaginary intersection between the basins on the opposite sides of the ridge of wax. In parts, only little
bits, in other parts, large portions of a rhombic plate
had been left between the opposed basins, but the work,
from the unnatural state of things, had not been neatly
performed. The bees must have worked at very nearly
the same rate on the opposite sides of the ridge of vermilion wax, as they circularly gnawed away and deep-

ened the basins on both sides, in order to have succeeded in thus leaving flat plates between the basins, by
stopping work along the intermediate planes or planes
of intersection.

Considering how flexible thin wax is, I do not see
that there is any difficulty in the bees, whilst at work on
the two sides of a strip of wax, perceiving when they
have gnawed the wax away to the proper thinness,
and then stopping their work. In ordinary combs it
has appeared to me that the bees do not always succeed
in working at exactly the same rate from the opposite
sides; for I have noticed half-completed rhombs at the
base of a just-commenced cell, which were slightly con-
cave on one side, where I suppose that the bees had excavated too quickly, and convex on the opposed side,
where the bees had worked less quickly. In one well-
marked instance, I put the comb back into the hive, and
allowed the bees to go on working for a short time, and
again examined the cell, and I found that the rhombic
plate had been completed, and had become *perfectly flat*:
it was absolutely impossible, from the extreme thinness
of the little rhombic plate, that they could have effected
this by gnawing away the convex side; and I suspect
that the bees in such cases stand in the opposed cells
and push and bend the ductile and warm wax (which
as I have tried is easily done) into its proper intermediate plane, and thus flatten it.

From the experiment of the ridge of vermilion wax
we can clearly see that if the bees were to build for
themselves a thin wall of wax, they could make their
cells of the proper shape, by standing at the proper distance from each other, by excavating at the same rate,
and by endeavouring to make equal spherical hollows,
but never allowing the spheres to break into each
other. Now bees, as may be clearly seen by examining

the edge of a growing comb, do make a rough, circumferential wall or rim all round the comb; and they gnaw into this from the opposite sides, always working circularly as they deepen each cell. They do not make the whole three-sided pyramidal base of any one cell at the same time, but only that one rhombic plate which stands on the extreme growing margin, or the two plates, as the case may be; and they never complete the upper edges of the rhombic plates, until the hexagonal walls are commenced. Some of these statements differ from those made by the justly celebrated elder Huber, but I am convinced of their accuracy; and if I had space, I could show that they are conformable with my theory.

Huber's statement that the very first cell is excavated out of a little parallel-sided wall of wax, is not, as far as I have seen, strictly correct; the first commencement having always been a little hood of wax; but I will not here enter on these details. We see how important a part excavation plays in the construction of the cells; but it would be a great error to suppose that the bees cannot build up a rough wall of wax in the proper position—that is, along the plane of intersection between two adjoining spheres. I have several specimens showing clearly that they can do this. Even in the rude circumferential rim or wall of wax round a growing comb, flexures may sometimes be observed, corresponding in position to the planes of the rhombic basal plates of future cells. But the rough wall of wax has in every case to be finished off, by being largely gnawed away on both sides. The manner in which the bees build is curious; they always make the first rough wall from ten to twenty times thicker than the excessively thin finished wall of the cell, which will ultimately be left. We shall understand how they work, by supposing masons first to

pile up a broad ridge of cement, and then to begin cutting
it away equally on both sides near the ground, till a
smooth, very thin wall is left in the middle; tho masons
always piling up the cut-away cement, and adding fresh
cement, on the summit of the ridge. We shall thus
have a thin wall steadily growing upward; but always
crowned by a gigantic coping. From all the cells,
both those just commenced and those completed, being
thus crowned by a strong coping of wax, the bees can
cluster and crawl over the comb without injuring the
delicate hexagonal walls. These walls, as Professor
Miller has kindly ascertained for me, vary greatly in
thickness; being, on an average of twelve measurements made near the border of the comb, $\frac{1}{353}$ of
an inch in thickness; whereas the basal rhomboidal
plates are thicker nearly in the proportion of three to
two, having a mean thickness, from twenty-one measurements, of $\frac{1}{229}$ of an inch. By the above singular manner
of building, strength is continually given to the comb,
with the utmost ultimate economy of wax.

It seems at first to add to the difficulty of understanding how the cells are made, that a multitude of bees all
work together; one bee after working a short time at
one cell going to another, so that, as Huber has stated,
a score of individuals work even at the commencement
of the first cell. I was able practically to show this
fact, by covering the edges of the hexagonal walls
of a single cell, or the extreme margin of the circumferential rim of a growing comb, with an extremely thin
layer of melted vermilion wax; and I invariably found
that the colour was most delicately diffused by the bees
—as delicately as a painter could have done with his
brush—by atoms of the coloured wax having been taken
from the spot on which it had been placed, and worked
into the growing edges of the cells all round. The work

of construction seems to be a sort of balance struck
between many bees, all instinctively standing at the
same relative distance from each other, all trying to
sweep equal spheres, and then building up, or leaving
ungnawed, the planes of intersection between these
spheres. It was really curious to note in cases of difficulty, as when two pieces of comb met at an angle, how
often the bees would pull down and rebuild in different
ways the same cell, sometimes recurring to a shape
which they had at first rejected.

When bees have a place on which they can stand in
their proper positions for working,—for instance, on a
slip of wood, placed directly under the middle of a comb
growing downwards so that the comb has to be built over
one face of the slip—in this case the bees can lay the
foundations of one wall of a new hexagon, in its strictly
proper place, projecting beyond the other completed
cells. It suffices that the bees should be enabled to
stand at their proper relative distances from each other
and from the walls of the last completed cells, and then,
by striking imaginary spheres, they can build up a wall
intermediate between two adjoining spheres; but, as far
as I have seen, they never gnaw away and finish off the
angles of a cell till a large part both of that cell and of
the adjoining cells has been built. This capacity in
bees of laying down under certain circumstances a
rough wall in its proper place between two just-commenced cells, is important, as it bears on a fact, which
seems at first quite subversive of the foregoing theory;
namely, that the cells on the extreme margin of wasp
combs are sometimes strictly hexagonal; but I have not
space here to enter on this subject. Nor does there
seem to me any great difficulty in a single insect (as in
the case of a queen-wasp) making hexagonal cells, if
she work alternately on the inside and outside of two

or three cells commenced at the same time, always
standing at the proper relative distance from the parts
of the cells just begun, sweeping spheres or cylinders,
and building up intermediate planes. It is even conceivable that an insect might, by fixing on a point at which
to commence a cell, and then moving outside, first to
one point, and then to five other points, at the proper
relative distances from the central point and from each
other, strike the planes of intersection, and so make an
isolated hexagon : but I am not aware that any such
case has been observed ; nor would any good be derived
from a single hexagon being built, as in its construction
more materials would be required than for a cylinder.

As natural selection acts only by the accumulation of
slight modifications of structure or instinct, each profitable to the individual under its conditions of life, it may
reasonably be asked, how a long and graduated succession
of modified architectural instincts, all tending towards the
present perfect plan of construction, could have profited
the progenitors of the hive-bee? I think the answer is
not difficult : it is known that bees are often hard pressed
to get sufficient nectar; and I am informed by Mr.
Tegetmeier that it has been experimentally found that
no less than from twelve to fifteen pounds of dry sugar
are consumed by a hive of bees for the secretion of each
pound of wax; so that a prodigious quantity of fluid
nectar must be collected and consumed by the bees in
a hive for the secretion of the wax necessary for the
construction of their combs. Moreover, many bees
have to remain idle for many days during the process of
secretion. A large store of honey is indispensable to
support a large stock of bees during the winter; and
the security of the hive is known mainly to depend on
a large number of bees being supported. Hence the
saving of wax by largely saving honey must be a most

important element of success in any family of bees. Of course the success of any species of bee may be dependent on the number of its parasites or other enemies, or on quite distinct causes, and so be altogether independent of the quantity of honey which the bees could collect. But let us suppose that this latter circumstance determined, as it probably often does determine, the numbers of a humble-bee which could exist in a country; and let us further suppose (differently to what really is the case) that the community lived throughout the winter, and consequently required a store of honey: there can in this case be no doubt that it would be an advantage to our humble-bee, if a slight modification of her instinct led her to make her waxen cells near together, so as to intersect a little; for a wall in common even to two adjoining cells, would save some little wax. Hence it would continually be more and more advantageous to our humble-bee, if she were to make her cells more and more regular, nearer together, and aggregated into a mass, like the cells of the Melipona; for in this case a large part of the bounding surface of each cell would serve to bound other cells, and much wax would be saved. Again, from the same cause, it would be advantageous to the Melipona, if she were to make her cells closer together, and more regular in every way than at present; for then, as we have seen, the spherical surfaces would wholly disappear, and would all be replaced by plane surfaces; and the Melipona would make a comb as perfect as that of the hive-bee. Beyond this stage of perfection in architecture, natural selection could not lead; for the comb of the hive-bee, as far as we can see, is absolutely perfect in economising wax.

Thus, as I believe, the most wonderful of all known instincts, that of the hive-bee, can be explained by natural selection having taken advantage of numerous, successive, slight modifications of simpler instincts; natural selection having by slow degrees, more and more perfectly, led the bees to sweep equal spheres at a given distance from each other in a double layer, and to build up and excavate the wax along the planes of intersection. The bees, of course, no more knowing that they swept their spheres at one particular distance from each other, than they know what are the several angles of the hexagonal prisms and of the basal rhombic plates. The motive power of the process of natural selection having been economy of wax, together with cells of due strength, and of the proper size and shape for the larvæ; that individual swarm which made the best cells, and wasted least honey in the secretion of wax, having succeeded best, and having transmitted by inheritance their newly acquired economical instincts to new swarms, which in their turn will have had the best chance of succeeding in the struggle for existence.

It has been objected to the foregoing view on the origin of instinct that "the variations of structure and of instinct must have been simultaneous and accurately adjusted to each other, as a modification in the one without an immediate corresponding change in the other would have been fatal." The force of this objection seems entirely to rest on the assumption that the changes in both instinct and structure are abrupt. To take as an illustration the case of the larger titmouse (Parus major) alluded to in the last chapter: this bird often holds the seeds of the yew between its feet on a branch, and hammers away till it gets into the kernel. Now what special difficulty would there be in natural selection preserving each slight variation of beak, better and better adapted to break open seeds, until a beak was formed, as well constructed for this purpose as that

of the nutbatch, at the same time that hereditary
habit, or compulsion from the want of other food, or the
preservation of chance variations of taste, made the bird
more and more of a seed-eater?   In this case the beak
is supposed to be slowly modified by natural selection,
subsequently to, but in accordance with, slowly changing
habit; but let the feet of the titmouse vary and grow
larger from correlation with the beak, or from any other
unknown cause, and is it very improbable that such larger
feet might lead the bird to climb more and more until
it acquired even the remarkable climbing instinct and
capacity of the nuthatch? In this case a gradual change
of structure is supposed to lead to changed instinctive
habits of life. To take one more case: few instincts
are more remarkable than that which leads the swift of
the Eastern Islands to make its nest wholly of inspissated saliva. Some birds build their nests of mud,
believed to be moistened with saliva; and one of the
swifts of North America makes its nest (as I have
seen) of sticks agglutinated with saliva, and even with
flakes of this substance. Is it then very improbable
that the natural selection of individual swifts, which
secreted more and more saliva, should at last produce
a species with instincts leading it to neglect other materials, and to make its nest exclusively of inspissated
saliva? And so in other cases. It must be admitted
that in many instances we cannot conjecture whether
instinct or structure has first slightly changed; nor can
we conjecture by what gradations many instincts have
been developed when they relate to organs (such as the
mammary glands) on the first origin of which we know
nothing.

No doubt many instincts of very difficult explanation
could be opposed to the theory of natural selection,—
cases, in which we cannot see how an instinct could

possibly have originated; cases, in which no intermediate gradutions are known to exist; cases of instinct
of apparently such trifling importance, that they could
hardly have been acted on by natural selection; cases
of instincts almost identically the same in animals so
remote in the scale of nature, that we cannot account
for their similarity by inheritance from a common
parent, and must therefore believe that they have been
acquired by independent acts of natural selection. I
will not here enter on these several cases, but will confine myself to one special difficulty, which at first appeared to me insuperable, and actually fatal to my
whole theory. I allude to the neuters or sterile females
in insect-communities: for these neuters often differ
widely in instinct and in structure from both the males
and fertile females, and yet, from being sterile, they
cannot propagate their kind.

The subject well deserves to be discussed at great
length, but I will here take only a single case, that of
working or sterile ants. How the workers have been
rendered sterile is a difficulty; but not much greater
than that of any other striking modification of structure;
for it can be shown that some insects and other articulate animals in a state of nature occasionally become
sterile; and if such insects had been social, and it had
been profitable to the community that a number should
have been annually born capable of work, but incapable
of procreation, I can see no very great difficulty in this
being effected by natural selection. But I must pass
over this preliminary difficulty. The great difficulty
lies in the working ants differing widely from both the
males and the fertile females in structure, as in the
shape of the thorax and in being destitute of wings and
sometimes of eyes, and in instinct. As far as instinct
alone is concerned, the prodigious difference in this

respect between the workers and the perfect females,
would have been better exemplified by the hive-bee. If
a working ant or other neuter insect had been an animal
in the ordinary state, I should have unhesitatingly assumed that all its characters had been slowly acquired
through natural selection; namely, by an individual
having been born with some slight profitable modification of structure, this being inherited by its offspring,
which again varied and were again selected, and so onwards. But with the working ant we have an insect
differing greatly from its parents, yet absolutely sterile ;
so that it could never have transmitted successively
acquired modifications of structure or instinct to its
progeny. It may well be asked how is it possible to re-
concile this case with the theory of natural selection?

First, let it be remembered that we have innumerable
instances, both in our domestic productions and in those
in a state of nature, of all sorts of differences of structure which have become correlated to certain ages, and
to either sex. We have differences correlated not only
to one sex, but to that short period alone when the reproductive system is active, as in the nuptial plumage of
many birds, and in the hooked jaws of the male salmon.
We have even slight differences in the horns of different
breeds of cattle in relation to an artificially imperfect
state of the male sex ; for oxen of certain breeds have
longer horns than in other breeds, in comparison with
the horns of the bulls or cows of these same breeds.
Hence I can see no real difficulty in any character
having become correlated with the sterile condition of
certain members of insect-communities : the difficulty
lies in understanding how such correlated modifications
of structure could have been slowly accumulated by
natural selection.

This difficulty, though appearing insuperable, is les-

senced, or, as I believe, disappears, when it is remembered
that selection may be applied to the family, as well as
to the individual, and may thus gain the desired end.
Thus, a well-flavoured vegetable is cooked, and the individual is destroyed; but the horticulturist sows seeds
of the same family, and confidently expects to get nearly
the same variety: breeders of cattle wish the flesh and
fat to be well marbled together; the animal has been
slaughtered, but the breeder goes with confidence to the
same stock. I have such faith in the powers of selection,
that I do not doubt that a breed of cattle, always
yielding oxen with extraordinarily long horns, could be
slowly formed by carefully watching which individual
bulls and cows, when matched, produced oxen with the
longest horns; and yet no one ox could ever have
propagated its kind. Thus I believe it has been with
social insects: a slight modification of structure, or
instinct, correlated with the sterile condition of certain
members of the community, has been advantageous to
the community: consequently the fertile males and
females of the same community flourished, and transmitted to their fertile offspring a tendency to produce
sterile members having the same modification. And I
believe that this process has been repeated, until that
prodigious amount of difference between the fertile and
sterile females of the same species has been produced,
which we see in many social insects.

But we have not as yet touched on the climax of the
difficulty; namely, the fact that the neuters of several
ants differ, not only from the fertile females and males,
but from each other, sometimes to an almost incredible
degree, and are thus divided into two or even three
castes. The castes, moreover, do not generally graduate into each other, but are perfectly well defined;
being as distinct from each other, as are any two species

of the same genus, or rather as any two genera of the
same family. Thus in Eciton, there are working and
soldier neuters, with jaws and instincts extraordinarily
different : in Cryptocerus, the workers of one caste alone
carry a wonderful sort of shield on their heads, the use
of which is quite unknown: in the Mexican Myrme-
cocystus, the workers of one caste never leave the nest;
they are fed by the workers of another caste, and they
have an enormously developed abdomen which secretes
a sort of honey, supplying the place of that excreted by
the aphides, or the domestic cattle as they may be called,
which our European ants guard or imprison.

It will indeed be thought that I have an overweening
confidence in the principle of natural selection, when I
do not admit that such wonderful and well-established
facts at once annihilate my theory. In the simpler case
of neuter insects all of one caste or of the same kind,
which have been rendered by natural selection, as I
believe to be quite possible, different from the fertile
males and females—in this case, we may safely conclude
from the analogy of ordinary variations, that each successive, slight, profitable modification did not probably
at first appear in all the individual neuters in the same
nest, but in a few alone ; and that by the long-continued
selection of the fertile parents which produced most
neuters with the profitable modification, all the neuters
ultimately came to have the desired character. On
this view we ought occasionally to find neuter insects of
the same species, in the same nest, presenting gradations of structure; and this we do find, even often,
considering how few neuter insects out of Europe have
been carefully examined. Mr. F. Smith has shown how
surprisingly the neuters of several British ants differ
from each other in size and sometimes in colour; and
that the extreme forms can sometimes be perfectly

linked together by individuals taken out of the same
nest : I have myself compared perfect gradations of this
kind. It often happens that the larger or the smaller
sized workers are the most numerous; or that both large
and small are numerous, with those of an intermediate
size scanty in numbers. Formica flava has larger and
smaller workers, with some of intermediate size ; and, in
this species, as Mr. F. Smith has observed, the larger
workers have simple eyes (ocelli), which though small
can be plainly distinguished, whereas the smaller
workers have their ocelli rudimentary. Having carefully dissected several specimens of these workers, I can
affirm that the eyes are far more rudimentary in the
smaller workers than can be accounted for merely by
their proportionally lesser size; and I fully believe,
though I dare not assert so positively, that the workers
of intermediate size have their ocelli in an exactly intermediate condition. So that we here have two bodies
of sterile workers in the same nest, differing not only
in size, but in their organs of vision, yet connected by
some few members in an intermediate condition. I may
digress by adding, that if the smaller workers had been
the most useful to the community, and those males and
females had been continually selected, which produced
more and more of the smaller workers, until all the
workers had come to be in this condition ; we should
then have had a species of ant with neuters very nearly
in the same condition with those of Myrmica. For the
workers of Myrmica have not even rudiments of ocelli,
though the male and female ants of this genus have
well-developed ocelli.

I may give one other case : so confidently did I expect
to find gradations in important points of structure between the different castes of neuters in the same species,
that I gladly availed myself of Mr. F. Smith's offer of

numerous specimens from the same nest of the driver
ant (Anomma) of West Africa. The reader will perhaps best appreciate the amount of difference in these
workers, by my giving not the actual measurements,
but a strictly accurate illustration: the difference was
the same as if we were to see a set of workmen building
a house of whom many were five feet four inches high,
and many sixteen feet high; but we must suppose that
the larger workmen had heads four instead of three
times as big as those of the smaller men, and jaws nearly
five times as big. The jaws, moreover, of the working
ants of the several sizes differed wonderfully in shape,
and in the form and number of the teeth. But the
important fact for us is, that though the workers can be
grouped into castes of different sizes, yet they graduate
insensibly into each other, as does the widely-different
structure of their jaws. I speak confidently on this latter
point, as Mr. Lubbock made drawings for me with the
camera lucida of the jaws which I had dissected from
the workers of the several sizes.

With these facts before me, I believe that natural
selection, by acting on the fertile parents, could form a
species which should regularly produce neuters, either
all of large size with one form of jaw, or all of small
size with jaws having a widely different structure; or
lastly, and this is our climax of difficulty, one set of
workers of one size and structure, and simultaneously
another set of workers of a different size and structure;
—a graduated series having been first formed, as in the
case of the driver ant, and then the extreme forms,
from being the most useful to the community, having
been produced in greater and greater numbers through
the natural selection of the parents which generated
them; until none with an intermediate structure were
produced.

Thus, as I believe, the wonderful fact of two distinctly defined castes of sterile workers existing in the same nest, both widely different from each other and from their parents, has originated. We can see how useful their production may have been to a social community of insects, on the same principle that the division of labour is useful to civilised man. As ants work by inherited instincts and by inherited organs or tools, and not by acquired knowledge and manufactured instruments, a perfect division of labour could be effected with them only by the workers being sterile; for had they been fertile, they would have intercrossed, and their instincts and structure would have become blended. And nature has, as I believe, effected this admirable division of labour in the communities of ants, by the means of natural selection. But I am bound to confess, that, with all my faith in this principle, I should never have anticipated that natural selection could have been efficient in so high a degree, had not the case of these neuter insects convinced me of the fact. I have, therefore, discussed this case, at some little but wholly insufficient length, in order to show the power of natural selection, and likewise because this is by far the most serious special difficulty, which my theory has encountered. The case, also, is very interesting, as it proves that with animals, as with plants, any amount of modification in structure can be effected by the accumulation of numerous, slight, and as we must call them accidental, variations, which are in any manner profitable, without exercise or habit having come into play. For no amount of exercise, or habit, or volition, in the utterly sterile members of a community could possibly affect the structure or instincts of the fertile members, which alone leave descendants. I am surprised that no one has advanced this demonstrative case of neuter insects, against the well-known doctrine of inherited habit advanced by Lamarck.

Summary.—I have endeavoured briefly in this chapter to show that the mental qualities of our domestic animals vary, and that the variations are inherited. Still more briefly I have attempted to show that instincts vary slightly in a state of nature. No one will dispute that instincts are of the highest importance to each animal. Therefore I can see no difficulty, under changing conditions of life, in natural selection accumulating slight modifications of instinct to any extent, in any useful direction. In some cases habit or use and disuse have probably come into play. I do not pretend that the facts given in this chapter strengthen in any great degree my theory; but none of the cases of difficulty, to the best of my judgment, annihilate it. On the other hand, the fact that instincts are not always absolutely perfect and are liable to mistakes;—that no instinct has been produced for the exclusive good of other animals, but that each animal takes advantage of the instincts of others;—that the canon in natural history, of "Natura non facit saltum," is applicable to instincts as well as to corporeal structure, and is plainly explicable on the foregoing views, but is otherwise inexplicable,—all tend to corroborate the theory of natural selection.

This theory is, also, strengthened by some few other facts in regard to instincts; as by that common case of closely allied, but distinct, species, when inhabiting distant parts of the world and living under considerably different conditions of life, yet often retaining nearly the same instincts. For instance, we can understand on the principle of inheritance, how it is that the thrush of South America lines its nest with mud, in the same

N

peculiar manner as does our British thrush : how it is
that the Hornbills of Africa and India have the same
extraordinary instinct of plastering up and imprisoning
the females in a hole in a tree, with only a small hole
left in the plaster through which the males feed them
and their young when hatched : how it is that the male
wrens (Troglodytes) of North America, build "cock-
nests," to roost in, like the males of our distinct Kitty-
wrens,—a habit wholly unlike that of any other known
bird. Finally, it may not be a logical deduction, but
to my imagination it is far more satisfactory to look at
such instincts as the young cuckoo ejecting its foster-
brothers,—ants making slaves,—the larvæ of ichneumo-
nidæ feeding within the live bodies of caterpillars,—
not as specially endowed or created instincts, but as
small consequences of one general law, leading to the
advancement of all organic beings, namely, multiply,
vary, let the strongest live and the weakest die.

CHAPTER VIIL

Hybridism.

Distinction between the sterility of first crosses and of hybrids — Sterility various in degree, not universal, affected by close interbreeding, removed by domestication — Laws governing the sterility of hybrids — Sterility not a special endowment, but incidental on other differences — Causes of the sterility of first crosses and of hybrids — Parallelism between the effects of changed conditions of life and of crossing — Fertility of varieties when crossed and of their mongrel offspring not universal — Hybrids and mongrels compared independently of their fertility — Summary.

THE view generally entertained by naturalists is that species, when intercrossed, have been specially endowed with the quality of sterility, in order to prevent the confusion of all organic forms. This view certainly seems at first probable, for species within the same country could hardly have kept distinct had they been capable of crossing freely. The importance of the fact that hybrids are very generally sterile, has, I think, been much underrated by some late writers. On the theory of natural selection the case is especially important, inasmuch as the sterility of hybrids could not possibly be of any advantage to them, and therefore could not have been acquired by the continued preservation of successive profitable degrees of sterility. I hope, however, to be able to show that sterility is not a specially acquired or endowed quality, but is incidental on other acquired and little known differences.

In treating this subject, two classes of facts, to a large extent fundamentally different, have generally been confounded together ; namely, the sterility of two species

N 2

when first crossed, and the sterility of the hybrids produced from them.

Pure species have of course their organs of reproduction in a perfect condition, yet when intercrossed they produce either few or no offspring. Hybrids, on the other hand, have their reproductive organs functionally impotent, as may be clearly seen in the state of the male element in both plants and animals; though the organs themselves are perfect in structure, as far as the microscope reveals. In the first case the two sexual elements which go to form the embryo are perfect; in the second case they are either not at all developed, or are imperfectly developed. This distinction is important, when the cause of the sterility, which is common to the two cases, has to be considered. The distinction has probably been slurred over, owing to the sterility in both cases being looked on as a special endowment, beyond the province of our reasoning powers.

The fertility of varieties, that is of the forms known or believed to have descended from common parents, when intercrossed, and likewise the fertility of their mongrel offspring, is, with reference to my theory, of equal importance with the sterility of species; for it seems to make a broad and clear distinction between varieties and species.

First, for the sterility of species when crossed and of their hybrid offspring. It is impossible to study the several memoirs and works of those two conscientious and admirable observers, Kölreuter and Gärtner, who almost devoted their lives to this subject, without being deeply impressed with the high generality of some degree of sterility. Kölreuter makes the rule universal; but then he cuts the knot, for in ten cases in which he found two forms, considered by most authors as distinct species, quite fertile together, he unhesitatingly ranks

them as varieties. Gärtner, also, makes the rule
equally universal; and he disputes the entire fertility
of Kölreuter's ten cases. But in these and in many
other cases, Gärtner is obliged carefully to count the
seeds, in order to show that there is any degree of
sterility. He always compares the maximum number
of seeds produced by two species when crossed and by
their hybrid offspring, with the average number produced by both pure parent species in a state of nature.
But a serious cause of error seems to me to be here
introduced : a plant to be hybridised must be castrated,
and, what is often more important, must be secluded
in order to prevent pollen being brought to it by insects from other plants. Nearly all the plants experimented on by Gärtner were potted, and apparently
were kept in a chamber in his house. That these processes are often injurious to the fertility of a plant cannot be doubted; for Gärtner gives in his table about a
score of cases of plants which he castrated, and artificially fertilised with their own pollen, and (excluding
all cases such as the Leguminosæ, in which there is an
acknowledged difficulty in the manipulation) half of
these twenty plants had their fertility in some degree
impaired. Moreover, as Gärtner during several years
repeatedly crossed the primrose and cowslip, which
we have such good reason to believe to be varieties,
and only once or twice succeeded in getting fertile
seed; as he found the common red and blue pim-
pernels (Anagallis arvensis and cœrulea), which the
best botanists rank as varieties, absolutely sterile together; and as he came to the same conclusion in
several other analogous cases, it seems to me that
we may well be permitted to doubt whether many
other species are really so sterile, when intercrossed,
as Gärtner believes.

It is certain, on the one hand, that the sterility of
various species when crossed is so different in degree
and graduates away so insensibly, and, on the other
hand, that the fertility of pure species is so easily
affected by various circumstances, that for all practical
purposes it is most difficult to say where perfect fertility ends and sterility begins. I think no better evidence of this can be required than that the two most
experienced observers who have ever lived, namely,
Kölreuter and Gärtner, should have arrived at diametrically opposite conclusions in regard to the very
same species. It is also most instructive to compare—
but I have not space here to enter on details—the evidence advanced by our best botanists on the question
whether certain doubtful forms should be ranked as
species or varieties, with the evidence from fertility adduced by different hybridisers, or by the same author,
from experiments made during different years. It can
thus be shown that neither sterility nor fertility affords
any clear distinction between species and varieties; but
that the evidence from this source graduates away,
and is doubtful in the same degree as is the evidence
derived from other constitutional and structural differences.

In regard to the sterility of hybrids in successive
generations; though Gärtner was enabled to rear some
hybrids, carefully guarding them from a cross with
either pure parent, for six or seven, and in one case for
ten generations, yet he asserts positively that their fertility never increased, but generally greatly decreased.
I do not doubt that this is usually the case, and that
the fertility often suddenly decreases in the first few
generations. Nevertheless I believe that in all these
experiments the fertility has been diminished by an
independent cause, namely, from close interbreeding.

I have collected so large a body of facts, showing that close interbreeding lessens fertility, and, on the other hand, that an occasional cross with a distinct individual or variety increases fertility, that I cannot doubt the correctness of this almost universal belief amongst breeders. Hybrids are seldom raised by experimentalists in great numbers; and as the parent-species, or other allied hybrids, generally grow in the same garden, the visits of insects must be carefully prevented during the flowering season: hence hybrids will generally be fertilised during each generation by their own individual pollen; and I am convinced that this would be injurious to their fertility, already lessened by their hybrid origin. I am strengthened in this conviction by a remarkable statement repeatedly made by Gärtner, namely, that if even the less fertile hybrids be artificially fertilised with hybrid pollen of the same kind, their fertility, notwithstanding the frequent ill effects of manipulation, sometimes decidedly increases, and goes on increasing. Now, in artificial fertilisation pollen is as often taken by chance (as I know from my own experience) from the anthers of another flower, as from the anthers of the flower itself which is to be fertilised; so that a cross between two flowers, though probably on the same plant, would be thus effected. Moreover, whenever complicated experiments are in progress, so careful an observer as Gärtner would have castrated his hybrids, and this would have insured in each generation a cross with a pollen from a distinct flower, either from the same plant or from another plant of the same hybrid nature. And thus, the strange fact of the increase of fertility in the successive generations of *artificially fertilised* hybrids may, I believe, be accounted for by close interbreeding having been avoided.

Now let us turn to the results arrived at by the third

most experienced hybridiser, namely, the Hon. and
Rev. W. Herbert. He is as emphatic in his conclusion
that some hybrids are perfectly fertile—as fertile as the
pure parent-species—as are Kölreuter and Gärtner that
some degree of sterility between distinct species is a
universal law of nature. He experimented on some of
the very same species as did Gärtner. The difference
in their results may, I think, be in part accounted
for by Herbert's great horticultural skill, and by his
having hothouses at his command. Of his many important statements I will here give only a single one as
an example, namely, that "every ovule in a pod of
Crinum capense fertilised by C. revolutum produced a
plant, which (he says) I never saw to occur in a case of
its natural fecundation." So that we here have perfect,
or even more than commonly perfect, fertility in a first
cross between two distinct species.

This case of the Crinum leads me to refer to a most
singular fact, namely, that there are individual plants
of certain species of Lobelia and of some other genera,
which can be far more easily fertilised by the pollen of
another and distinct species, than by their own pollen;
and all the individuals of nearly all the species of Hippeastrum seem to be in this predicament. For these
plants have been found to yield seed to the pollen
of a distinct species, though quite sterile with their
own pollen, notwithstanding that their own pollen
was found to be perfectly good, for it fertilised distinct species. So that certain individual plants and
all the individuals of certain species can actually be
hybridised much more readily than they can be self-
fertilised ! For instance, a bulb of Hippeastrum au-
licum produced four flowers ; three were fertilised by
Herbert with their own pollen, and the fourth was
subsequently fertilised by the pollen of a compound

hybrid descended from three other and distinct species : the result was that "the ovaries of the three first flowers soon ceased to grow, and after a few days perished entirely, whereas the pod impregnated by the pollen of the hybrid made vigorous growth and rapid progress to maturity, and bore good seed, which vegetated freely." In a letter written in 1839, Mr. Herbert told me that he had then tried the experiment during five years, and he continued to try it during several subsequent years, and always with the same result. This result has, also, been confirmed by other observers in the case of Hippeastrum with its subgenera, and in the case of some other genera, as Lobelia, Passiflora, and Verbascum. Although the plants in these experiments appeared perfectly healthy, and although both the ovules and pollen of the same flower were perfectly good with respect to other species, yet as they were functionally imperfect in their mutual self-action, we must infer that the plants were in an unnatural state. Nevertheless these facts show on what slight and mysterious causes the lesser or greater fertility of species when crossed, in comparison with the same species when self-fertilised, sometimes depends.

The practical experiments of horticulturists, though not made with scientific precision, deserve some notice. It is notorious in how complicated a manner the species of Pelargonium, Fuchsia, Calceolaria, Petunia, Rhododendron, &c., have been crossed, yet many of these hybrids seed freely. For instance, Herbert asserts that a hybrid from Calceolaria integrifolia and plantaginea, species most widely dissimilar in general habit, "reproduced itself as perfectly as if it had been a natural species from the mountains of Chile." I have taken some pains to ascertain the degree of fertility of some of the complex crosses of

Rhododendrons, and I am assured that many of them
are perfectly fertile. Mr. C. Noble, for instance, informs
me that he raises stocks for grafting from a hybrid
between Rhod. Ponticum and Catawbiense, and that
this hybrid "seeds as freely as it is possible to imagine." Had hybrids, when fairly treated, gone on decreasing in fertility in each successive generation, as
Gärtner believes to be the case, the fact would have
been notorious to nurserymen. Horticulturists raise
large beds of the same hybrids, and such alone are
fairly treated, for by insect agency the several individuals of the same hybrid variety are allowed to freely
cross with each other, and the injurious influence of
close interbreeding is thus prevented. Any one may
readily convince himself of the efficiency of insect-
agency by examining the flowers of the more sterile
kinds of hybrid Rhododendrons, which produce no pollen, for he will find on their stigmas plenty of pollen
brought from other flowers.

In regard to animals, much fewer experiments have
been carefully tried than with plants. If our systematic
arrangements can be trusted, that is, if the genera of
animals are as distinct from each other as are the genera
of plants, then we may infer that animals more widely
separated in the scale of nature can be more easily
crossed than in the case of plants; but the hybrids
themselves are, I think, more sterile. I doubt whether
any case of a perfectly fertile hybrid animal can be
considered as thoroughly well authenticated. It should,
however, be borne in mind that, owing to few animals
breeding freely under confinement, few experiments
have been fairly tried: for instance, the canary-bird
has been crossed with nine other finches, but as not
one of these nine species breeds freely in confinement,
we have no right to expect that the first crosses be-

tween them and the canary, or that their hybrids, should be perfectly fertile. Again, with respect to the fertility in successive generations of the more fertile hybrid animals, I hardly know of an instance in which two families of the same hybrid have been raised at the same time from different parents, so as to avoid the ill effects of close interbreeding. On the contrary, brothers and sisters have usually been crossed in each successive generation, in opposition to the constantly repeated admonition of every breeder. And in this case, it is not at all surprising that the inherent sterility in the hybrids should have gone on increasing. If we were to act thus, and pair brothers and sisters in the case of any pure animal, which from any cause had the least tendency to sterility, the breed would assuredly be lost in a few generations.

Although I do not know of any thoroughly well-authenticated cases of perfectly fertile hybrid animals, I have reason to believe that the hybrids from Cervulus vaginalis and Reevesii, and from Phasianus colchicus with P. torquatus and with P. versicolor, are perfectly fertile. There is no doubt that these three pheasants, namely, the common, the true ring-necked, and the Japan, intercross, and are becoming blended together in the woods of several parts of England. The hybrids from the common and Chinese geese (A. cignoides), species which are so different that they are generally ranked in distinct genera, have often bred in this country with either pure parent, and in one single instance they have bred *inter se*. This was effected by Mr. Eyton, who raised two hybrids from the same parents but from different hatches; and from these two birds he raised no less than eight hybrids (grandchildren of the pure geese) from one nest. In India, however, these cross-bred geese must be far more fertile; for

I am assured by two eminently capable judges, namely
Mr. Blyth and Capt. Hutton, that whole flocks of these
crossed geese are kept in various parts of the country;
and as they are kept for profit, where neither pure
parent-species exists, they must certainly be highly
fertile.

A doctrine which originated with Pallas, has been
largely accepted by modern naturalists; namely, that
most of our domestic animals have descended from two
or more wild species, since commingled by intercrossing. On this view, the aboriginal species must
either at first have produced quite fertile hybrids, or
the hybrids must have become in subsequent generations quite fertile under domestication. This latter
alternative seems the most probable, and I am inclined
to believe in its truth, although it rests on no direct
evidence. I believe, for instance, that our dogs have
descended from several wild stocks; yet, with perhaps the exception of certain indigenous domestic dogs
of South America, all are quite fertile together; and
analogy makes me greatly doubt, whether the several
aboriginal species would at first have freely bred together and have produced quite fertile hybrids. So
again there is reason to believe that our European and
the humped Indian cattle are quite fertile together;
but from facts communicated to me by Mr. Blyth, I
think they must be considered as distinct species. On
this view of the origin of many of our domestic animals,
we must either give up the belief of the almost universal sterility of distinct species of animals when
crossed; or we must look at sterility, not as an in-
delible characteristic, but as one capable of being
removed by domestication.

Finally, looking to all the ascertained facts on the
intercrossing of plants and animals, it may be con-

cluded that some degree of sterility, both in first crosses and in hybrids, is an extremely general result; but that it cannot, under our present state of knowledge, be considered as absolutely universal.

Laws governing the Sterility of first Crosses and of Hybrids.—We will now consider a little more in detail the circumstances and rules governing the sterility of first crosses and of hybrids.   Our chief object will be to see whether or not the rules indicate that species have specially been endowed with this quality, in order to prevent their crossing and blending together in utter confusion.   The following rules and conclusions are chiefly drawn up from Gärtner's admirable work on the hybridisation of plants.   I have taken much pains to ascertain how far the rules apply to animals, and considering how scanty our knowledge is in regard to hybrid animals, I have been surprised to find how generally the same rules apply to both kingdoms.

It has been already remarked, that the degree of fertility, both of first crosses and of hybrids, graduates from zero to perfect fertility.   It is surprising in how many curious ways this gradation can be shown to exist; but only the barest outline of the facts can here be given.   When pollen from a plant of one family is placed on the stigma of a plant of a distinct family, it exerts no more influence than so much inorganic dust.   From this absolute zero of fertility, the pollen of different species of the same genus applied to the stigma of some one species, yields a perfect gradation in the number of seeds produced, up to nearly complete or even quite complete fertility; and, as we have seen, in certain abnormal cases, even to an excess of fertility, beyond that which the plant's own pollen will produce.   So in hybrids themselves, there are some which never

have produced, and probably never would produce, even
with the pollen of either pure parent, a single fertile seed :
but in some of these cases a first trace of fertility may
be detected, by the pollen of one of the pure parent-
species causing the flower of the hybrid to wither
earlier than it otherwise would have done; and the
early withering of the flower is well known to be a sign
of incipient fertilization. From this extreme degree
of sterility we have self-fertilised hybrids producing a
greater and greater number of seeds up to perfect
fertility.

Hybrids from two species which are very difficult to
cross, and which rarely produce any offspring, are generally very sterile ; but the parallelism between the difficulty of making a first cross, and the sterility of the
hybrids thus produced—two classes of facts which are
generally confounded together—is by no means strict.
There are many cases, in which two pure species can be
united with unusual facility, and produce numerous
hybrid-offspring, yet these hybrids are remarkably
sterile. On the other hand, there are species which
can be crossed very rarely, or with extreme difficulty,
but the hybrids, when at last produced, are very fertile.
Even within the limits of the same genus, for instance
in Dianthus, these two opposite cases occur.

The fertility, both of first crosses and of hybrids, is
more easily affected by unfavourable conditions, than
is the fertility of pure species. But the degree of
fertility is likewise innately variable ; for it is not always
the same when the same two species are crossed under
the same circumstances, but depends in part upon the
constitution of the individuals which happen to have
been chosen for the experiment. So it is with hybrids,
for their degree of fertility is often found to differ
greatly in the several individuals raised from seed out

of the same capsule and exposed to the same conditions.

By the term systematic affinity is meant, the resemblance between species in structure and in constitution, more especially in the structure of parts which are of high physiological importance and which differ little in the allied species. Now the fertility of first crosses between species, and of the hybrids produced from them, is largely governed by their systematic affinity. This is clearly shown by hybrids never having been raised between species ranked by systematists in distinct families; and on the other hand, by very closely allied species generally uniting with facility. But the correspondence between systematic affinity and the facility of crossing is by no means strict. A multitude of cases could be given of very closely allied species which will not unite, or only with extreme difficulty; and on the other hand of very distinct species which unite with the utmost facility. In the same family there may be a genus, as Dianthus, in which very many species can most readily be crossed; and another genus, as Silene, in which the most persevering efforts have failed to produce between extremely close species a single hybrid. Even within the limits of the same genus, we meet with this same difference; for instance, the many species of Nicotiana have been more largely crossed than the species of almost any other genus; but Gärtner found that N. acuminata, which is not a particularly distinct species, obstinately failed to fertilise, or to be fertilised by, no less than eight other species of Nicotiana. Very many analogous facts could be given.

No one has been able to point out what kind, or what amount, of difference in any recognisable character is sufficient to prevent two species crossing. It can be shown that plants most widely different in habit and

general appearance, and having strongly marked differences in every part of the flower, even in the pollen, in
the fruit, and in the cotyledons, can be crossed. Annual
and perennial plants, deciduous and evergreen trees,
plants inhabiting different stations and fitted for extremely different climates, can often be crossed with ease.
By a reciprocal cross between two species, I mean
the case, for instance, of a stallion-horse being first
crossed with a female-ass, and then a male-ass with a
mare : these two species may then be said to have been
reciprocally crossed. There is often the widest possible
difference in the facility of making reciprocal crosses.
Such cases are highly important, for they prove that
the capacity in any two species to cross is often completely independent of their systematic affinity, or of
any recognisable difference in their whole organisation.
On the other hand, these cases clearly show that the
capacity for crossing is connected with constitutional
differences imperceptible by us, and confined to the
reproductive system. This difference in the result of
reciprocal crosses between the same two species was
long ago observed by Kölreuter. To give an instance :
Mirabilis jalapa can easily be fertilised by the pollen
of M. longiflora, and the hybrids thus produced are
sufficiently fertile ; but Kölreuter tried more than two
hundred times, during eight following years, to fertilise
reciprocally M. longiflora with the pollen of M. jalapa,
and utterly failed. Several other equally striking cases
could be given. Thuret has observed the same fact
with certain sea-weeds or Fuci. Gärtner, moreover,
found that this difference of facility in making reciprocal crosses is extremely common in a lesser degree.
He has observed it even between forms so closely related
(as Matthiola annua and glabra) that many botanists
rank them only as varieties. It is also a remarkable

fact, that hybrids raised from reciprocal crosses, though
of course compounded of the very same two species,
the one species having first been used as the father and
then as the mother, generally differ in fertility in a
small, and occasionally in a high degree.

·Several other singular rules could be given from
Gärtner: for instance, some species have a remarkable
power of crossing with other species; other species of
the same genus have a remarkable power of impressing
their likeness on their hybrid offspring; but these
two powers do not at all necessarily go together.
There are certain hybrids which instead of having, as
is usual, an intermediate character between their two
parents, always closely resemble one of them; and
such hybrids, though externally so like one of their
pure parent-species, are with rare exceptions extremely
sterile. So again amongst hybrids which are usually
intermediate in structure between their parents, exceptional and abnormal individuals sometimes are born,
which closely resemble one of their pure parents; and
these hybrids are almost always utterly sterile, even
when the other hybrids raised from seed from the same
capsule have a considerable degree of fertility. These
facts show how completely fertility in the hybrid is
independent of its external resemblance to either pure
parent.

Considering the several rules now given, which
govern the fertility of first crosses and of hybrids, we
see that when forms, which must be considered as
good and distinct species, are united, their fertility
graduates from zero to perfect fertility, or even to
fertility under certain conditions in excess. That
their fertility, besides being eminently susceptible to
favourable and unfavourable conditions, is innately
variable. That it is by no means always the same in

degree in the first cross and in the hybrids produced
from this cross. That the fertility of hybrids is not
related to the degree in which they resemble in external appearance either parent. And lastly, that the
facility of making a first cross between any two species
is not always governed by their systematic affinity or
degree of resemblance to each other. This latter statement is clearly proved by the difference in the result of
reciprocal crosses between the same two species, for
according as the one species or the other is used as the
father or the mother, there is generally some difference,
and occasionally the widest possible difference, in the
facility of effecting an union. The hybrids, moreover,
produced from reciprocal crosses often differ in fertility.
    Now do these complex and singular rules indicate
that species have been endowed with sterility simply
to prevent their becoming confounded in nature? I
think not. For why should the sterility be so extremely
different in degree, when various species are crossed, all
of which we must suppose it would be equally important
to keep from blending together? Why should the
degree of sterility be innately variable in the individuals
of the same species? Why should some species cross
with facility, and yet produce very sterile hybrids;
and other species cross with extreme difficulty, and yet
produce fairly fertile hybrids? Why should there often
be so great a difference in the result of a reciprocal
cross between the same two species? Why, it may
even be asked, has the production of hybrids been permitted? To grant to species the special power of producing hybrids, and then to stop their further propaga-
tion by different degrees of sterility, not strictly related
to the facility of the first union between their parents,
seems to be a strange arrangement.
    The foregoing rules and facts, on the other hand, ap-

pear to me clearly to indicate that the sterility both of
first crosses and of hybrids is simply incidental or dependent on unknown differences, chiefly in the reproductive systems, of the species which are crossed.   The
differences being of so peculiar and limited a nature,
that, in reciprocal crosses between two species, the male
sexual element of the one will often freely act on the
female sexual element of the other, but not in a reversed direction.   It will be advisable to explain a little
more fully by an example what I mean by sterility
being incidental on other differences, and not a specially endowed quality.   As the capacity of one plant
to be grafted or budded on another is so unimportant
for its welfare in a state of nature, I presume that
no one will suppose that this capacity is a *specially*
endowed quality, but will admit that it is incidental
on differences in the laws of growth of the two plants.
We can sometimes see the reason why one tree will
not take on another, from differences in their rate of
growth, in the hardness of their wood, in the period of
the flow or nature of their sap, &c.; but in a multitude
of cases we can assign no reason whatever.   Great diversity in the size of two plants, one being woody and
the other herbaceous, one being evergreen and the
other deciduous, and adaptation to widely different
climates, do not always prevent the two grafting together.   As in hybridisation, so with grafting, the
capacity is limited by systematic affinity, for no one
has been able to graft trees together belonging to quite
distinct families; and, on the other hand, closely allied
species, and varieties of the same species, can usually,
but not invariably, be grafted with ease.   But this
capacity, as in hybridisation, is by no means absolutely
governed by systematic affinity.   Although many distinct genera within the same family have been grafted

together, in other cases species of the same genus will not take on each other. The pear can be grafted far more readily on the quince, which is ranked as a distinct genus, than on the apple, which is a member of the same genus. Even different varieties of the pear take with different degrees of facility on the quince; so do different varieties of the apricot and peach on certain varieties of the plum.

As Gärtner found that there was sometimes an innate difference in different *individuals* of the same two species in crossing; so Sagaret believes this to be the case with different individuals of the same two species in being grafted together. As in reciprocal crosses, the facility of effecting an union is often very far from equal, so it sometimes is in grafting; the common gooseberry, for instance, cannot be grafted on the currant, whereas the currant will take, though with difficulty, on the gooseberry.

We have seen that the sterility of hybrids, which have their reproductive organs in an imperfect condition, is a very different case from the difficulty of uniting two pure species, which have their reproductive organs perfect; yet these two distinct cases run to a certain extent parallel. Something analogous occurs in grafting; for Thouin found that three species of Robinia, which seeded freely on their own roots, and which could be grafted with no great difficulty on another species, when thus grafted were rendered barren. On the other hand, certain species of Sorbus, when grafted on other species, yielded twice as much fruit as when on their own roots. We are reminded by this latter fact of the extraordinary case of Hippeastrum, Lobelia, &c., which seeded much more freely when fertilised with the pollen of distinct species, than when fertilised with their own pollen.

We thus see, that although there is a clear and fundamental difference between the mere adhesion of grafted stocks, and the union of the male and female elements in the act of reproduction, yet that there is a rude degree of parallelism in the results of grafting and of crossing distinct species. And as we must look at the curious and complex laws governing the facility with which trees can be grafted on each other as incidental on unknown differences in their vegetative systems, so I believe that the still more complex laws governing the facility of first crosses are incidental on unknown differences, chiefly in their reproductive systems. These differences, in both cases, follow to a certain extent, as might have been expected, systematic affinity, by which every kind of resemblance and dissimilarity between organic beings is attempted to be expressed. The facts by no means seem to me to indicate that the greater or lesser difficulty of either grafting or crossing together various species has been a special endowment; although in the case of crossing, the difficulty is as important for the endurance and stability of specific forms, as in the case of grafting it is unimportant for their welfare.

Causes of the Sterility of first Crosses and of Hybrids. — We may now look a little closer at the probable causes of the sterility of first crosses and of hybrids. These two cases are fundamentally different, for, as just remarked, in the union of two pure species the male and female sexual elements are perfect, whereas in hybrids they are imperfect. Even in first crosses, the greater or lesser difficulty in effecting an union apparently depends on several distinct causes. There must sometimes be a physical impossibility in the male element reaching the ovule, as would be the case with a

plant having a pistil too long for the pollen-tubes to reach
the ovarium. It has also been observed that when pollen
of one species is placed on the stigma of a distantly
allied species, though the pollen-tubes protrude, they
do not penetrate the stigmatic surface. Again, the
male element may reach the female element, but be
incapable of causing an embryo to be developed, as
seems to have been the case with some of Thuret's
experiments on Fuci. No explanation can be given
of these facts, any more than why certain trees cannot
be grafted on others. Lastly, an embryo may be
developed, and then perish at an early period. This
latter alternative has not been sufficiently attended to ;
but I believe, from observations communicated to me by
Mr. Hewitt, who has had great experience in hybridising
gallinaceous birds, that the early death of the embryo
is a very frequent cause of sterility in first crosses. I
was at first very unwilling to believe in this view ;
as hybrids, when once born, are generally healthy and
long-lived, as we see in the case of the common mule.
Hybrids, however, are differently circumstanced before
and after birth: when born and living in a country
where their two parents can live, they are generally
placed under suitable conditions of life. But a hybrid
partakes of only half of the nature and constitution
of its mother, and therefore before birth, as long as
it is nourished within its mother's womb, or within
the egg or seed produced by the mother, it may be
exposed to conditions in some degree unsuitable,
and consequently be liable to perish at an early
period; more especially as all very young beings seem
eminently sensitive to injurious or unnatural conditions
of life.

In regard to the sterility of hybrids, in which the
sexual elements are imperfectly developed, the case is

very different. I have more than once alluded to a
large body of facts, which I have collected, showing
that when animals and plants are removed from their
natural conditions, they are extremely liable to have
their reproductive systems seriously affected. This, in
fact, is the great bar to the domestication of animals.
Between the sterility thus superinduced and that of
hybrids, there are many points of similarity. In both
cases the sterility is independent of general health, and is
often accompanied by excess of size or great luxuriance.
In both cases the sterility occurs in various degrees ; in
both, the male element is the most liable to be affected ;
but sometimes the female more than the male. In both,
the tendency goes to a certain extent with systematic
affinity, for whole groups of animals and plants are rendered impotent by the same unnatural conditions ; and
whole groups of species tend to produce sterile hybrids.
On the other hand, one species in a group will sometimes resist great changes of conditions with unimpaired
fertility ; and certain species in a group will produce
unusually fertile hybrids. No one can tell, till he tries,
whether any particular animal will breed under confinement or any exotic plant seed freely under culture ; nor
can he tell, till he tries, whether any two species of a
genus will produce more or less sterile hybrids. Lastly,
when organic beings are placed during several generations under conditions not natural to them, they are
extremely liable to vary, which is due, as I believe, to
their reproductive systems having been specially affected,
though in a lesser degree than when sterility ensues.
So it is with hybrids, for hybrids in successive generations are eminently liable to vary, as every experi-
mentalist has observed.

Thus we see that when organic beings are placed
under new and unnatural conditions, and when hybrids

are produced by the unnatural crossing of two species, the reproductive system, independently of the general state of health, is affected by sterility in a very similar manner. In the one case, the conditions of life have been disturbed, though often in so slight a degree as to be inappreciable by us ; in the other case, or that of hybrids, the external conditions have remained the same, but the organisation has been disturbed by two different structures and constitutions having been blended into one. For it is scarcely possible that two organisations should be compounded into one, without some disturbance occurring in the development, or periodical action, or mutual relation of the different parts and organs one to another, or to the conditions of life. When hybrids are able to breed *inter se*, they transmit to their offspring from generation to generation the same compounded organisation, and hence we need not be surprised that their sterility, though in some degree variable, rarely diminishes.

It must, however, be confessed that we cannot understand, except on vague hypotheses, several facts with respect to the sterility of hybrids ; for instance, the unequal fertility of hybrids produced from reciprocal crosses ; or the increased sterility in those hybrids which occasionally and exceptionally resemble closely either pure parent. Nor do I pretend that the foregoing remarks go to the root of the matter : no explanation is offered why an organism, when placed under unnatural conditions, is rendered sterile. All that I have attempted to show, is that in two cases, in some respects allied, sterility is the common result,—in the one case from the conditions of life having been disturbed, in the other case from the organisation having been disturbed by two organisations having been compounded into one.

It may seem fanciful, but I suspect that a similar parallelism extends to an allied yet very different class of facts. It is an old and almost universal belief, founded, I think, on a considerable body of evidence, that slight changes in the conditions of life are beneficial to all living things. We see this acted on by farmers and gardeners in their frequent exchanges of seed, tubers, &c., from one soil or climate to another, and back again. During the convalescence of animals, we plainly see that great benefit is derived from almost any change in the habits of life. Again, both with plants and animals, there is abundant evidence that a cross between very distinct individuals of the same species, that is between members of different strains or sub-breeds, gives vigour and fertility to the offspring. I believe, indeed, from the facts alluded to in our fourth chapter, that a certain amount of crossing is indispensable even with hermaphrodites; and that close interbreeding continued during several generations between the nearest relations, especially if these be kept under the same conditions of life, always induces weakness and sterility in the progeny.

Hence it seems that, on the one hand, slight changes in the conditions of life benefit all organic beings, and on the other hand, that slight crosses, that is crosses between the males and females of the same species which have varied and become slightly different, give vigour and fertility to the offspring. But we have seen that greater changes, or changes of a particular nature, often render organic beings in some degree sterile; and that greater crosses, that is crosses between males and females which have become widely or specifically different, produce hybrids which are generally sterile in some degree. I cannot persuade myself that this parallelism is an accident or an illusion. Both series

O

of facts seem to be connected together by some common but unknown bond, which is essentially related to the principle of life.

Fertility of Varieties when crossed, and of their Mongrel offspring.—It may be urged, as a most forcible argument, that there must be some essential distinction between species and varieties, and that there must be some error in all the foregoing remarks, inasmuch as varieties, however much they may differ from each other in external appearance, cross with perfect facility, and yield perfectly fertile offspring. With some exceptions, presently to be given, I fully admit that this is very generally the rule. But the subject is surrounded by difficulties, for looking to varieties produced under nature, if two forms hitherto reputed to be varieties be found in any degree sterile together, they are at once ranged by most naturalists as species. For instance, the blue and red pimpernel, the primrose and cowslip, which are considered by many of our best botanists as varieties, are said by Gärtner not to be quite fertile when crossed, and he consequently ranks them as undoubted species. If we thus argue in a circle, the fertility of all varieties produced under nature will assuredly have to be granted.

If we turn to varieties produced, or supposed to have been produced, under domestication, we are still involved in doubt. For when it is stated, for instance, that the German Spitz dog unites more easily than other dogs with foxes, or that certain South American indigenous domestic dogs do not readily cross with European dogs, the explanation which will occur to every one, and probably the true one, is that these dogs have descended from several aboriginally distinct species. Nevertheless the perfect fertility of so many domestic varieties, differing widely from each other in appearance, for instance of the pigeon or of the cabbage, is a remarkable fact; more especially when we reflect how many species there are, which, though resembling each other most closely, are utterly sterile when intercrossed. Several considerations, however, render this fertility of domestic varieties less remarkable. In the first place we must remember how ignorant we are regarding the precise cause of sterility, both when species are crossed and when species are removed from their natural conditions. On this latter head I have not had space to adduce the many remarkable facts which could have been given; with respect to sterility from crossing, reflect on the difference in the result of reciprocal crosses,—reflect on the singular cases in which a plant can be more easily fertilised by foreign pollen than by its own. When we think over such cases, and on that of the differently coloured varieties of Verbascum presently to be given, we must feel how ignorant we are, and how little likely it is that we should understand why certain forms are fertile and other forms are sterile when crossed. It can, in the second place, be clearly shown that mere external dissimilarity between two species does not determine their greater or lesser degree of sterility when crossed; and we may apply the same rule to domestic varieties. In the third place, some eminent naturalists believe that a long course of domestication tends to eliminate sterility in the successive generations of hybrids which were at first only slightly sterile; and if this be so, we surely ought not to expect to find sterility both appearing and disappearing under nearly the same domestic conditions of life. Lastly, and this seems to me the most important consideration, new races of animals and plants are produced under domestication chiefly by man's methodical and unconscious

power of selection, for his own use and pleasure : he
neither wishes to select, nor could select, slight differences in the reproductive system, or other constitutional
differences correlated with the reproductive system.
Domestic productions are less closely adapted to climate
and to the other physical conditions of the countries which
they inhabit than are those in a state of nature, for
they can generally be removed to other and differently
constituted countries with entire impunity. Man sup-
plies his several varieties with the same food ; he treats
them in nearly the same manner, and does not wish to
alter their general habits of life. Nature acts uniformly
and slowly during vast periods of time on the whole
organisation, in any way which may be for each creature's own good ; and thus she may, either directly, or
more probably indirectly, through correlation, modify
the reproductive systems of the several descendants
from any one species. Seeing this difference in the
process of selection, as carried on by man and nature,
we need not be surprised at some difference in the
result.

I have as yet spoken as if the varieties of the same
species were almost invariably fertile when intercrossed.
But it is impossible to resist the evidence of the existence of a certain amount of sterility in the few
following cases, which I will briefly abstract. The evidence is at least as good as that from which we believe
in the sterility of a multitude of species. The evidence
is, also, derived from hostile witnesses, who in all other
cases consider fertility and sterility as safe criterions
of specific distinction. Gärtner kept during several
years a dwarf kind of maize with yellow seeds, and a
tall variety with red seeds, growing near each other in
his garden ; and although these plants have separated
sexes, they never naturally crossed. He then fertilised

thirteen flowers of the one with the pollen of the other; but only a single head produced any seed, and this one head produced only five grains. Manipulation in this case could not have been injurious, as the plants have separated sexes. No one, I believe, has suspected that these varieties of maize are distinct species; and it is important to notice that the hybrid plants thus raised were themselves *perfectly fertile*; so that even Gärtner did not venture to consider the two varieties as specifically distinct.

Girou de Buzareingues crossed three varieties of gourd, which like the maize has separated sexes, and he asserts that their mutual fertilisation is by so much the less easy as their differences are greater. How far these experiments may be trusted, I know not; but the forms experimented on are ranked by Sageret, who mainly founds his classification by the test of infertility, as varieties.

The following case is far more remarkable, and seems at first quite incredible; but it is the result of an astonishing number of experiments made during many years on nine species of Verbascum, by so good an observer and so hostile a witness as Gärtner: namely, that yellow and white varieties of the same species of Verbascum when intercrossed produce less seed, than do either coloured varieties when fertilised with pollen from their own coloured flowers. Moreover, he asserts that when yellow and white varieties of one species are crossed with yellow and white varieties of a *distinct* species, more seed is produced by the crosses between the similarly coloured flowers, than between those which are differently coloured. Yet these varieties of Verbascum present no other difference besides the mere colour of the flower; and one variety can sometimes be raised from the seed of the other.

From experiments which I have made on certain varieties of hollyhock, I am inclined to suspect that they present analogous facts.

Kölreuter, whose accuracy has been confirmed by every subsequent observer, has proved the remarkable fact, that one variety of the common tobacco is more fertile, when crossed with a widely distinct species, than are the other varieties. He experimented on five forms, which are commonly reputed to be varieties, and which he tested by the severest trial, namely, by reciprocal crosses, and he found their mongrel offspring perfectly fertile. But one of these five varieties, when used either as father or mother, and crossed with the Nicotiana glutinosa, always yielded hybrids not so sterile as those which were produced from the four other varieties when crossed with N. glutinosa. Hence the reproductive system of this one variety must have been in some manner and in some degree modified.

From these facts; from the great difficulty of ascertaining the infertility of varieties in a state of nature, for a supposed variety if infertile in any degree would generally be ranked as species; from man selecting only external characters in the production of the most distinct domestic varieties, and from not wishing or being able to produce recondite and functional differences in the reproductive system; from these several considerations and facts, I do not think that the very general fertility of varieties can be proved either to be of universal occurrence, or to form a fundamental distinction between varieties and species. The general fertility of varieties, considering how entirely ignorant we are on the causes of both fertility and sterility, does not seem to me sufficient to overthrow the view taken with respect to the very general, but not invariable, sterility of first crosses between species and of their hybrids, namely, that it is not a special endowment, but is incidental on slowly acquired modifications, more especially in the reproductive systems of the forms which are crossed.

Hybrids and Mongrels compared, independently of their fertility.—Independently of the question of fertility, the offspring of species when crossed and of varieties when crossed may be compared in several other respects. Gärtner, whose strong wish it was to draw a distinct line between species and varieties, could find very few, and, as it seems to me, quite unimportant differences between the so-called hybrid offspring of species, and the so-called mongrel offspring of varieties. And, on the other hand, they agree most closely in very many important respects.

I shall here discuss this subject with extreme brevity. The most important distinction is, that in the first generation mongrels are more variable than hybrids; but Gärtner admits that hybrids from species which have long been cultivated are often variable in the first generation; and I have myself seen striking instances of this fact. Gärtner further admits that hybrids between very closely allied species are more variable than those from very distinct species; and this shows that the difference in the degree of variability graduates away. When mongrels and the more fertile hybrids are propagated for several generations an extreme amount of variability in their offspring is notorious; but some few cases both of hybrids and mongrels long retaining uniformity of character could be given. The variability, however, in the successive generations of mongrels is perhaps, greater than in hybrids.

This greater variability of mongrels than of hybrids does not seem to me at all surprising. For the parents of mongrels are varieties, and mostly domestic varieties (very few experiments having been tried on natural

varieties), and this implies in most cases that there has
been recent variability ; and therefore we might expect
that such variability would often continue and be super-
added to that arising from the mere act of crossing.
The slight degree of variability in hybrids from the first
cross or in the first generation, in contrast with their
extreme variability in the succeeding generations, is a
curious fact and deserves attention. For it bears on
and corroborates the view which I have taken on the
cause of ordinary variability; namely, that it is due to
the reproductive system being eminently sensitive to
any change in the conditions of life, being thus often
rendered either impotent or at least incapable of its
proper function of producing offspring identical with
the parent-form. Now hybrids in the first generation
are descended from species (excluding those long cultivated) which have not had their reproductive systems
in any way affected, and they are not variable ; but
hybrids themselves have their reproductive systems
seriously affected, and their descendants are highly
variable.

But to return to our comparison of mongrels and hybrids : Gärtner states that mongrels are more liable than
hybrids to revert to either parent-form ; but this, if it
be true, is certainly only a difference in degree. Gärtner
further insists that when any two species, although most
closely allied to each other, are crossed with a third
species, the hybrids are widely different from each other ;
whereas if two very distinct varieties of one species are
crossed with another species, the hybrids do not differ
much. But this conclusion, as far as I can make out,
is founded on a single experiment ; and seems directly
opposed to the results of several experiments made by
Kölreuter.

These alone are the unimportant differences, which

Gärtner is able to point out between hybrid and mongrel
plants. On the other hand, the resemblance in mongrels and in hybrids to their respective parents, more
especially in hybrids produced from nearly related
species, follows according to Gärtner the same laws.
When two species are crossed, one has sometimes a prepotent power of impressing its likeness on the hybrid;
and so I believe it to be with varieties of plants. With
animals one variety certainly often has this prepotent
power over another variety. Hybrid plants produced
from a reciprocal cross, generally resemble each other
closely; and so it is with mongrels from a reciprocal
cross. Both hybrids and mongrels can be reduced to
either pure parent form, by repeated crosses in successive generations with either parent.

These several remarks are apparently applicable to
animals; but the subject is here excessively complicated, partly owing to the existence of secondary sexual
characters; but more especially owing to prepotency in
transmitting likeness running more strongly in one sex
than in the other, both when one species is crossed with
another, and when one variety is crossed with another
variety. For instance, I think those authors are right,
who maintain that the ass has a prepotent power over
the horse, so that both the mule and the hinny more
resemble the ass than the horse; but that the prepotency runs more strongly in the male-ass than in the
female, so that the mule, which is the offspring of the
male-ass and mare, is more like an ass, than is the
hinny, which is the offspring of the female-ass and
stallion.

Much stress has been laid by some authors on the
supposed fact, that mongrel animals alone are born
closely like one of their parents; but it can be shown
that this does sometimes occur with hybrids; yet I grant

much less frequently with hybrids than with mongrels. Looking to the cases which I have collected of cross-bred animals closely resembling one parent, the resemblances seem chiefly confined to characters almost monstrous in their nature, and which have suddenly appeared—such as albinism, melanism, deficiency of tail or horns, or additional fingers and toes; and do not relate to characters which have been slowly acquired by selection. Consequently, sudden reversions to the perfect character of either parent would be more likely to occur with mongrels, which are descended from varieties often suddenly produced and semi-monstrous in character, than with hybrids, which are descended from species slowly and naturally produced. On the whole I entirely agree with Dr. Prosper Lucas, who, after arranging an enormous body of facts with respect to animals, comes to the conclusion, that the laws of resemblance of the child to its parents are the same, whether the two parents differ much or little from each other, namely in the union of individuals of the same variety, or of different varieties, or of distinct species.

Laying aside the question of fertility and sterility, in all other respects there seems to be a general and close similarity in the offspring of crossed species, and of crossed varieties. If we look at species as having been specially created, and at varieties as having been produced by secondary laws, this similarity would be an astonishing fact. But it harmonises perfectly with the view that there is no essential distinction between species and varieties.

Summary of Chapter.—First crosses between forms sufficiently distinct to be ranked as species, and their hybrids, are very generally, but not universally, sterile.

The sterility is of all degrees, and is often so slight that the two most careful experimentalists who have ever lived, have come to diametrically opposite conclusions in ranking forms by this test. The sterility is innately variable in individuals of the same species, and is eminently susceptible of favourable and unfavourable conditions. The degree of sterility does not strictly follow systematic affinity, but is governed by several curious and complex laws. It is generally different, and sometimes widely different, in reciprocal crosses between the same two species. It is not always equal in degree in a first cross and in the hybrid produced from this cross.

In the same manner as in grafting trees, the capacity of one species or variety to take on another, is incidental on generally unknown differences in their vegetative systems, so in crossing, the greater or less facility of one species to unite with another, is incidental on unknown differences in their reproductive systems. There is no more reason to think that species have been specially endowed with various degrees of sterility to prevent them crossing and blending in nature, than to think that trees have been specially endowed with various and somewhat analogous degrees of difficulty in being grafted together in order to prevent them becoming inarched in our forests.

The sterility of first crosses between pure species, which have their reproductive systems perfect, seems to depend on several circumstances; in some cases largely on the early death of the embryo. The sterility of hybrids, which have their reproductive systems imperfect, and which have had this system and their whole organisation disturbed by being compounded of two distinct species, seems closely allied to that sterility which so

frequently affects pure species, when their natural conditions of life have been disturbed. This view is supported by a parallelism of another kind :—namely, that
the crossing of forms only slightly different is favourable
to the vigour and fertility of their offspring; and that
slight changes in the conditions of life are apparently
favourable to the vigour and fertility of all organic beings.
It is not surprising that the degree of difficulty in uniting
two species, and the degree of sterility of their hybrid-
offspring, should generally correspond, though due to
distinct causes; for both depend on the amount of
difference of some kind between the species which are
crossed.   Nor is it surprising that the facility of effecting
a first cross, the fertility of the hybrids produced from
it, and the capacity of being grafted together—though
this latter capacity evidently depends on widely different circumstances—should all run, to a certain extent,
parallel with the systematic affinity of the forms which
are subjected to experiment ; for systematic affinity
attempts to express all kinds of resemblance between all
species.

First crosses between forms known to be varieties, or
sufficiently alike to be considered as varieties, and their
mongrel offspring, are very generally, but not, as is so
often falsely stated, universally fertile.   Nor is this
nearly general and perfect fertility surprising, when we
remember how liable we are to argue in a circle with
respect to varieties in a state of nature ; and when we
remember that the greater number of varieties have
been produced under domestication by the selection of
mere external differences, and not of differences in the
reproductive system.   In all other respects, excluding
fertility, there is a close general resemblance between
hybrids and mongrels.   Finally, then, although we are

profoundly ignorant in every case of the precise cause
of sterility, the facts briefly given in this chapter do not
seem to me opposed to, but even rather to support in
some respects the view, that there is no fundamental
distinction between species and varieties.

CHAPTER IX.

ON THE IMPERFECTION OF THE GEOLOGICAL RECORD.

On the absence of intermediate varieties at the present day—On the nature of extinct intermediate varieties; on their number— On the vast lapse of time, as inferred from the rate of deposition and of denudation — On the poorness of our palæontological collections—On the denudation of granitic areas—On the inter- mittence of geological formations—On the absence of intermediate varieties in any one formation—On the sudden appearance of groups of species—On their sudden appearance in the lowest known fossiliferous strata.

IN the sixth chapter I enumerated the chief objections which might be justly urged against the views main- tained in this volume. Most of them have now been discussed. One, namely the distinctness of specific forms, and their not being blended together by innu- merable transitional links, is a very obvious difficulty. I assigned reasons why such links do not commonly occur at the present day, under the circumstances ap- parently most favourable for their presence, namely on an extensive and continuous area with graduated physical conditions. I endeavoured to show, that the life of each species depends in a more important manner on the presence of other already defined organic forms, than on climate; and, therefore, that the really go- verning conditions of life do not graduate away quite insensibly like heat or moisture. I endeavoured, also, to show that intermediate varieties, from existing in lesser numbers than the forms which they connect, will generally be beaten out and exterminated during the course of further modification and improvement. The

main cause, however, of innumerable intermediate links
not now occurring everywhere throughout nature depends on the very process of natural selection, through
which new varieties continually take the places of and
exterminate their parent-forms. But just in proportion
as this process of extermination has acted on an
enormous scale, so must the number of intermediate
varieties, which have formerly existed on the earth,
be truly enormous. Why then is not every geological formation and every stratum full of such intermediate links? Geology assuredly does not reveal any
such finely-graduated organic chain; and this, perhaps,
is the most obvious and serious objection which can be
urged against my theory. The explanation lies, as I
believe, in the extreme imperfection of the geological
record.

In the first place it should always be borne in mind
what sort of intermediate forms must, on my theory,
have formerly existed. I have found it difficult, when
looking at any two species, to avoid picturing to myself,
forms *directly* intermediate between them. But this
is a wholly false view; we should always look for forms
intermediate between each species and a common but
unknown progenitor; and the progenitor will generally
have differed in some respects from all its modified descendants. To give a simple illustration: the fantail and
pouter pigeons have both descended from the rock-pigeon;
if we possessed all the intermediate varieties which have
ever existed, we should have an extremely close series
between both and the rock-pigeon; but we should have
no varieties directly intermediate between the fantail
and pouter; none, for instance, combining a tail somewhat expanded with a crop somewhat enlarged, the
characteristic features of these two breeds. These two
breeds, moreover, have become so much modified, that

if we had no historical or indirect evidence regarding
their origin, it would not have been possible to have
determined from a mere comparison of their structure
with that of the rock-pigeon (C. livia), whether they
had descended from this species or from some other
allied species, such as C. oenas.

So with natural species, if we look to forms very
distinct, for instance to the horse and tapir, we have
no reason to suppose that links ever existed directly
intermediate between them, but between each and an
unknown common parent. The common parent will
have had in its whole organisation much general resemblance to the tapir and to the horse; but in some points
of structure may have differed considerably from both,
even perhaps more than they differ from each other.
Hence in all such cases, we should be unable to recognise the parent-form of any two or more species, even
if we closely compared the structure of the parent with
that of its modified descendants, unless at the same
time we had a nearly perfect chain of the intermediate
links.

It is just possible by my theory, that one of two living
forms might have descended from the other; for instance, a horse from a tapir; and in this case direct
intermediate links will have existed between them.
But such a case would imply that one form had remained for a very long period unaltered, whilst its
descendants had undergone a vast amount of change;
and the principle of competition between organism and
organism, between child and parent, will render this a
very rare event; for in all cases the new and improved
forms of life tend to supplant the old and unimproved
forms.

By the theory of natural selection all living species
have been connected with the parent-species of each

genus, by differences not greater than we see between the varieties of the same species at the present
day; and these parent-species, now generally extinct,
have in their turn been similarly connected with more
ancient species; and so on backwards, always converging to the common ancestor of each great class.
So that the number of intermediate and transitional
links, between all living and extinct species, must have
been inconceivably great. But assuredly, if this theory
be true, such have lived upon the earth.

On the lapse of Time.—Independently of our not
finding fossil remains of such infinitely numerous connecting links, it may be objected, that time will not
have sufficed for so great an amount of organic change,
all changes having been effected very slowly through
natural selection. It is hardly possible for me even to
recall to the reader, who may not be a practical geologist, the facts leading the mind feebly to comprehend
the lapse of time. He who can read Sir Charles Lyell's
grand work on the Principles of Geology, which the
future historian will recognise as having produced a
revolution in natural science, yet does not admit how
incomprehensibly vast have been the past periods of
time, may at once close this volume. Not that it suffices
to study the Principles of Geology, or to read special
treatises by different observers on separate formations,
and to mark how each author attempts to give an inadequate idea of the duration of each formation or even
each stratum. A man must for years examine for
himself great piles of superimposed strata, and watch
the sea at work grinding down old rocks and making
fresh sediment, before he can hope to comprehend anything of the lapse of time, the monuments of which we
see around us.

It is good to wander along lines of sea-coast, when formed of moderately hard rocks, and mark the process of degradation. The tides in most cases reach the cliffs only for a short time twice a day, and the waves eat into them only when they are charged with sand or pebbles; for there is good evidence that pure water effects nothing in wearing away rock. At last the base of the cliff is undermined, huge fragments fall down, and these remaining fixed, have to be worn away atom by atom, until reduced in size they can be rolled about by the waves, and then are more quickly ground into pebbles, sand, or mud. But how often do we see along the bases of retreating cliffs rounded boulders, all thickly clothed by marine productions, showing how little they are abraded and how seldom they are rolled about! Moreover, if we follow for a few miles any line of rocky cliff, which is undergoing degradation, we find that it is only here and there, along a short length or round a promontory, that the cliffs are at the present time suffering. The appearance of the surface and the vegetation show that elsewhere years have elapsed since the waters washed their base.

He who most closely studies the action of the sea on our shores, will, I believe, be most deeply impressed with the slowness with which rocky coasts are worn away. The observations on this head by Hugh Miller, and by that excellent observer Mr. Smith of Jordan Hill, are most impressive. With the mind thus impressed, let any one examine beds of conglomerate many thousand feet in thickness, which, though probably formed at a quicker rate than many other deposits, yet, from being formed of worn and rounded pebbles, each of which bears the stamp of time, are good to show how slowly the mass has been accumulated. In the Cordillera I estimated one pile of conglomerate at ten thousand feet in thickness. Let the observer remember Lyell's profound remark that the thickness and extent of sedimentary formations are the result and measure of the degradation which the earth's crust has elsewhere suffered. And what an amount of degradation is implied by the sedimentary deposits of many countries! Professor Ramsay has given me the maximum thickness, in most cases from actual measurement, in a few cases from estimate, of each formation in different parts of Great Britain; and this is the result :—

	Feet
Palæozoic strata (not including igneous beds) ..	57,154
Secondary strata	2,190
Tertiary strata	2,240

making altogether 72,584 feet; that is, very nearly thirteen and three-quarters British miles. Some of the formations, which are represented in England by thin beds, are thousands of feet in thickness on the Continent. Moreover, between each successive formation, we have, in the opinion of most geologists, enormously long blank periods. So that the lofty pile of sedimentary rocks in Britain, gives but an inadequate idea of the time which has elapsed during their accumulation; yet what time this must have consumed! Good observers have estimated that sediment is deposited by the great Mississippi river at the rate of only 600 feet in a hundred thousand years. This estimate has no pretension to strict exactness; yet, considering over what wide spaces very fine sediment is transported by the currents of the sea, the process of accumulation over any one extensive area must be extremely slow.

But the amount of denudation which the strata have in many places suffered, independently of the rate of accumulation of the degraded matter, probably offers the best evidence of the lapse of time. I remember having been much struck with the evidence of denudation, when viewing volcanic islands, which have been worn by the waves and pared all round into perpendicular cliffs of one or two thousand feet in height; for the gentle slope of the lava-streams, due to their formerly liquid state, showed at a glance how far the hard, rocky beds had once extended into the open ocean. The same story is still more plainly told by faults,—those great cracks along which the strata have been upheaved on one side, or thrown down on the other, to the height or depth of thousands of feet; for since the crust cracked (it makes no great difference whether the upheaval was sudden, or, as most geologists now believe, was very slow and effected by many starts), the surface of the land has been so completely planed down by the action of the sea, that no trace of these vast dislocations is externally visible.

The Craven fault, for instance, extends for upwards of 30 miles, and along this line the vertical displacement of the strata has varied from 600 to 3000 feet. Prof. Ramsay has published an account of a downthrow in Anglesea of 2300 feet; and he informs me that he fully believes there is one in Merionethshire of 12,000 feet; yet in these cases there is nothing on the surface of the land to show such prodigious movements; the pile of rocks on the one or other side having been smoothly swept away. The consideration of these facts impresses the mind almost in the same manner as does the vain endeavour to grapple with the idea of eternity.

I have made these few remarks because it is highly important for us to gain some notion, however imperfect, of the lapse of time. During each year, over the whole world, the land and the water have been peopled by hosts of living forms. What an infinite number of

generations, which the mind cannot grasp, must have
succeeded each other in the long roll of years! Now
turn to our richest geological museums, and what a
paltry display we behold!

On the poorness of our Palæontological collections.
That our palæontological collections are very imperfect,
is admitted by every one. The remark of that admirable palæontologist, the late Edward Forbes, should not
be forgotten, namely, that numbers of our fossil species
are known and named from single and often broken
specimens, or from a few specimens collected on some
one spot. Only a small portion of the surface of the
earth has been geologically explored, and no part with
sufficient care, as the important discoveries made every
year in Europe prove. No organism wholly soft can be
preserved. Shells and bones will decay and disappear
when left on the bottom of the sea, where sediment is
not accumulating. I believe we are continually taking
a most erroneous view, when we tacitly admit to ourselves that sediment is being deposited over nearly the
whole bed of the sea, at a rate sufficiently quick to
embed and preserve fossil remains. Throughout an
enormously large proportion of the ocean, the bright
blue tint of the water bespeaks its purity. The many
cases on record of a formation conformably covered,
after an enormous interval of time, by another and
later formation, without the underlying bed having
suffered in the interval any wear and tear, seem explicable only on the view of the bottom of the sea not
rarely lying for ages in an unaltered condition. The remains which do become embedded, if in sand or gravel,
will when the beds are upraised generally be dissolved
by the percolation of rain-water charged with carbonic
acid. Some of the many kinds of animals which live

at the beach between high and low water mark were
to be rarely preserved. For instance, the several
species of the Chthamalinæ (a sub-family of sessile
cirripedes) coat the rocks all over the world in infinite
numbers: they are all strictly littoral, with the exception of a single Mediterranean species, which inhabits
deep water, and has been found fossil in Sicily, whereas
not one other species has hitherto been found in any
tertiary formation: yet it is now known that the genus
Chthamalus existed during the chalk period. The mol-
luscan genus Chiton offers a partially analogous case.

With respect to the terrestrial productions which
lived during the Secondary and Palæozoic periods, it is
superfluous to state that our evidence from fossil
remains is fragmentary in an extreme degree. For
instance, not a land shell is known belonging to either
of these vast periods, with the exception of one species
discovered by Sir C. Lyell and Dr. Dawson in the carboniferous strata of North America, of which shell above
a hundred specimens have now been collected. In
regard to mammiferous remains, a single glance at the
historical table published in the Supplement to Lyell's
Manual, will bring home the truth, how accidental and
rare is their preservation, far better than pages of detail.
Nor is their rarity surprising, when we remember how
large a proportion of the bones of tertiary mammals
have been discovered either in caves or in lacustrine
deposits; and that not a cave or true lacustrine bed is
known belonging to the age of our secondary or palæozoic formations.

But the imperfection in the geological record largely
results from another and more important cause than any
of the foregoing; namely, from the several formations
being separated from each other by wide intervals of
time. This doctrine has been most emphatically

admitted by many geologists and palæontologists, who,
like E. Forbes, entirely disbelieve in the change of
species. When we see the formations tabulated in
written works, or when we follow them in nature, it is
difficult to avoid believing that they are closely consecutive. But we know, for instance, from Sir R.
Murchison's great work on Russia, what wide gaps
there are in that country between the superimposed
formations; so it is in North America, and in many
other parts of the world. The most skilful geologist, if
his attention had been exclusively confined to these
large territories, would never have suspected that during
the periods which were blank and barren in his own
country, great piles of sediment, charged with new and
peculiar forms of life, had elsewhere been accumulated. And if in each separate territory, hardly any
idea can be formed of the length of time which has
elapsed between the consecutive formations, we may infer
that this could nowhere be ascertained. The frequent
and great changes in the mineralogical composition of
consecutive formations, generally implying great changes
in the geography of the surrounding lands, whence the
sediment has been derived, accords with the belief of
vast intervals of time having elapsed between each
formation.

But we can, I think, see why the geological formations of each region are almost invariably intermittent; that is, have not followed each other in close
sequence. Scarcely any fact struck me more when
examining many hundred miles of the South American
coasts, which have been upraised several hundred feet
within the recent period, than the absence of any recent
deposits sufficiently extensive to last for even a short
geological period. Along the whole west coast, which
is inhabited by a peculiar marine fauna, tertiary beds

are so poorly developed, that no record of several successive and peculiar marine faunas will probably be preserved to a distant age.   A little reflection will explain why along the rising coast of the western side of South America, no extensive formations with recent or tertiary remains can anywhere be found, though the supply of sediment must for ages have been great, from the enormous degradation of the coast-rocks and from muddy streams entering the sea.   The explanation, no doubt, is, that the littoral and sub-littoral deposits are continually worn away, as soon as they are brought up by the slow and gradual rising of the land within the grinding action of the coast-waves.

We may, I think, safely conclude that sediment must be accumulated in extremely thick, solid, or extensive masses, in order to withstand the incessant action of the waves, when first upraised and during subsequent oscillations of level.   Such thick and extensive accumulations of sediment may be formed in two ways; either in profound depths of the sea, in which case, judging from the researches of E. Forbes, we may conclude that the bottom will be inhabited by few animals, but it will not be, as we at last know from the telegraphic soundings, barren of life; consequently the mass when upraised will give a most imperfect record of the forms of life which existed during the period of deposition.   Or, sediment may be accumulated to any thickness and extent over a shallow bottom, if it continue slowly to subside.   In this latter case, as long as the rate of subsidence and supply of sediment nearly balance each other, the sea will remain shallow and favourable for life, and thus a rich fossiliferous formation thick enough, when upraised, to resist any amount of degradation, may be formed.

I am convinced that nearly all our ancient formations,

which are throughout the greater part of their thickness
rich in *fossils*, have thus been formed during subsidence.
Since publishing my views on this subject in 1845, I
have watched the progress of Geology, and have been
surprised to note how author after author, in treating
of this or that great formation, has come to the conclusion that it was accumulated during subsidence. I
may add, that the only ancient tertiary formation on
the west coast of South America, which has been bulky
enough to resist such degradation as it has as yet suffered, but which will hardly last to a distant geological
age, was certainly deposited during a downward oscilla-
tion of level, and thus gained considerable thickness.

All geological facts tell us plainly that each area
has undergone numerous slow oscillations of level, and
apparently these oscillations have affected wide spaces.
Consequently formations rich in fossils and sufficiently
thick and extensive to resist subsequent degradation,
may have been formed over wide spaces during periods
of subsidence, but only where the supply of sediment
was sufficient to keep the sea shallow and to embed and
preserve the remains before they had time to decay.
On the other hand, as long as the bed of the sea remained stationary, thick deposits could not have been
accumulated in the shallow parts, which are the most
favourable to life. Still less could this have happened
during the alternate periods of elevation; or, to speak
more accurately, the beds which were then accumulated
will generally have been destroyed by being upraised
and brought within the limits of the coast-action.

These remarks apply chiefly to littoral and sub-littoral
deposits. In the case of an extensive and shallow sea,
such as that within a large part of the Malay Archipelago, where the depth varies from 30 or 40 to 60
fathoms, a widely extended formation might be formed

z

during a period of elevation and yet not suffer excessively from denudation during its slow upheaval; but the thickness of the formation could not be great, for owing to the elevatory movement it would be less than the depth, supposed to be shallow; the deposit would not generally be much consolidated, nor would it be capped by overlying formations, so that it would run a good chance of being worn away during subsequent oscillations of level. It has been suggested by Mr. Hopkins, that if one part of the area, after rising and before being denuded, subsided, the deposit formed during the rising movement, though not thick, might become protected by fresh accumulations, and thus be preserved for an extremely long period,—a consideration which I formerly overlooked.

Mr. Hopkins, in commenting on this subject, states that he believes the entire destruction of any sedimentary bed of considerable horizontal extent to have been of rare occurrence. My remarks apply solely to beds rich in fossils: I have admitted that sediment accumulated in extremely thick, solid, or extensive masses would escape denudation. The point in question is, whether widely extended formations, rich in fossils, and of sufficient thickness to last for a long period, would be formed except during periods of subsidence? My impression is that this has rarely been the case. As the subject of complete denudation has been broached by Mr. Hopkins, I may remark that all geologists, -excepting the few who believe that they see in the metamorphic schists and plutonic rocks the heated primordial nucleus of the globe, will probably admit that rocks of this nature must have been largely denuded. For it is scarcely possible that these rocks should have been solidified and crystallized in a naked condition; but if the metamorphic action occurred at

profound depths of the ocean, the former mantle may
not have been thick. Admitting then that such rocks
as gneiss, mica-schist, granite, diorite, &c., were once
necessarily covered up, how can we account for the
extensive and naked areas of such rocks in many parts
of the world, except on the belief that they have subsequently been completely denuded of all overlying
strata? That such extensive areas do exist cannot
be doubted: the granitic region of Parime is described by Humboldt as being at least nineteen times
as large as Switzerland. South of the Amazon Boué
colours an area composed of such rocks as equal to that
of Spain, France, Italy, part of Germany, and the
British Islands all conjoined. This region has not been
carefully explored, but from the concurrent testimony
of travellers the granitic area must be very large:
thus, Von Eschwege gives a detailed section of these
rocks, stretching from Rio de Janeiro for 260 geographical miles inland in a straight line; and I travelled for 150 miles in another direction, and saw
nothing but granitic rocks. Numerous specimens, collected along the whole coast from near Rio Janeiro to
the mouth of the Plata, a distance of 1100 geographical
miles, were shown me, and they all belonged to this
class. Inland, along the whole northern bank of the
Plata I saw, besides modern tertiary beds, only one
small patch of slightly metamorphosed rock, which alone
could have formed a part of the original capping of the
granitic series. Turning to a well-known region,
namely, to the United States and Canada, as shown
in Professor H. D. Rogers's beautiful map, I have estimated the areas by cutting out and weighing the
paper, and I find that the metamorphic (excluding "the
semi metamorphic") and granitic rocks exceed, in the
proportion of 19 to 125, not only the true coal measures,

well known to be here developed in extraordinary force, but likewise the Cumbral series, which together compose the whole newer Palaeozoic formation. In many regions the metamorphic and granitic areas would be greatly increased in size, if we could remove all the sedimentary beds which rest unconformably on them, and which at the line of junction have not been metamorphosed, showing that they could not have formed part of the original mantle under which the granitic rocks were crystallized. Hence it is probable that in some parts of the world whole formations, representing at least some stages in the several successive geological epochs, have been completely denuded, with not a wreck left behind.

One remark is here worth a passing notice. During periods of elevation the area of the land and of the adjoining shoal parts of the sea will be increased, and new stations will often be formed;—all circumstances favourable, as previously explained, for the formation of new varieties and species; but during such periods there will generally be a blank in the geological record. On the other hand, during subsidence, the inhabited area and number of inhabitants will decrease (excepting the productions on the shores of a continent when first broken up into an archipelago), and consequently during subsidence, though there will be much extinction, few new varieties or species will be formed; and it is during these very periods of subsidence, that our great deposits rich in fossils have been accumulated. Nature may almost be said to have guarded against the frequent discovery of her true transitional or linking forms.

From these several considerations it cannot be doubted that the geological record, viewed as a whole, is extremely imperfect; but if we confine our attention to any one formation, it becomes much more difficult to understand, why we do not therein find closely graduated varieties between the allied species which lived at its commencement and at its close. Some cases are on record of the same species presenting distinct varieties in the upper and lower parts of the same formation, but, as they are rare, they may be here passed over. Although each formation has indisputably required a vast number of years for its deposition, I can see several reasons why each should not include a graduated series of links between the species which then lived; but I can by no means pretend to assign due proportional weight to the following considerations.

Although each formation may mark a very long lapse of years, each perhaps is short compared with the period requisite to change one species into another. I am aware that two palæontologists, whose opinions are worthy of much deference, namely Bronn and Woodward, have concluded that the average duration of each formation is twice or thrice as long as the average duration of specific forms. But insuperable difficulties, as it seems to me, prevent us coming to any just conclusion on this head. When we see a species first appearing in the middle of any formation, it would be rash in the extreme to infer that it had not elsewhere previously existed. So again when we find a species disappearing before the uppermost layers have been deposited, it would be equally rash to suppose that it then became wholly extinct. We forget how small the area of Europe is compared with the rest of the world; nor have the several stages of the same formation throughout Europe been correlated with perfect accuracy.

With marine animals of all kinds, we may safely infer a large amount of migration during climatal and other changes; and when we see a species first appearing in any formation, the probability is that it only they first immigrated into that area. It is well known, for instance, that several species appeared somewhat earlier in the palæozoic beds of North America than in those of Europe; time having apparently been required for their migration from the American to the European seas. In examining the latest deposits of various quarters of the world, it has everywhere been noted, that some few still existing species are common in the deposit, but have become extinct in the immediately surrounding sea; or, conversely, that some are now abundant in the neighbouring sea, but are rare or absent in this particular deposit. It is an excellent lesson to reflect on the ascertained amount of migration of the inhabitants of Europe during the glacial period, which forms only a part of one whole geological period; and likewise to reflect on the great changes of level, on the inordinately great change of climate, on the prodigious lapse of time, all included within this same glacial period. Yet it may be doubted whether in any quarter of the world, sedimentary deposits, *including fossil remains*, have gone on accumulating within the same area during the whole of this period. It is not, for instance, probable that sediment was deposited during the whole of the glacial period near the mouth of the Mississippi, within that limit of depth at which marine animals can flourish; for we know what vast geographical changes occurred in other parts of America during this space of time. When such beds as were deposited in shallow water near the mouth of the Mississippi during some part of the glacial period shall have been upraised, organic remains will probably first appear and disappear at different levels, owing to the migration of species and to geographical changes. And in the distant future, a geologist examining these beds, might be tempted to conclude that the average duration of life of the embedded fossils had been less than that fauna of the glacial period, instead of having been really far greater, that is, extending from before the glacial epoch to the present day.

In order to get a perfect gradation between two forms in the upper and lower parts of the same formation, the deposit must have gone on accumulating for a very long period, so that there may have been time sufficient for the slow process of variation; hence the deposit will generally have to be a very thick one; and the species undergoing modification will have had to live on the same area throughout this whole time. But we have seen that a thick formation, fossiliferous throughout its thickness, can only be accumulated during a period of subsidence; and to keep the depth approximately the same, which is necessary that the same species may live on the same space, the supply of sediment must nearly have counterbalanced the amount of subsidence. But this same movement of subsidence will often tend to sink the area whence the sediment is derived, and thus diminish the supply whilst the downward movement continues. In fact, this nearly exact balancing between the supply of sediment and the amount of subsidence is probably a rare contingency; for it has been observed by more than one palaeontologist, that very thick deposits are usually barren of organic remains, except near their upper or lower limits.

It would seem that each separate formation, like the whole pile of formations in any country, has generally been intermittent in its accumulation. When we see, as is so often the case, a formation composed of beds of different mineralogical composition, we may reasonably suspect that the process of deposition has been much interrupted, as a change in the currents of the sea and a supply of sediment of a different nature will generally have been due to geographical changes requiring much time. Nor will the closest inspection of a formation give any idea of the time which its deposition has consumed. Many instances could be given of beds only a few feet in thickness, representing formations, elsewhere thousands of feet in thickness, and which must have required an enormous period for their accumulation; yet no one ignorant of this fact would have suspected the vast lapse of time represented by the thinner formation. Many cases could be given of the lower beds of a formation having been upraised, denuded, submerged, and then re-covered by the upper beds of the same formation,—facts, showing what wide, yet easily overlooked, intervals have occurred in its accumulation. In other cases we have the plainest evidence in great fossilized trees, still standing upright as they grew, of many long intervals of time and changes of level during the process of deposition, which would never even have been suspected, had not the trees chanced to have been preserved: thus Messrs. Lyell and Dawson found carboniferous beds 1400 feet thick in Nova Scotia, with ancient root-bearing strata, one above the other, at no less than sixty-eight different levels. Hence, when the same species occur at the bottom, middle, and top of a formation, the probability is that they have not lived on the same spot during the whole period of deposition, but have disappeared and reappeared, perhaps many times, during the same geological period. So that if such species were to undergo a considerable amount of modification during any one geological period, a section would not probably include all the fine intermediate gradations which must on my theory have existed between them, but abrupt, though perhaps slight, changes of form.

It is all-important to remember that naturalists have

no golden rule by which to distinguish species and
varieties; they grant some little variability to each
species, but when they meet with a somewhat greater
amount of difference between any two forms, they rank
both as species, unless they are enabled to connect them
together by close intermediate gradations. And this
from the reasons just assigned we can seldom hope to
effect in any one geological section. Supposing B and
C to be two species, and a third, A, to be found in an
underlying and older bed; even if A were strictly intermediate between B and C, it would simply be ranked
as a third and distinct species, unless at the same time
it could be most closely connected with either one or
both forms by intermediate varieties. Nor should it be
forgotten, as before explained, that A might be the
actual progenitor of B and C, and yet might not at all
necessarily be strictly intermediate between them in all
points of structure. So that we might obtain the parent-
species and its several modified descendants from the
lower and upper beds of a formation, and unless we
obtained numerous transitional gradations, we should
not recognise their blood-relationship, and should consequently be compelled to rank them all as distinct
species.

It is notorious on what excessively slight differences
many palæontologists have founded their species; and
they do this the more readily if the specimens come
from different sub-stages of the same formation. Some
experienced conchologists are now sinking many of the
very fine species of D'Orbigny and others into the rank
of varieties; and on this view we do find the kind of
evidence of change which on my theory we ought to
find. Look again at the later tertiary deposits, which
include many shells believed by the majority of naturalists to be identical with existing species; but some

excellent naturalists, as Agassiz and Pictet, maintain that all these tertiary species are specifically distinct, though the distinction is admitted to be very slight; so that here, unless we believe that these eminent naturalists have been misled by their imaginations, and that these late tertiary species really present no difference whatever from their living representatives, or unless we believe that the great majority of naturalists are wrong, and that these tertiary species are all truly distinct from the recent, we have evidence of a very general slight modification of form of the kind required. If we look to rather wider intervals of time, namely, to distinct but consecutive stages of the same great formation, we find that the embedded fossils, though almost universally ranked as specifically different, yet are far more closely related to each other than are the species found in more widely separated formations; so that here again we have undoubted evidence of change, though not strictly of variation, in the direction required by my theory; but to this latter subject I shall have to return in the following chapter.

One other consideration is worth notice: with animals and plants that can propagate rapidly and are not highly locomotive, there is reason to suspect, as we have formerly seen, that their varieties are generally at first local; and that such local varieties do not spread widely and supplant their parent-forms until they have been modified and perfected in some considerable degree. According to this view, the chance of discovering in a formation in any one country all the early stages of transition between any two forms, is small, for the successive changes are supposed to have been local or confined to some one spot. Most marine animals have a wide range; and we have seen that with plants it is those which have the widest range, that oftenest present varieties; so that with shells and other marine animals, it is highly probable that those which have had the widest ranges, far exceeding the limits of the known geological formations of Europe, have oftenest given rise, first to local varieties and ultimately to new species; and this again would greatly lessen the chance of our being able to trace the stages of transition in any one geological formation.

It should not be forgotten that at the present day, with perfect specimens for examination, two forms can seldom be connected by intermediate varieties, and thus proved to be the same species, until many specimens have been collected from many places; and in the case of fossil species this could rarely be effected by palæontologists. We shall, perhaps, best perceive the improbability of our being enabled to connect species by numerous, fine, intermediate fossil links, by asking ourselves whether, for instance, geologists at some future period will be able to prove that our different breeds of cattle, sheep, horses, and dogs have descended from a single stock or from several aboriginal stocks; or, again, whether certain sea-shells inhabiting the shores of North America, which are ranked by some conchologists as distinct species from their European representatives, and by other conchologists as only varieties, are really varieties, or are, as it is called, specifically distinct. This could be effected only by the future geologist discovering in a fossil state numerous intermediate gradations; and such success is improbable in the highest degree.

It has been asserted over and over again by writers who believe in the immutability of species, that geology has yielded no linking forms. This assertion is entirely erroneous. As Mr. Lubbock has recently remarked, "Every species is a link between other allied forms." We clearly see this if we take a genus having

a score of recent and extinct species and destroy four-
fifths of them; for in this case no one will doubt that
the remainder will stand much more distinct from each
other. If the extreme forms in the genus happen to
have been thus destroyed, the genus itself in most cases
will stand more distinct from other allied genera. The
camel and the pig, or the horse and the tapir, are now
obviously very distinct forms; but if we add the several
fossil quadrupeds which have already been discovered
to the families including the camel and pig, these forms
become joined by links not extremely wide apart. The
chain of breeding forms does not, however, in these
cases, or in any case, run straight from the one living
form to the other, but takes a circuitous sweep through
the forms which lived during long past ages. What
geological research has not revealed is the former
existence of infinitely numerous gradations, as fine as
existing varieties, connecting all known species. And
this not having been effected by geology is the most
obvious of the many objections which may be urged
against my views.

Hence it will be worth while to sum up the foregoing
remarks on the causes of the imperfection of the geological record under an imaginary illustration. The
Malay Archipelago is about the size of Europe from
the North Cape to the Mediterranean, and from Britain
to Russia; and therefore equals all the geological formations which have been examined with any accuracy,
excepting those of the United States of America. I
fully agree with Mr. Godwin-Austen, that the present
condition of the Malay Archipelago, with its numerous
large islands separated by wide and shallow seas, probably represents the former state of Europe, whilst most
of our formations were accumulating. The Malay Archipelago is one of the richest regions of the whole world

in organic beings; yet if all the species were to be collected which have ever lived there, how imperfectly would they represent the natural history of the world!

But we have every reason to believe that the terrestrial productions of the archipelago would be preserved in an excessively imperfect manner in the formations which we suppose to be there accumulating. I suspect that not many of the strictly littoral animals, or of those which lived on naked submarine rocks, would be embedded; and those embedded in gravel or sand, would not endure to a distant epoch. Wherever sediment did not accumulate on the bed of the sea, or where it did not accumulate at a sufficient rate to protect organic bodies from decay, no remains could be preserved.

As the occasional rule formations rich in fossils would be formed in the archipelago, of thickness sufficient to last to an age as distant in futurity as the secondary formations lie in the past, only during periods of subsidence. These periods of subsidence would be separated from each other by enormous intervals, during which the area would be either stationary or rising; whilst rising, the fossiliferous formations on the steeper shores would be destroyed, almost as soon as accumulated, by the incessant coast-action, as we now see on the shores of South America; even throughout the extensive and shallow seas within the archipelago sedimentary beds could not, during the periods of elevation, be accumulated of great thickness, or become capped and protected by subsequent deposits, so as to have a good chance of enduring to an extremely distant future. During the periods of subsidence, there would probably be much extinction of life; during the periods of elevation, there would be much variation, but the geological record would then be less perfect.

It may be doubted whether the duration of any one

great period of subsidence over the whole or part of the
archipelago, together with a contemporaneous accumulation of sediment, would exceed the average duration
of the same specific forms; and these contingencies are
indispensable for the preservation of all the transitional
gradations between any two or more species. If such
gradations were not all fully preserved, transitional
varieties would merely appear as so many distinct
species. It is also probable that each great period of
subsidence would be interrupted by oscillations of level,
and that slight climatal changes would intervene during
such lengthy periods; and in these cases the inhabitants
of the archipelago would migrate, and no closely consecutive record of their modifications could be preserved
in any one formation.

Very many of the marine inhabitants of the archipelago now range thousands of miles beyond its confines;
and analogy plainly leads to the belief that it would be
chiefly these far-ranging species which would oftenest
produce new varieties; and the varieties would at first
generally be local or confined to one place, but if possessed of any decided advantage, or when further modified and improved, they would slowly spread and supplant their parent-forms. When such varieties returned to their ancient homes, as they would differ
from their former state in a nearly uniform, though
perhaps extremely slight degree, and as they would be
found embedded in slightly different sub-stages of the
same formation, they would, according to the principles
followed by many palæontologists, be ranked as new
and distinct species.

If then there be some degree of truth in these
remarks, we have no right to expect to find in our
geological formations, an infinite number of those fine
transitional forms which, on my theory, have connected

all the past and present species of the same group into one long and branching chain of life. We ought only to look for a few links, and such assuredly we do find — some more distantly, some more closely, related to each other; and these links, let them be ever so close, if found in different stages of the same formation, would, by many palæontologists, be ranked as distinct species. But I do not pretend that I should ever have suspected how poor a record of the mutations of life the best preserved geological section presented, had not the difficulty of our not discovering innumerable transitional links between the species which appeared at the commencement and close of each formation, pressed so hardly on my theory.

On the sudden appearance of whole groups of allied Species. — The abrupt manner in which whole groups of species suddenly appear in certain formations, has been urged by several palæontologists — for instance, by Agassiz, Pictet, and Sedgwick — as a fatal objection to the belief in the transmutation of species. If numerous species, belonging to the same genera or families, have really started into life at once, the fact would be fatal to the theory of descent with slow modification through natural selection. For the development of a group of forms, all of which have descended from some one progenitor, must have been an extremely slow process; and the progenitors must have lived long ages before their modified descendants. But we continually overrate the perfection of the geological record, and falsely infer, because certain genera or families have not been found beneath a certain stage, that they did not exist before that stage. In all cases positive palæontological evidence may be implicitly trusted; negative evidence is worthless, as experience has so often shown. We continually forget how large the world is, compared with the area over which our geological formations have been carefully examined; we forget that groups of species may elsewhere have long existed, and have slowly multiplied, before they invaded the ancient archipelagoes of Europe and the United States. We do not make due allowance for the enormous intervals of time which have elapsed between our consecutive formations,—longer perhaps in many cases than the time required for the accumulation of each formation. These intervals will have given time for the multiplication of species from some one or more few parent-forms; and in the succeeding formation such groups of species will appear as if suddenly created.

I may here recall a remark formerly made, namely, that it might require a long succession of ages to adapt an organism to some new and peculiar line of life, for instance, to fly through the air; and consequently that the transitional forms would often long remain confined to some one region; but that, when this adaptation had once been effected, and a few species had thus acquired a great advantage over other organisms, a comparatively short time would be necessary to produce many divergent forms, which would be able to spread rapidly and widely throughout the world. Professor Pictet, in his excellent Review of this work, in commenting on early transitional forms, and taking birds as an illustration, cannot see how the successive modifications of the anterior limbs of a supposed prototype could possibly have been of any advantage. But look at the penguins of the Southern Ocean; have not these birds their front limbs in this precise intermediate state of "neither true arms nor true wings"? Yet these birds hold their place victoriously in the battle for life; for they exist in infinite numbers and of many kinds. I do not sup-

suppose that we here see the real transitional grade through which the wings of birds have passed; but what special difficulty is there in believing that it might profit the modified descendants of the penguin, first to become enabled to flap along the surface of the sea like the logger-headed duck, and ultimately to rise from its surface and glide through the air?

I will now give a few examples to illustrate the foregoing remarks, and to show how liable we are to error in supposing that whole groups of species have suddenly been produced. Even in so short an interval as that between the first and second editions of Pictet's great work on Palæontology, published in 1844-46 and in 1853-57, the conclusions on the first appearance and disappearance of several groups of animals have been considerably modified; and a third edition would require still further changes. I may recall the well-known fact that in geological treatises, published not many years ago, the whole class of mammals was always spoken of as having abruptly come in at the commencement of the tertiary series. And now one of the richest known accumulations of fossil mammals, for its thickness, belongs to the middle of the secondary series; and one true mammal has been discovered in the new red sandstone at nearly the commencement of this great series. Cuvier used to urge that no monkey occurred in any tertiary stratum; but now extinct species have been discovered in India, South America, and in Europe even as far back as the eocene stage. Had it not been for the rare accident of the preservation of footsteps in the new red sandstone of the United States, who would have ventured to suppose that, besides reptiles, no less than at least thirty kinds of birds, some of gigantic size, existed during that period? Not a

fragment of bone has been discovered in these beds.
Notwithstanding that the number of joints shown in the
fossil impressions correspond with the number in the
several toes of living birds' feet, some authors doubt
whether the animals which left the impressions were
really birds. Until quite recently these authors might
have maintained, and some have maintained, that the
whole class of birds came suddenly into existence
during an early tertiary period; but now we know, on
the authority of Professor Owen, that a bird certainly
lived during the deposition of the upper greensand.

I may give another instance, which from having passed
under my own eyes, has much struck me. In a memoir
on Fossil Sessile Cirripedes, I have stated that, from the
number of existing and extinct tertiary species; from
the extraordinary abundance of the individuals of many
species all over the world, from the Arctic regions to
the equator, inhabiting various zones of depths from
the upper tidal limits to 50 fathoms; from the perfect
manner in which specimens are preserved in the oldest
tertiary beds; from the ease with which even a fragment of a valve can be recognised: from all these circumstances, I inferred that had sessile cirripedes existed
during the secondary periods, they would certainly have
been preserved and discovered; and as not one species
had then been discovered in beds of this age, I concluded
that this great group had been suddenly developed at
the commencement of the tertiary series. This was a
sore trouble to me, adding as I thought one more instance of the abrupt appearance of a great group of
species. But my work had hardly been published, when
a skilful palæontologist, M. Bosquet, sent me a drawing
of a perfect specimen of an unmistakeable sessile cirripede, which he had himself extracted from the chalk

of Belgium. And, as if to make the case as striking as possible, this sessile cirripede was a Chthamalus, a very common, large, and ubiquitous genus, of which not one specimen has as yet been found even in any tertiary stratum. Hence we now positively know that sessile cirripedes existed during the secondary period; and these cirripedes might have been the progenitors of our many tertiary and existing species.

The case most frequently insisted on by palæontologists of the apparently sudden appearance of a whole group of species, is that of the teleostean fishes, low down in the Chalk period. This group includes the large majority of existing species. Lately, Professor Pictet has carried their existence one sub-stage further back; and some palæontologists believe that certain much older fishes, of which the affinities are as yet imperfectly known, are really teleostean. Assuming, however, that the whole of them did appear, as Agassiz believes, at the commencement of the chalk formation, the fact would certainly be highly remarkable; but I cannot see that it would be an insuperable difficulty on my theory, unless it could likewise be shown that the species of this group appeared suddenly and simultaneously throughout the world at this same period. It is almost superfluous to remark that hardly any fossil-fish are known from south of the equator; and by running through Pictet's Palæontology it will be seen that very few species are known from several formations in Europe. Some few families of fish now have a confined range; the teleostean fish might formerly have had a similarly confined range, and after having been largely developed in some one sea, might have spread widely. Nor have we any right to suppose that the seas of the world have always been so freely open from south to north as they are at present.

Even at this day, if the Malay Archipelago were converted into land, the tropical parts of the Indian Ocean would form a large and perfectly enclosed basin, in which any great group of marine animals might be multiplied; and here they would remain confined, until some of the species became adapted to a cooler climate, and were enabled to double the southern capes of Africa or Australia, and thus reach other and distant seas.

From these and similar considerations, but chiefly from our ignorance of the geology of other countries beyond the confines of Europe and the United States; and from the revolution in our palæontological ideas on many points, which the discoveries of even the last dozen years have effected, it seems to me to be about as rash in us to dogmatize on the succession of organic beings throughout the world, as it would be for a naturalist to land for five minutes on some one barren point in Australia, and then to discuss the number and range of its productions.

On the sudden appearance of groups of allied Species in the lowest known fossiliferous strata.—There is another and allied difficulty, which is much graver. I allude to the manner in which numbers of species of the same group, suddenly appear in the lowest known fossiliferous rocks. Most of the arguments which have convinced me that all the existing species of the same group have descended from one progenitor, apply with nearly equal force to the earliest known species. For instance, I cannot doubt that all the Silurian trilobites have descended from some one crustacean, which must have lived long before the Silurian age, and which probably differed greatly from any known animal. Some of the most ancient Silurian animals, as the Nautilus, Lingula, &c., do not differ much from living species;

and it cannot on my theory be supposed, that these old species were the progenitors of all the species of the orders to which they belong, for they do not present characters in any degree intermediate between them. If, moreover, they had been the progenitors of these orders, they would almost certainly have been long ago supplanted and exterminated by their numerous and improved descendants.

Consequently, if my theory be true, it is indisputable that before the lowest Silurian stratum was deposited, long periods elapsed, as long as, or probably far longer than, the whole interval from the Silurian age to the present day; and that during these vast, yet quite unknown, periods of time, the world swarmed with living creatures.

To the question why we do not find records of these vast primordial periods, I can give no satisfactory answer. Several of the most eminent geologists, with Sir R. Murchison at their head, are convinced that we see in the organic remains of the lowest Silurian stratum the dawn of life on this planet. Other highly competent judges, as Lyell and the late E. Forbes, dispute this conclusion. We should not forget that only a small portion of the world is known with accuracy. M. Barrande has lately added another and lower stage to the Silurian system, abounding with new and peculiar species. Traces of life have been detected in the Longmynd beds beneath Barrande's so-called primordial zone. The presence of phosphatic nodules and bituminous matter in some of the lowest azoic rocks, probably indicates the former existence of life at these periods. But the difficulty of understanding the absence of vast piles of fossiliferous strata, which on my theory no doubt were somewhere accumulated before the Silurian epoch, is very great. If these most ancient beds had been wholly worn away by denudation, or obliterated by metamorphic action, we ought to find only small remnants of the formations next succeeding them in age, and these ought to be very generally in a metamorphosed condition. But the descriptions which we now possess of the Silurian deposits over immense territories in Russia and in North America, do not support the view, that the older a formation is the more it has always suffered the extremity of derangement and metamorphism.

The case at present must remain inexplicable; and may be truly urged as a valid argument against the views here entertained. To show that it may hereafter receive some explanation, I will give the following hypothesis. From the nature of the organic remains which do not appear to have inhabited profound depths, in the several formations of Europe and of the United States; and from the amount of sediment, miles in thickness, of which the formations are composed, we may infer that from first to last large islands or tracts of land, whence the sediment was derived, occurred in the neighbourhood of the existing continents of Europe and North America. But we do not know what was the state of things in the intervals between the successive formations; whether Europe and the United States during these intervals existed as dry land, or as a submarine surface near land, on which sediment was not deposited, or as the bed of an open and unfathomable sea.

Looking to the existing oceans, which are thrice as extensive as the land, we see them studded with many islands; but not one oceanic island is as yet known to afford even a remnant of any palæozoic or secondary formation. Hence we may perhaps infer, that during the palæozoic and secondary periods, neither continents nor continental islands existed where our oceans now extend; for had they existed there, palæozoic and secondary formations would in all probability have been accumulated from sediment derived from their wear and tear; and would have been at least partially upheaved by the oscillations of level, which we may fairly conclude must have intervened during these enormously long periods. If then we may infer anything from these facts, we may infer that where our oceans now extend, oceans have extended from the remotest period of which we have any record; and on the other hand, that where continents now exist, large tracts of land have existed, subjected no doubt to great oscillations of level, since the earliest Silurian period. The coloured map appended to my volume on Coral Reefs, led me to conclude that the great oceans are still mainly areas of subsidence, the great archipelagoes still areas of oscillations of level, and the continents areas of elevation. But have we any right to assume that things have thus remained from the beginning of this world? Our continents seem to have been formed by a preponderance, during many oscillations of level, of the force of elevation; but may not the areas of preponderant movement have changed in the lapse of ages? At a period immeasurably antecedent to the Silurian epoch, continents may have existed where oceans are now spread out; and clear and open oceans may have existed where our continents now stand. Nor should we be justified in assuming that if, for instance, the bed of the Pacific Ocean were now converted into a continent, we should there find formations older than the Silurian strata, supposing such to have been formerly deposited; for it might well happen that strata which had subsided some miles nearer to the centre of the earth, and which had been pressed on by an enormous weight of superincumbent water, might have undergone far more metamorphic action than strata which have always remained nearer to the surface. The immense areas in some parts of the world, for instance in North America, of bare metamorphic rocks, which must have been heated under great pressure, have always seemed to me to require some special explanation; and we may perhaps believe that we see in these large areas the many formations long anterior to the Silurian epoch in a completely metamorphosed but likewise denuded condition.

The several difficulties here discussed, namely—that though we find in our geological formations many links between the species which now exist and have existed, we do not find infinitely numerous fine transitional forms closely joining them all together;—the sudden manner in which several whole groups of species first appeared in our European formations;—the almost entire absence, as at present known, of fossiliferous formations beneath the Silurian strata,—are all undoubtedly of the most serious nature. We see this in the fact that the most eminent palæontologists, namely Cuvier, Agassiz, Barrande, Pictet, Falconer, E. Forbes, &c., and all our greatest geologists, as Lyell, Murchison, Sedgwick, &c., have unanimously, often vehemently, maintained the immutability of species. But I have reason to believe that one great authority, Sir Charles Lyell, from further reflection entertains grave doubts on this subject. I feel how rash it is to differ from these authorities, to whom, with others, we owe all our knowledge. Those who think the natural geological record in any degree perfect, and who do not attach much weight to the facts and arguments of other kinds given in this volume, will undoubtedly at once reject my theory. For my part, following out Lyell's metaphor, I look at the natural geological record, as a history of the world imperfectly kept, and written in a changing dialect; of this history we possess the last volume alone, relating only to two or three countries. Of this volume, only here and there a short chapter has been preserved; and of each page, only here and there a few lines. Each word of the slowly-changing language, in which the history is written, being more or less different in the successive chapters, may represent the apparently abruptly changed forms of life, entombed in our consecutive, but widely separated, formations. On this view, the difficulties above discussed are greatly diminished, or even disappear.

CHAPTER X.

ON THE GEOLOGICAL SUCCESSION OF ORGANIC BEINGS.

On the slow and successive appearance of new species — On their
different rates of change — Species once lost do not reappear —
Groups of species follow the same general rules in their appearance and disappearance as do single species — On Extinction —
On simultaneous changes in the forms of life throughout the
world — On the affinities of extinct species to each other and to
living species — On the state of development of ancient forms —.
On the succession of the same types within the same areas —
Summary of preceding and present chapter.

LET us now see whether the several facts and rules
relating to the geological succession of organic beings,
better accord with the common view of the immutability
of species, or with that of their slow and gradual modification, through descent and natural selection.

New species have appeared very slowly, one after
another, both on the land and in the waters. Lyell has
shown that it is hardly possible to resist the evidence
on this head in the case of the several tertiary stages ;
and every year tends to fill up the blanks between them,
and to make the percentage system of lost and new
forms more gradual. In some of the most recent beds,
though undoubtedly of high antiquity if measured by
years, only one or two species are lost forms, and only
one or two are new forms, having here appeared for
the first time, either locally, or, as far as we know, on
the face of the earth. If we may trust the observations
of Philippi in Sicily, the successive changes in the marine
inhabitants of that island have been many and most
gradual. The secondary formations are more broken ;
but, as Bronn has remarked, neither the appearance

the disappearance of their many now extinct species has been almost universal in each separate formation.

Species of different genera and classes have not changed at the same rate or in the same degree. In the older tertiary beds a few living shells may still be found in the midst of a multitude of extinct forms. Falconer has given a striking instance of a similar fact, in an existing crocodile associated with many strange lost mammals and reptiles in the sub-Himalayan deposits. The Silurian Lingula differs but little from the living species of this genus; whereas most of the other Silurian Molluscs and all the Crustaceans have changed greatly. The productions of the land seem to change at a quicker rate than those of the sea, of which a striking instance has lately been observed in Switzerland. There is some reason to believe that organisms, considered high in the scale of nature, change more quickly than those that are low: though there are exceptions to this rule. The amount of organic change, as Pictet has remarked, does not strictly correspond with the succession of our geological formations; so that between each two consecutive formations, the forms of life have seldom changed in exactly the same degree. Yet if we compare any but the most closely related formations, all the species will be found to have undergone some change. When a species has once disappeared from the face of the earth, we have no reason to believe that the same identical form ever reappears. The strongest apparent exception to this latter rule, is that of the so-called "colonies" of M. Barrande, which intrude for a period in the midst of an older formation, and then allow the pre-existing fauna to reappear: but Lyell's explanation, namely, that it is a case of temporary migration from a distinct geographical province, seems to me satisfactory.

These several facts accord well with my theory. I believe in no fixed law of development, causing all the inhabitants of a country to change abruptly, or simultaneously, or to an equal degree. The process of modification must be extremely slow. The variability of each species is quite independent of that of all others. Whether such variability be taken advantage of by natural selection, and whether the variations be accumulated to a greater or lesser amount, thus causing a greater or lesser amount of modification in the varying species, depends on many complex contingencies, —on the variability being of a beneficial nature, on the power of intercrossing, on the rate of breeding, on the slowly changing physical conditions of the country, and more especially on the nature of the other inhabitants with which the varying species comes into competition. Hence it is by no means surprising that one species should retain the same identical form much longer than others ; or, if changing, that it should change less. We see the same fact in geographical distribution ; for instance, in the land-shells and coleopterous insects of Madeira having come to differ considerably from their nearest allies on the continent of Europe, whereas the marine shells and birds have remained unaltered. We can perhaps understand the apparently quicker rate of change in terrestrial and in more highly organised productions compared with marine and lower productions, by the more complex relations of the higher beings to their organic and inorganic conditions of life, as explained in a former chapter. When many of the inhabitants of a country have become modified and improved, we can understand, on the principle of competition, and on that of the many all-important relations of organism to organism, that any form which does not become in some degree modified and improved, will be liable to be exterminated. Hence we can see why all the species in the same region do at last, if we look to wide enough intervals of time, become modified, for those which do not change will become extinct.

In members of the same class the average amount of change, during long and equal periods of time, may, perhaps, be nearly the same; but as the accumulation of long-enduring fossiliferous formations depends on great masses of sediment having been deposited on areas whilst subsiding, our formations have been almost necessarily accumulated at wide and irregularly intermittent intervals; consequently the amount of organic change exhibited by the fossils embedded in consecutive formations is not equal. Each formation, on this view, does not mark a new and complete act of creation, but only an occasional scene, taken almost at hazard, in a slowly changing drama.

We can clearly understand why a species when once lost should never reappear, even if the very same conditions of life, organic and inorganic, should recur. For though the offspring of one species might be adapted (and no doubt this has occurred in innumerable instances) to fill the exact place of another species in the economy of nature, and thus supplant it; yet the two forms—the old and the new—would not be identically the same; for both would almost certainly inherit different characters from their distinct progenitors. For instance, it is just possible, if our fantail-pigeons were all destroyed, that fanciers, by striving during long ages for the same object, might make a new breed hardly distinguishable from our present fantail; but if the parent rock-pigeon were also destroyed, and in nature we have every reason to believe that the parent-form will generally be supplanted and exterminated by its improved offspring, it is quite in-

credible that a fantail, identical with the existing breed,
could be raised from any other species of pigeon, or
even from the other well-established races of the domestic pigeon, for the newly-formed fantail would be
almost sure to inherit from its new progenitor some
slight characteristic differences.

Groups of species, that is, genera and families, follow
the same general rules in their appearance and disappearance as do single species, changing more or less
quickly, and in a greater or lesser degree. A group
does not reappear after it has once disappeared; or
its existence, as long as it lasts, is continuous. I am
aware that there are some apparent exceptions to this
rule, but the exceptions are surprisingly few, so few
that E. Forbes, Pictet, and Woodward (though all
strongly opposed to such views as I maintain) admit
its truth; and the rule strictly accords with my theory.
For as all the species of the same group have descended
from some one species, it is clear that as long as any
species of the group have appeared in the long succession
of ages, so long must its members have continuously
existed, in order to have generated either new and
modified or the same old and unmodified forms. Species
of the genus Lingula, for instance, must have continuously existed by an unbroken succession of generations,
from the lowest Silurian stratum to the present day.

We have seen in the last chapter that the species
of a group sometimes falsely appear to have come in
abruptly in a body; and I have attempted to give an
explanation of this fact, which if true would have been
fatal to my views. But such cases are certainly exceptional; the general rule being a gradual increase
in number, till the group reaches its maximum, and
then, sooner or later, it gradually decreases. If the
number of the species of a genus, or the number of

the genera of a family, be represented by a vertical line of varying thickness, according through the successive geological formations in which the species are found, the line will sometimes falsely appear to begin at its lower end, not in a sharp point, but abruptly; it then gradually thickens upwards, often keeping for a space of equal thickness, and ultimately thins out in the upper beds, marking the decrease and final extinction of the species. This gradual increase in number of the species of a group is strictly conformable with my theory; as the species of the same genus, and the genera of the same family, can increase only slowly and progressively; for the process of modification and the production of a number of allied forms must be slow and gradual,—one species giving rise first to two or three varieties, these being slowly converted into species, which in their turn produce by equally slow steps other species, and so on, like the branching of a great tree from a single stem, till the group becomes large.

On Extinction.—We have as yet spoken only incidentally of the disappearance of species and of groups of species. On the theory of natural selection the extinction of old forms and the production of new and improved forms are intimately connected together. The old notion of all the inhabitants of the earth having been swept away at successive periods by catastrophes, is very generally given up, even by those geologists, as Elie de Beaumont, Murchison, Barrande, &c., whose general views would naturally lead them to this conclusion. On the contrary, we have every reason to believe, from the study of the tertiary formations, that species and groups of species gradually disappear, one after another, first from one spot, then from another, and finally from the world; in some few cases, however, as by the breaking of an isthmus and the consequent irruption of a multitude of new inhabitants, or by the final subsidence of an island, the extinction may have been comparatively rapid. Both single species and whole groups of species last for very unequal periods; some groups, as we have seen, having endured from the earliest known dawn of life to the present day; some having disappeared before the close of the palæozoic period. No fixed law seems to determine the length of time during which any single species or any single genus endures. There is reason to believe that the complete extinction of the species of a group is generally a slower process than their production: if the appearance and disappearance of a group of species be represented, as before, by a vertical line of varying thickness, the line is found to taper more gradually at its upper end, which marks the progress of extermination, than at its lower end, which marks the first appearance and increase in numbers of the species. In some cases, however, the extermination of whole groups of beings, as of ammonites towards the close of the secondary period, has been wonderfully sudden relatively to that of most other groups.

The whole subject of the extinction of species has been involved in the most gratuitous mystery. Some authors have even supposed that as the individual has a definite length of life, so have species a definite duration. No one I think can have marvelled more at the extinction of species, than I have done. When I found in La Plata the tooth of a horse embedded with the remains of Mastodon, Megatherium, Toxodon, and other extinct monsters, which all co-existed with still living shells at a very late geological period, I was filled with astonishment; for seeing that the horse, since its introduction by the Spaniards into South America, has run

wild over the whole country and has increased in
numbers at an unparalleled rate, I asked myself what
could so recently have exterminated the former horse
under conditions of life apparently so favourable. But
how utterly groundless was my astonishment! Professor Owen soon perceived that the tooth, though so
like that of the existing horse, belonged to an extinct
species. Had this horse been still living, but in some
degree rare, no naturalist would have felt the least surprise at its rarity; for rarity is the attribute of a vast
number of species of all classes, in all countries. If
we ask ourselves why this or that species is rare, we
answer that something is unfavourable in its conditions
of life; but what that something is, we can hardly ever
tell. On the supposition of the fossil horse still existing as a rare species, we might have felt certain from
the analogy of all other mammals, even of the slow-
breeding elephant, and from the history of the naturalisation of the domestic horse in South America, that
under more favourable conditions it would in a very few
years have stocked the whole continent. But we could
not have told what the unfavourable conditions were
which checked its increase, whether some one or
several contingencies, and at what period of the horse's
life, and in what degree, they severally acted. If
the conditions had gone on, however slowly, becoming less and less favourable, we assuredly should not
have perceived the fact, yet the fossil horse would certainly have become rarer and rarer, and finally extinct;
—its place being seized on by some more successful
competitor.

It is most difficult always to remember that the
increase of every living being is constantly being
checked by unperceived injurious agencies; and that
these same unperceived agencies are amply sufficient to

cause rarity, and finally extinction. So little is this subject understood, that I have heard surprise repeatedly expressed at such great monsters as the Mastodon and the more ancient Dinosaurians having become extinct; as if mere bodily strength gave victory in the battle of life. Mere size, on the contrary, would in some cases determine quicker extermination from the greater amount of requisite food. Before man inhabited India or Africa, some cause must have checked the continued increase of the existing elephant. A highly capable judge believes that at the present day insects (as Bruce has likewise described in Abyssinia), from incessantly harassing and weakening the elephant, are one chief check to its increase. It is certain that insects of different kinds, and blood-sucking bats, determine the existence of the larger naturalised quadrupeds in several parts of S. America.

We see in many cases in the more recent tertiary formations, that rarity precedes extinction; and we know that this has been the progress of events with those animals which have been exterminated, either locally or wholly, through man's agency. I may repeat what I published in 1845, namely, that to admit that species generally become rare before they become extinct—to feel no surprise at the rarity of a species, and yet to marvel greatly when it ceases to exist, is much the same as to admit that sickness in the individual is the forerunner of death—to feel no surprise at sickness, but when the sick man dies, to wonder and to suspect that he died by some unknown deed of violence.

The theory of natural selection is grounded on the belief that each new variety, and ultimately each new species, is produced and maintained by having some advantage over those with which it comes into competion; and the consequent extinction of less-favoured forms almost inevitably follows. It is the same with our domestic productions: when a new and slightly improved variety has been raised, it at first supplants the less improved varieties in the same neighbourhood; when much improved it is transported far and near, like our short-horn cattle, and takes the place of other breeds in other countries. Thus the appearance of new forms and the disappearance of old forms, both natural and artificial, are bound together. In certain flourishing groups, the number of new specific forms which have been produced within a given time is probably greater than that of the old specific forms which have been exterminated; but we know that the number of species has not gone on indefinitely increasing, at least during the later geological periods, so that looking to later times we may believe that the production of new forms has caused the extinction of about the same number of old forms.

The competition will generally be most severe, as formerly explained and illustrated by examples, between the forms which are most like each other in all respects. Hence the improved and modified descendants of a species will generally cause the extermination of the parent-species; and if many new forms have been developed from any one species, the nearest allies of that species, i. e. the species of the same genus, will be the most liable to extermination. Thus, as I believe, a number of new species descended from one species, that is a new genus, comes to supplant an old genus, belonging to the same family. But it must often have happened that a new species belonging to some one group will have seized on the place occupied by a species belonging to a distinct group, and thus cause its extermination; and if many allied forms be developed from the success-

ful inhabitants many will have to yield their places; and
it will generally be allied forms, which will suffer
from some inherited inferiority in common. But whether it be species belonging to the same or to a
distinct class, which yield their places to other species
which have been modified and improved, a few of the
sufferers may often long be preserved, from being
fitted to some peculiar line of life, or from inhabiting
some distant and isolated station, where they have
escaped severe competition. For instance, a single
species of Trigonia, a great genus of shells in the
secondary formations, survives in the Australian seas;
and a few members of the great and almost extinct
group of Ganoid fishes still inhabit our fresh waters.
Therefore the utter extinction of a group is generally, as we have seen, a slower process than its production.

With respect to the apparently sudden extermination
of whole families or orders, as of Trilobites at the close
of the palæozoic period and of Ammonites at the close
of the secondary period, we must remember what has
been already said on the probable wide intervals of time
between our consecutive formations; and in these intervals there may have been much slow extermination.
Moreover, when by sudden immigration or by unusually
rapid development, many species of a new group have
taken possession of a new area, they will have exterminated in a correspondingly rapid manner many of the
old inhabitants; and the forms which thus yield their
places will commonly be allied, for they will partake of
some inferiority in common.

Thus, as it seems to me, the manner in which single
species and whole groups of species become extinct,
accords well with the theory of natural selection. We
need not marvel at extinction; if we must marvel, let

it be our presumption in imagining for a moment that we understand the many complex contingencies, on which the existence of each species depends. If we forget for an instant, that each species tends to increase inordinately, and that some check is always in action, yet seldom perceived by us, the whole economy of nature will be utterly obscured. Whenever we can precisely say why this species is more abundant in individuals than that; why this species and not another can be naturalised in a given country; then, and not till then, we may justly feel surprise why we cannot account for the extinction of this particular species or group of species.

On the Forms of Life changing almost simultaneously throughout the World.—Scarcely any palæontological discovery is more striking than the fact that the forms of life change almost simultaneously throughout the world. Thus our European Chalk formation can be recognised in many distant parts of the world, under the most different climates, where not a fragment of the mineral chalk itself can be found; namely, in North America, in equatorial South America, in Tierra del Fuego, at the Cape of Good Hope, and in the peninsula of India. For at these distant points, the organic remains in certain beds present an unmistakeable degree of resemblance to those of the Chalk. It is not that the same species are met with; for in some cases not one species is identically the same, but they belong to the same families, genera, and sections of genera, and sometimes are similarly characterised in such trifling points as mere superficial sculpture. Moreover other forms, which are not found in the Chalk of Europe, but which occur in the formations either above or below, are similarly absent at these distant points of the world. In

the several successive palæozoic formations of Russia,
Western Europe and North America, a similar parallelism in the forms of life has been observed by several
authors: so it is, according to Lyell, with the several
European and North American tertiary deposits. Even
if the few fossil species which are common to the Old
and New Worlds be kept wholly out of view, the general parallelism in the successive forms of life, in the
stages of the palæozoic and tertiary periods, would still
be manifest, and the several formations could be easily
correlated.

These observations, however, relate to the marine
inhabitants of distant parts of the world: we have not
sufficient data to judge whether the productions of the
land and of fresh water change at distant points in the
same parallel manner. We may doubt whether they
have thus changed: if the Megatherium, Mylodon,
Macrauchenia, and Toxodon had been brought to Europe
from La Plata, without any information in regard to
their geological position, no one would have suspected
that they had coexisted with still living sea-shells; but
as these anomalous monsters coexisted with the Mastodon and Horse, it might at least have been inferred
that they had lived during one of the later tertiary
stages.

When the marine forms of life are spoken of as
having changed simultaneously throughout the world,
it must not be supposed that this expression relates to
the same thousandth or hundred-thousandth year, or
even that it has a very strict geological sense; for if
all the marine animals which live at the present day in
Europe, and all those that lived in Europe during the
pleistocene period (an enormously remote period as
measured by years, including the whole glacial epoch),
were to be compared with those now living in South

America or in Australia, the most skilful naturalist
would hardly be able to say whether the existing or the
pleistocene inhabitants of Europe resembled most closely
those of the southern hemisphere. So, again, several
highly competent observers believe that the existing
productions of the United States are more closely related
to those which lived in Europe during certain later tertiary stages, than to those which now live here; and
if this be so, it is evident that fossiliferous beds deposited at the present day on the shores of North
America would hereafter be liable to be classed with
somewhat older European beds. Nevertheless, looking
to a remotely future epoch, there can, I think, be little
doubt that all the more modern *marine* formations,
namely, the upper pliocene, the pleistocene and strictly
modern beds, of Europe, North and South America, and
Australia, from containing fossil remains in some degree
allied, and from not including those forms which are
only found in the older underlying deposits, would be
correctly ranked as simultaneous in a geological sense.

The fact of the forms of life changing simultaneously,
in the above large sense, at distant parts of the world,
has greatly struck those admirable observers, MM.
de Verneuil and d'Archiac. After referring to the
parallelism of the palæozoic forms of life in various
parts of Europe, they add, "If struck by this strange
sequence, we turn our attention to North America, and
there discover a series of analogous phenomena, it will
appear certain that all these modifications of species,
their extinction, and the introduction of new ones, cannot be owing to mere changes in marine currents or
other causes more or less local and temporary, but depend on general laws which govern the whole animal
kingdom." M. Barrande has made forcible remarks to
precisely the same effect. It is, indeed, quite futile to

look to changes of currents, climate, or other physical
conditions, as the cause of these great mutations in the
forms of life throughout the world, under the most different climates. We must, as Barrande has remarked,
look to some special law. We shall see this more clearly
when we treat of the present distribution of organic
beings, and find how slight is the relation between the
physical conditions of various countries, and the nature
of their inhabitants.

The great fact of the parallel succession of the forms
of life throughout the world, is explicable on the theory
of natural selection. New species are formed by new
varieties arising, which have some advantage over
older forms; and those forms, which are already dominant, or have some advantage over the other forms in
their own country, would naturally oftenest give rise to
new varieties or incipient species; for these latter must
be victorious in a still higher degree in order to be preserved and to survive. We have distinct evidence on
this head, in the plants which are dominant, that is,
which are commonest and most widely diffused, compared with other plants within their own homes, having
produced the greatest number of new varieties. It is
also natural that the dominant, varying, and far-spread-
ing species, which already have invaded to a certain
extent the territories of other species, should be those
which would have the best chance of spreading still
further, and of giving rise in new countries to new
varieties and species. The process of diffusion may
often be very slow, being dependent on climatal and
geographical changes, or on strange accidents, or on
the gradual acclimatisation of new species to the various
climates through which they must pass, but in the long
run the dominant forms will generally succeed in
spreading. The diffusion would, it is probable, be

 klowr with the terrestrial inhabitants of distinct continents than with the marine inhabitants of the continuous
sea. We might therefore expect to find, as we apparently do find, a less strict degree of parallel succession
in the productions of the land than of the sea.

Dominant species spreading from any region might
encounter still more dominant species, and then their
triumphant course, or even their existence, would cease.
We know not at all precisely what are all the conditions
most favourable for the multiplication of new and dominant species; but we can, I think, clearly see that a
number of individuals, from giving a better chance of
the appearance of favourable variations, and that severe
competition with many already existing forms, would be
highly favourable, as would be the power of spreading
into new territories. A certain amount of isolation,
recurring at long intervals of time, would probably be
also favourable, as before explained. One quarter of
the world may have been most favourable for the production of new and dominant species on the land, and
another for those in the waters of the sea. If two great
regions had been for a long period favourably circumstanced in an equal degree, whenever their inhabitants
met, the battle would be prolonged and severe; and
some from one birthplace and some from the other
might be victorious. But in the course of time, the
forms dominant in the highest degree, wherever produced, would tend everywhere to prevail. As they
prevailed, they would cause the extinction of other and
inferior forms; and as these inferior forms would be
allied in groups by inheritance, whole groups would
tend slowly to disappear; though here and there a
single member might long be enabled to survive.

Thus, as it seems to me, the parallel, and, taken in a

large areas, simultaneous, succession of the same forms of life throughout the world, accords well with the principle of new species having been formed by dominant species spreading widely and varying; the new species thus produced being themselves dominant owing to inheritance, and to having already had some advantage over their parents or over other species; these again spreading, varying, and producing new species. The forms which are beaten and which yield their places to the new and victorious forms, will generally be allied in groups, from inheriting some inferiority in common; and therefore, as new and improved groups spread throughout the world, old groups will disappear from the world; and the succession of forms in both ways will everywhere tend to correspond.

There is one other remark connected with this subject worth making. I have given my reasons for believing that most of our greater formations, rich in fossils, were deposited during periods of subsidence; and that blank intervals of vast duration occurred during the periods when the bed of the sea was either stationary or rising, and likewise when sediment was not thrown down quickly enough to embed and preserve organic remains. During these long and blank intervals I suppose that the inhabitants of each region underwent a considerable amount of modification and extinction, and that there was much migration from other parts of the world. As we have reason to believe that large areas are affected by the same movement, it is probable that strictly contemporaneous formations have often been accumulated over very wide spaces in the same quarter of the world; but we are far from having any right to conclude that this has invariably been the case, and that large areas have invarably have been affected by the same movements. When two formations have been deposited in two regions during nearly, but not exactly the same period, we should find in both, from the causes explained in the foregoing paragraphs, the same general succession in the forms of life; but the species would not exactly correspond; for there will have been a little more time in the one region than in the other for modification, extinction, and immigration.

I suspect that cases of this nature occur in Europe. Mr. Prestwich, in his admirable Memoirs on the eocene deposits of England and France, is able to draw a close general parallelism between the successive stages in the two countries; but when he compares certain stages in England with those in France, although he finds in both a curious accordance in the numbers of the species belonging to the same genera, yet the species themselves differ in a manner very difficult to account for, considering the proximity of the two areas,—unless, indeed, it be assumed that an isthmus separated two seas inhabited by distinct, but contemporaneous, faunas. Lyell has made similar observations on some of the later tertiary formations. Barrande, also, shows that there is a striking general parallelism in the successive Silurian deposits of Bohemia and Scandinavia; nevertheless he finds a surprising amount of difference in the species. If the several formations in these regions have not been deposited during the same exact periods,—a formation in one region often corresponding with a blank interval in the other, and if in both regions the species have gone on slowly changing during the accumulation of the several formations and during the long intervals of time between them; in this case, the several formations in the two regions could be arranged in the same order, in accordance with the general succession of the forms of life, and the order would falsely appear to be strictly parallel; nevertheless the species would not be all the same in the apparently corresponding stages in the two regions.

On the Affinities of extinct Species to each other, and to living forms.—Let us now look to the mutual affinities of extinct and living species. They all fall into one grand natural system, and this fact is at once explained on the principle of descent. The more ancient any form is, the more, as a general rule, it differs from living forms. But, as Buckland long ago remarked, all fossils can be classed either in still existing groups, or between them. That the extinct forms of life help to fill up the wide intervals between existing genera, families, and orders, cannot be disputed. For if we confine our attention either to the living or to the extinct alone, the series is far less perfect than if we combine both into one general system. With respect to the Vertebrata, whole pages could be filled with striking illustrations from our great palæontologist, Owen, showing how extinct animals fall in between existing groups. Cuvier ranked the Ruminants and Pachyderms, as the two most distinct orders of mammals; but Owen has discovered so many fossil links, that he has had to alter the whole classification of these two orders; and has placed certain pachyderms in the same sub-order with ruminants: for example, he dissolves by fine gradations the apparently wide difference between the pig and the camel. In regard to the Invertebrata, Barrande, and a higher authority could not be named, asserts that he is every day taught that Palæozoic animals, though belonging to the same orders, families, or genera with those living at the present day, were not at this early epoch limited in such distinct groups as they now are.

Some writers have objected to any extinct species or group of species being considered as intermediate between living species or groups. If by this term it is meant that an extinct form is directly intermediate in all its characters between two living forms, the objection is valid. But I apprehend that in a perfectly natural classification many fossil species would have to stand between living species, and some extinct genera between living genera, even between genera belonging to distinct families. The most common case, especially with respect to very distinct groups, such as fish and reptiles, seems to be, that supposing them to be distinguished at the present day from each other by a dozen characters, the ancient members of the same two groups would be distinguished by a somewhat lesser number of characters, so that the two groups, though formerly quite distinct, at that period made some small approach to each other.

It is a common belief that the more ancient a form is, by so much the more it tends to connect by some of its characters groups now widely separated from each other. This remark no doubt must be restricted to those groups which have undergone much change in the course of geological ages; and it would be difficult to prove the truth of the proposition, for every now and then even a living animal, as the Lepidosiren, is discovered having affinities directed towards very distinct groups. Yet if we compare the older Reptiles and Batrachians, the older Fish, the older Cephalopods, and the eocene Mammals, with the more recent members of the same classes, we must admit that there is some truth in the remark.

Let us see how far these several facts and inferences accord with the theory of descent with modification. As the subject is somewhat complex, I must request the reader to turn to the diagram in the fourth chapter. We may suppose that the numbered letters represent genera, and the dotted lines diverging from them the species in each genus. The diagram is much too simple, too few genera and too few species being given, but this is unimportant for us. The horizontal lines may represent successive geological formations, and all the forms beneath the uppermost line may be considered as extinct. The three existing genera a^{14}, q^{14}, p^{14}, will form a small family; b^{14} and f^{14} a closely allied family or sub-family; and o^{14}, e^{14}, m^{14}, a third family. These three families, together with the many extinct genera on the several lines of descent diverging from the parent-form (A), will form an order; for all will have inherited something in common from their ancient and common progenitor. On the principle of the continued tendency to divergence of character, which was formerly illustrated by this diagram, the more recent any form is, the more it will generally differ from its ancient progenitor. Hence we can understand the rule that the most ancient fossils differ most from existing forms. We must not, however, assume that divergence of character is a necessary contingency; it depends solely on the descendants from a species being thus enabled to seize on many and different places in the economy of nature. Therefore it is quite possible, as we have seen in the case of some Silurian forms, that a species might go on being slightly modified in relation to its slightly altered conditions of life, and yet retain throughout a vast period the same generic characteristics. This is represented in the diagram by the letter F^{14}.

All the many forms, extinct and recent, descended from (A), make, as before remarked, one order; and this order, from the continued effects of extinction and

divergence of character, has become divided into several
sub-families and families, some of which are supposed
to have perished at different periods, and some to have
endured to the present day.

By looking at the diagram we can see that if many
of the extinct forms, supposed to be embedded in the
successive formations, were discovered at several points
low down in the series, the three existing families on the
uppermost line would be rendered less distinct from each
other. If, for instance, the genera a^1, a^5, a^{10}, f^8, m^3, m^6, m^9,
were disinterred, these three families would be so closely
linked together that they probably would have to be
united into one great family, in nearly the same manner
as has occurred with ruminants and certain pachyderms.
Yet he who objected to call the extinct genera, which
thus linked the living genera of three families together,
intermediate in character, would be justified, as they
are intermediate, not directly, but only by a long and
circuitous course through many widely different forms.
If many extinct forms were to be discovered above one
of the middle horizontal lines or geological formations
—for instance, above No. VI.—but none from beneath
this line, then only the two families on the left hand
(namely, a^{14}, &c., and b^{14}, &c.) would have to be united
into one family; and the two other families (namely,
a^{14} to f^{14} now including five genera, and o^{14} to m^{14}) would
yet remain distinct. These two families, however, would
be less distinct from each other than they were before the
discovery of the fossils. If, for instance, we suppose the
existing genera of the two families to differ from each
other by a dozen characters, in this case the genera, at
the early period marked VI., would differ by a lesser
number of characters; for at this early stage of descent
they have not diverged in character from the common
progenitor of the order, nearly so much as they did

sequently diverged. Thus it comes that ancient and
extinct genera are often in some slight degree intermediate in character between their modified descendants,
or between their collateral relations.

In nature the case will be far more complicated than
is represented in the diagram; for the groups will have
been more numerous, they will have endured for extremely unequal lengths of time, and will have been
modified in various degrees. As we possess only the last
volume of the geological record, and that in a very broken
condition, we have no right to expect, except in very
rare cases, to fill up wide intervals in the natural system,
and thus unite distinct families or orders. All that
we have a right to expect, is that those groups, which
have within known geological periods undergone much
modification, should in the older formations make some
slight approach to each other; so that the older members should differ less from each other in some of their
characters than do the existing members of the same
groups; and this by the concurrent evidence of our best
palæontologists seems frequently to be the case.

Thus, on the theory of descent with modification, the
main facts with respect to the mutual affinities of the
extinct forms of life to each other and to living forms,
seem to me explained in a satisfactory manner. And
they are wholly inexplicable on any other view.

On this same theory, it is evident that the fauna of
any great period in the earth's history will be intermediate in general character between that which pre-
ceded and that which succeeded it. Thus, the species
which lived at the sixth great stage of descent in the
diagram are the modified offspring of those which lived
at the fifth stage, and are the parents of those which
became still more modified at the seventh stage; hence
they could hardly fail to be nearly intermediate in

character between the forms of life above and below.
We must, however, allow for the entire extinction of
some preceding forms, and in any one region for the
immigration of new forms from other regions, and for a
large amount of modification, during the long and blank
intervals between the successive formations. Subject to
these allowances, the fauna of each geological period
undoubtedly is intermediate in character, between the
preceding and succeeding faunas. I need give only one
instance, namely, the manner in which the fossils of the
Devonian system, when this system was first discovered,
were at once recognised by palæontologists as intermediate in character between those of the overlying
carboniferous, and underlying Silurian system. But
each fauna is not necessarily exactly intermediate, as
unequal intervals of time have elapsed between consecutive formations.

It is no real objection to the truth of the statement,
that the fauna of each period as a whole is nearly intermediate in character between the preceding and
succeeding faunas, that certain genera offer exceptions
to the rule. For instance, mastodons and elephants,
when arranged by Dr. Falconer in two series, first
according to their mutual affinities and then according
to their periods of existence, do not accord in arrangement. The species extreme in character are not the
oldest, or the most recent; nor are those which are
intermediate in character, intermediate in age. But
supposing for an instant, in this and other such cases,
that the record of the first appearance and disappearance
of the species was perfect, we have no reason to believe
that forms successively produced necessarily endure for
corresponding lengths of time: a very ancient form
might occasionally last much longer than a form elsewhere subsequently produced, especially in the case of

R

terrestrial productions inhabiting separated districts.
To compare small things with great : if the principal
living and extinct races of the domestic pigeon were
arranged as well as they could be in serial affinity, this
arrangement would not closely accord with the order in
time of their production, and still less with the order of
their disappearance; for the parent rock-pigeon now
lives; and many varieties between the rock-pigeon and
the carrier have become extinct; and carriers which are
extreme in the important character of length of beak
originated earlier than short-beaked tumblers, which are
at the opposite end of the series in this respect.

Closely connected with the statement, that the organic remains from an intermediate formation are in
some degree intermediate in character, is the fact,
insisted on by all palæontologists, that fossils from two
consecutive formations are far more closely related to
each other, than are the fossils from two remote formations. Pictet gives as a well-known instance, the
general resemblance of the organic remains from the
several stages of the Chalk formation, though the species
are distinct in each stage. This fact alone, from its
generality, seems to have shaken Professor Pictet in his
firm belief in the immutability of species. He who is
acquainted with the distribution of existing species over
the globe, will not attempt to account for the close
resemblance of the distinct species in closely consecutive
formations, by the physical conditions of the ancient
areas having remained nearly the same. Let it be
remembered that the forms of life, at least those inhabiting the sea, have changed almost simultaneously
throughout the world, and therefore under the most
different climates and conditions. Consider the prodigious vicissitudes of climate during the pleistocene
period, which includes the whole glacial period, and

note how little the specific forms of the inhabitants of the sea have been affected.

On the theory of descent, the full meaning of the fact of fossil remains from closely consecutive formations, though ranked as distinct species, being closely related, is obvious. As the accumulation of each formation has often been interrupted, and as long blank intervals have intervened between successive formations, we ought not to expect to find, as I attempted to show in the last chapter, in any one or two formations all the intermediate varieties between the species which appeared at the commencement and close of these periods; but we ought to find after intervals, very long as measured by years, but only moderately long as measured geologically, closely allied forms, or, as they have been called by some authors, representative species; and these we assuredly do find. We find, in short, such evidence of the slow and scarcely sensible mutation of specific forms, as we have a just right to expect to find.

On the State of Development of Ancient compared with Living Forms. — We have seen in the fourth chapter that the degree of differentiation and specialisation of the parts of all organic beings, when come to maturity, is the best standard, as yet suggested, of their degree of perfection or highness. We have also seen that as the specialisation of parts and organs is an advantage to each being, so natural selection will constantly tend thus to render the organisation of each being more specialised and perfect, and in this sense higher; not but that it may and will leave many creatures with simple and unimproved structures fitted for simple conditions of life, and in some cases will even degrade or simplify the organisation, yet leaving such degraded beings better fitted for their new walks of life. In another and more

B 2

general manner we can see that on the theory of
natural selection the more recent forms will tend to be
higher than their progenitors; for each new species is
formed by having had some advantage in the struggle
for life over other and preceding forms. If under a
nearly similar climate the eocene inhabitants of one
quarter of the world were put into competition with the
existing inhabitants of the same or some other quarter,
the eocene fauna or flora would certainly be beaten
and exterminated, as would a secondary fauna by an
eocene, and a palæozoic fauna by a secondary fauna.
So that by this fundamental test of victory in the battle
for life, as well as by the standard of the specialisation
of organs, modern forms ought on the theory of natural
selection to stand higher than ancient forms. Is this
the case? A large majority of palæontologists would
certainly answer in the affirmative; but in my judgment I can, after having read the discussions on this
subject by Lyell, and Hooker's views in regard to
plants, concur only to a limited extent. Nevertheless
it may be anticipated that the evidence will be rendered
more decisive by future geological research.

The problem is in many ways excessively intricate.
The geological record, at all times imperfect, does not
extend far enough back, as I believe, to show with
unmistakeable clearness that within the known history
of the world organisation has largely advanced. Even
at the present day, looking to members of the same
class, naturalists are not unanimous which forms are
highest: thus, some look at the selaceans or sharks
from their approach in some important points of structure to reptiles as the highest fish; others look at the
teleosteans as the highest. The ganoids stand intermediate between the selaceans and teleosteans; the
latter at the present day are largely preponderant in

number; but formerly ganoids and placoids alone existed; and in this case, according to the standard of highness chosen, so will it be said that fishes have advanced or retrograded in organisation. To attempt to compare in the scale of highness members of distinct types seems hopeless: who will decide whether a cuttle-fish be higher than a bee—that insect which the great Von Baer believed to be " in fact more highly organised than a fish, although upon another type"? In the complex struggle for life it is quite credible that crustaceans, for instance, not very high in their own class, might beat the cephalopods or highest mollusks; and such crustaceans, though not highly developed, would stand very high in the scale of invertebrate animals if judged by the most decisive of all trials—the law of battle. Besides these inherent difficulties in deciding which forms are the most advanced in organisation, we ought not solely to compare the highest members of a class at any two distant periods—though undoubtedly this is one and perhaps the most important element in striking a balance—but we ought to compare all the members, high and low, at the two periods. At an ancient epoch the highest and lowest mollusks, namely, cephalopods and brachiopods, swarmed in numbers: at the present time both these orders have been greatly reduced, whereas other orders, intermediate in grade of organisation, have largely increased; consequently some naturalists have maintained that mollusks were formerly more highly developed than at present; but a stronger case can be made out on the other side, by considering the vast reduction at the present day of the lowest mollusks, more especially as the existing cephalopods, though so few in number, are more highly organised than their ancient representatives. We ought also to

consider the relative proportional numbers of the high
and low classes in the population of the world at the
two periods: if, for instance, at the present day there
be fifty thousand kinds of vertebrate animals, and if
we knew that at some former period only ten thousand
kinds had existed, we ought to look at this increase in
number of the highest class, which implies a great displacement of lower forms, as a decided advance in the
organisation of the world, whether the higher or the
lower vertebrata had thus largely increased. We can
thus see how hopelessly difficult it will apparently for
ever be to compare with perfect fairness, under such
extremely complex relations, the standard of organisation of the imperfectly-known faunas of successive
periods of the earth's history.

We shall appreciate under one important point of
view this difficulty the more clearly, by looking to the
case of certain existing faunas and floras. From the
extraordinary manner in which European productions
have recently spread over New Zealand, and have seized
on places which must have been previously occupied, we
may believe, if all the animals and plants of Great
Britain were set free in New Zealand, that in the course
of time a multitude of British forms would become
thoroughly naturalized there, and would exterminate
many of the natives. On the other hand, from what we
see now occurring in New Zealand, and from hardly a
single inhabitant of the southern hemisphere having
become wild in any part of Europe, we may doubt, if all
the productions of New Zealand were set free in Great
Britain, whether any considerable number would be
enabled to seize on places now occupied by our native
plants and animals. Under this point of view, the productions of Great Britain may be said to be higher than

those of New Zealand. Yet the most skilful naturalist from an examination of the species of the two countries could not have foreseen this result.

Agassiz insists that ancient animals resemble to a certain extent the embryos of recent animals of the same classes; or that the geological succession of extinct forms is in some degree parallel to the embryological development of recent forms. I must follow Pictet and Huxley in thinking that the truth of this doctrine is very far from proved. Yet I fully expect to see it hereafter confirmed, at least in regard to subordinate groups, which have branched off from each other within comparatively recent times. For this doctrine of Agassiz accords well with the theory of natural selection. In a future chapter I shall attempt to show that the adult differs from its embryo, owing to variations supervening at a not early age, and being inherited at a corresponding age. This process, whilst it leaves the embryo almost unaltered, continually adds, in the course of successive generations, more and more difference to the adult.

Thus the embryo comes to be left as a sort of picture, preserved by nature, of the ancient and less modified condition of each animal. This view may be true, and yet it may never be capable of full proof. Seeing, for instance, that the oldest known mammals, reptiles, and fish strictly belong to their own proper classes, though some of these old forms are in a slight degree less distinct from each other than are the typical members of the same groups at the present day, it would be vain to look for animals having the common embryological character of the Vertebrata, until beds far beneath the lowest Silurian strata are discovered—a discovery of which the chance is very small.

On the Succession of the same Types within the same areas, during the later tertiary periods.—Mr. Clift many years ago showed that the fossil mammals from the Australian caves were closely allied to the living marsupials of that continent. In South America, a similar relationship is manifest, even to an uneducated eye, in the gigantic pieces of armour like those of the armadillo, found in several parts of La Plata; and Professor Owen has shown in the most striking manner that most of the fossil mammals, buried there in such numbers, are related to South American types. This relationship is even more clearly seen in the wonderful collection of fossil bones made by MM. Lund and Clausen in the caves of Brazil. I was so much impressed with these facts that I strongly insisted, in 1839 and 1845, on this "law of the succession of types,"—on "this wonderful relationship in the same continent between the dead and the living." Professor Owen has subsequently extended the same generalisation to the mammals of the Old World. We see the same law in this author's restorations of the extinct and gigantic birds of New Zealand. We see it also in the birds of the caves of Brazil. Mr. Woodward has shown that the same law holds good with sea-shells, but from the wide distribution of most genera of molluscs, it is not well displayed by them. Other cases could be added, as the relation between the extinct and living land-shells of Madeira; and between the extinct and living brackish-water shells of the Aralo-Caspian Sea.

Now what does this remarkable law of the succession of the same types within the same areas mean? He would be a bold man, who after comparing the present climate of Australia and of parts of South America under the same latitude, would attempt to account, on the one hand, by dissimilar physical conditions for the dissimilarity of the inhabitants of these two continents, and, on the other hand, by similarity of conditions, for the uniformity of the same types in each during the later tertiary periods. Nor can it be pretended that it is an immutable law that marsupials should have been chiefly or solely produced in Australia ; or that Edentata and other American types should have been solely produced in South America. For we know that Europe in ancient times was peopled by numerous marsupials ; and I have shown in the publications above alluded to, that in America the law of distribution of terrestrial mammals was formerly different from what it now is. North America formerly partook strongly of the present character of the southern half of the continent ; and the southern half was formerly more closely allied, than it is at present, to the northern half. In a similar manner we know from Falconer and Cautley's discoveries, that northern India was formerly more closely related in its mammals to Africa than it is at the present time. Analogous facts could be given in relation to the distribution of marine animals.

On the theory of descent with modification, the great law of the long enduring, but not immutable, succession of the same types within the same areas, is at once explained ; for the inhabitants of each quarter of the world will obviously tend to leave in that quarter, during the next succeeding period of time, closely allied though in some degree modified descendants. If the inhabitants of one continent formerly differed greatly from those of another continent, so will their modified descendants still differ in nearly the same manner and degree. But after very long intervals of time and after great geographical changes, permitting much inter-migration, the feebler will yield to the more dominant forms, and there will be nothing immutable in the laws of past and present distribution.

It may be asked in ridicule, whether I suppose that the megatherium and other allied huge monsters have left behind them in South America, the sloth, armadillo, and anteater, as their degenerate descendants. This cannot for an instant be admitted. These huge animals have become wholly extinct, and have left no progeny. But in the caves of Brazil, there are many extinct species which are closely allied in size and in other characters to the species still living in South America; and some of these fossils may be the actual progenitors of living species. It must not be forgotten that, on my theory, all the species of the same genus have descended from some one species; so that if six genera, each having eight species, be found in one geological formation, and in the next succeeding formation there be six other allied or representative genera with the same number of species, then we may conclude that only one species of each of the six older genera has left modified descendants, constituting the six new genera. The other seven species of the old genera have all died out and have left no progeny. Or, which would probably be a far commoner case, two or three species of two or three alone of the six older genera will have been the parents of the six new genera; the other old species and the other whole old genera having become utterly extinct. In failing orders, with the genera and species decreasing in numbers, as apparently is the case of the Edentata of South America, still fewer genera and species will have left modified blood-descendants.

Summary of the preceding and present Chapter. — I

have attempted to show that the geological record is extremely imperfect; that only a small portion of the globe has been geologically explored with care; that only certain classes of organic beings have been largely preserved in a fossil state; that the number both of specimens and of species, preserved in our museums, is absolutely as nothing compared with the incalculable number of generations which must have passed away even during a single formation; that, owing to subsidence being almost necessary for the accumulation of deposits rich in fossils and thick enough to resist future degradation, enormous intervals of time have elapsed between most of our successive formations; that there has probably been more extinction during the periods of subsidence, and more variation during the periods of elevation, and during the latter the record will have been least perfectly kept; that each single formation has not been continuously deposited; that the duration of each formation is, perhaps, short compared with the average duration of specific forms; that migration has played an important part in the first appearance of new forms in any one area and formation; that widely ranging species are those which have varied most, and have oftenest given rise to new species; and that varieties have at first often been local. These causes taken conjointly, must have made the geological record extremely imperfect, and will to a large extent explain why—though we do find many links between the members of the same group—we do not find interminable varieties, connecting together all extinct and existing forms by the finest graduated steps.

He who rejects these views on the nature of the geological record, will rightly reject my whole theory. For he may ask in vain where are the numberless transitional links which must formerly have connected the closely allied or representative species, found in the several stages of the same great formation? He may disbelieve in the enormous intervals of time which have elapsed between our consecutive formations; he may overlook how important a part migration must have played, when the formations of any one great region alone, as that of Europe, are considered; he may urge the apparent, but often falsely apparent, sudden coming in of whole groups of species. He may ask where are the remains of those infinitely numerous organisms which must have existed long before the first bed of the Silurian system was deposited? I can answer this latter question only hypothetically, by saying that as far as we can see, where our oceans now extend they have for an enormous period extended, and where our oscillating continents now stand they have stood ever since the Silurian epoch; but that long before that period, the world may have presented a wholly different aspect; and that the older continents, formed of formations older than any known to us, may now all be in a metamorphosed condition, or may lie buried under the ocean.

Passing from these difficulties, the other great leading facts in palæontology seem to me simply to follow on the theory of descent with modification through natural selection. We can thus understand how it is that new species come in slowly and successively; how species of different classes do not necessarily change together, or at the same rate, or in the same degree; yet in the long run that all undergo modification to some extent. The extinction of old forms is the almost inevitable consequence of the production of new forms. We can understand why when a species has once disappeared it never reappears. Groups of species increase in numbers slowly, and endure for unequal periods of time; for the process of modification is necessarily slow, and depends on many complex contingencies. The dominant species of the larger dominant groups tend to leave many modified descendants, and thus new sub-groups and groups are formed. As these are formed, the species of the less vigorous groups, from their inferiority inherited from a common progenitor, tend to become extinct together, and to leave no modified offspring on the face of the earth. But the utter extinction of a whole group of species may often be a very slow process, from the survival of a few descendants, lingering in protected and isolated situations. When a group has once wholly disappeared, it does not reappear; for the link of generation has been broken.

We can understand how the spreading of the dominant forms of life, which are those that oftenest vary, will in the long run tend to people the world with allied, but modified, descendants; and these will generally succeed in taking the places of those groups of species which are their inferiors in the struggle for existence. Hence, after long intervals of time, the productions of the world will appear to have changed simultaneously.

We can understand how it is that all the forms of life, ancient and recent, make together one grand system; for all are connected by generation. We can understand, from the continued tendency to divergence of character, why the more ancient a form is, the more it generally differs from those now living. Why ancient and extinct forms often tend to fill up gaps between existing forms, sometimes blending two groups previ-

nerely classed as distinct into one; but more commonly
only bringing them a little closer together. The more
ancient a form is, the more often, apparently, it dis-
plays characters in some degree intermediate between
groups now distinct; for the more ancient a form is,
the more nearly it will be related to, and consequently
resemble, the common progenitor of groups, since become widely divergent. Extinct forms are seldom
directly intermediate between existing forms; but are
intermediate only by a long and circuitous course
through many other extinct and different forms. We
can clearly see why the organic remains of closely consecutive formations are more closely allied to each
other, than are those of remote formations; for the
forms are more closely linked together by generation:
we can clearly see why the remains of an intermediate
formation are intermediate in character.

The inhabitants of each successive period in the
world's history must have beaten their predecessors in
the race for life, and are, in so far, higher in the scale
of nature, and their structure will generally have become more specialised; and this may account for that
vague yet common sentiment, felt by many palæontologists, that organisation on the whole has progressed.
If it should hereafter be proved that ancient animals
resemble to a certain extent the embryos of more
recent animals of the same class, the fact will be intelligible. The succession of the same types of structure within the same areas during the later geological
periods ceases to be mysterious, and is simply explained
by inheritance.

If then the geological record be as imperfect as I
believe it to be, and it may at least be asserted that
the record cannot be proved to be much more perfect,

the main objections to the theory of natural selection
are greatly diminished or disappear. On the other
hand, all the chief laws of palæontology plainly proclaim, as it seems to me, that species have been produced by ordinary generation: old forms having been
supplanted by new and improved forms of life, produced
by the laws of variation still acting round us, and preserved by Natural Selection.

CHAPTER XI.

GEOGRAPHICAL DISTRIBUTION.

Present distribution cannot be accounted for by differences in physical conditions—Importance of barriers—Affinity of the productions of the same continent—Centres of creation—Means of dispersal, by changes of climate and of the level of the land, and by occasional means—Dispersal during the Glacial period co-extensive with the world.

In considering the distribution of organic beings over the face of the globe, the first great fact which strikes us is, that neither the similarity nor the dissimilarity of the inhabitants of various regions can be accounted for by their climatal and other physical conditions. Of late, almost every author who has studied the subject has come to this conclusion. The case of America alone would almost suffice to prove its truth: for if we exclude the northern parts where the circumpolar land is almost continuous, all authors agree that one of the most fundamental divisions in geographical distribution is that between the New and Old Worlds; yet if we travel over the vast American continent, from the central parts of the United States to its extreme southern point, we meet with the most diversified conditions; the most humid districts, arid deserts, lofty mountains, grassy plains, forests, marshes, lakes, and great rivers, under almost every temperature. There is hardly a climate or condition in the Old World which cannot be paralleled in the New—at least as closely as the same species generally require; for it is a most rare case to find a group of organisms confined to any small spot, having conditions peculiar in only a slight degree; for instance, small areas in the Old World can be pointed out hotter than any in the New World, yet these are not inhabited by a peculiar fauna or flora. Notwithstanding this parallelism in the conditions of the Old and New Worlds, how widely different are their living productions!

In the southern hemisphere, if we compare large tracts of land in Australia, South Africa, and western South America, between latitudes 25° and 35°, we shall find parts extremely similar in all their conditions, yet it would not be possible to point out three faunas and floras more utterly dissimilar. Or again we may compare the productions of South America south of lat. 35° with those north of 25°, which consequently inhabit a considerably different climate, and they will be found incomparably more closely related to each other than they are to the productions of Australia or Africa under nearly the same climate. Analogous facts could be given with respect to the inhabitants of the sea.

A second great fact which strikes us in our general review is, that barriers of any kind, or obstacles to free migration, are related in a close and important manner to the differences between the productions of various regions. We see this in the great difference of nearly all the terrestrial productions of the New and Old Worlds, excepting in the northern parts, where the land almost joins, and where, under a slightly different climate, there might have been free migration for the northern temperate forms, as there now is for the strictly arctic productions. We see the same fact in the great difference between the inhabitants of Australia, Africa, and South America under the same latitude: for these countries are almost as much isolated from each other as is possible. On each continent, also, we see the same fact; for on the opposite sides of lofty and continuous mountain-ranges, and of great deserts, and sometimes even of large rivers, we find different productions; though as mountain-chains, deserts, &c., are not as impassable, or likely to have endured so long as the oceans separating continents, the differences are very inferior in degree to those characteristic of distinct continents.

Turning to the sea, we find the same law. No two marine faunas are more distinct, with hardly a fish, shell, or crab in common, than those of the eastern and western shores of South and Central America; yet these great faunas are separated only by the narrow, but impassable, isthmus of Panama. Westward of the shores of America, a wide space of open ocean extends, with not an island as a halting-place for emigrants; here we have a barrier of another kind, and as soon as this is passed we meet in the eastern islands of the Pacific with another and totally distinct fauna. So that here three marine faunas range far northward and southward, in parallel lines not far from each other, under corresponding climates; but from being separated from each other by impassable barriers, either of land or open sea, they are wholly distinct. On the other hand, proceeding still farther westward from the eastern islands of the tropical parts of the Pacific, we encounter no impassable barriers, and we have innumerable islands as halting-places, or continuous coasts, until after travelling over a hemisphere we come to the shores of Africa; and over this vast space we meet with no well-defined and distinct marine faunas. Although hardly one shell, crab or fish is common to the above-named three approximate faunas of Eastern and Western America and the eastern Pacific islands, yet many fish range from the Pacific into the Indian Ocean, and many shells are common to the eastern islands of the Pacific

and the eastern shores of Africa, on almost exactly
opposite meridians of longitude.

A third great fact, partly included in the foregoing
statements, is the affinity of the productions of the
same continent or sea, though the species themselves
are distinct at different points and stations. It is
a law of the widest generality, and every continent
offers innumerable instances. Nevertheless the naturalist in travelling, for instance, from north to south
never fails to be struck by the manner in which successive groups of beings, specifically distinct, yet clearly
related, replace each other. He hears from closely
allied, yet distinct kinds of birds, notes nearly similar,
and sees their nests similarly constructed, but not quite
alike, with eggs coloured in nearly the same manner.
The plains near the Straits of Magellan are inhabited
by one species of Rhea (American ostrich), and northward the plains of La Plata by another species of the
same genus; and not by a true ostrich or emu, like
those found in Africa and Australia under the same
latitude. On these same plains of La Plata, we see
the agouti and bizcacha, animals having nearly the
same habits as our hares and rabbits and belonging to
the same order of Rodents, but they plainly display
an American type of structure. We ascend the lofty
peaks of the Cordillera and we find an alpine species
of bizcacha; we look to the waters, and we do not find
the beaver or musk-rat, but the coypu and capybara,
rodents of the American type. Innumerable other instances could be given. If we look to the islands off
the American shore, however much they may differ in
geological structure, the inhabitants, though they may
be all peculiar species, are essentially American. We
may look back to past ages, as shown in the last
chapter, and we find American types then prevalent on

the American continent and in the American seas. We see in these facts some deep organic bond, prevailing throughout space and time, over the same areas of land and water, and independent of their physical conditions. The naturalist must feel little curiosity, who is not led to inquire what this bond is.

This bond, on my theory, is simply inheritance, that cause which alone, as far as we positively know, produces organisms quite like, or, as we see in the case of varieties, nearly like each other. The dissimilarity of the inhabitants of different regions may be attributed to modification through natural selection, and in a quite subordinate degree to the direct influence of different physical conditions. The degree of dissimilarity will depend on the migration of the more dominant forms of life from one region into another having been effected with more or less ease, at periods more or less remote;—on the nature and number of the former immigrants;—and on their action and reaction, in their mutual struggles for life;—the relation of organism to organism being, as I have already often remarked, the most important of all relations. Thus the high importance of barriers comes into play by checking migration; as does time for the slow process of modification through natural selection. Widely-ranging species, abounding in individuals, which have already triumphed over many competitors in their own widely-extended homes, will have the best chance of seizing on new places, when they spread into new countries. In their new homes they will be exposed to new conditions, and will frequently undergo further modification and improvement; and thus they will become still further victorious, and will produce groups of modified descendants. On this principle of inheritance with modification, we can understand how it is that sections of genera, whole genera, and even families, are confined to the same areas, as is so commonly and notoriously the case.

I believe, as was remarked in the last chapter, in no law of necessary development. As the variability of each species is an independent property, and will be taken advantage of by natural selection, only so far as it profits the individual in its complex struggle for life, so the degree of modification in different species will be no uniform quantity. If, for instance, a number of species, which stand in direct competition with each other, migrate in a body into a new and afterwards isolated country, they will be little liable to modification; for neither migration nor isolation in themselves can do anything. These principles come into play only by bringing organisms into new relations with each other, and in a lesser degree with the surrounding physical conditions. As we have seen in the last chapter that some forms have retained nearly the same character from an enormously remote geological period, so certain species have migrated over vast spaces, and have not become greatly or at all modified.

On these views, it is obvious that the several species of the same genus, though inhabiting the most distant quarters of the world, must originally have proceeded from the same source, as they have descended from the same progenitor. In the case of those species which have undergone during whole geological periods but little modification, there is not much difficulty in believing that they may have migrated from the same region; for during the vast geographical and climatal changes which will have supervened since ancient times, almost any amount of migration is possible. But in many other cases, in which we have reason to believe that the species of a genus have been produced within comparatively recent times, there is great difficulty on this head. It is also obvious that the individuals of the same species, though now inhabiting distant and isolated regions, must have proceeded from one spot, where their parents were first produced; for, as explained in the last chapter, it is incredible that individuals identically the same should have been produced through natural selection from parents specifically distinct.

We are thus brought to the question which has been largely discussed by naturalists, namely, whether species have been created at one or more points of the earth's surface. Undoubtedly there are very many cases of extreme difficulty in understanding how the same species could possibly have migrated from some one point to the several distant and isolated points, where now found. Nevertheless the simplicity of the view that each species was first produced within a single region captivates the mind. He who rejects it, rejects the vera causa of ordinary generation with subsequent migration, and calls in the agency of a miracle. It is universally admitted, that in most cases the area inhabited by a species is continuous; and when a plant or animal inhabits two points so distant from each other, or with an interval of such a nature, that the space could not be easily passed over by migration, the fact is given as something remarkable and exceptional. The capacity of migrating across the sea is more distinctly limited in terrestrial mammals, than perhaps in any other organic beings; and, accordingly, we find no inexplicable cases of the same mammal inhabiting distant points of the world. No geologist will feel any difficulty in such cases as Great Britain having been formerly united to Europe, and consequently possessing the same quadrupeds. But if the same species can be produced at two separate points, why do we not find a single mammal common to Europe and Australia or North America?

The conditions of life are nearly the same, so that a multitude of European animals and plants have become naturalised in America and Australia; and some of the aboriginal plants are identically the same at these distant points of the northern and southern hemispheres? The answer, as I believe, is, that mammals have not been able to migrate, whereas some plants, from their varied means of dispersal, have migrated across the vast and broken interspace. The great and striking influence which barriers of every kind have had on distribution, is intelligible only on the view that the great majority of species have been produced on one side alone, and have not been able to migrate to the other side. Some few families, many sub-families, very many genera, and a still greater number of sections of genera are confined to a single region; and it has been observed by several naturalists, that the most natural genera, or those genera in which the species are most closely related to each other, are generally local, or if they have a wide range that their range is continuous. What a strange anomaly it would be, if, when coming one step lower in the series, to the individuals of the same species, a directly opposite rule prevailed; and species were not local, but had been produced in two or more quite distinct areas!

Hence it seems to me, as it has to many other naturalists, that the view of each species having been produced in one area alone, and having subsequently migrated from that area as far as its powers of migration and subsistence under past and present conditions permitted, is the most probable. Undoubtedly many cases occur, in which we cannot explain how the same species could have passed from one point to the other. But the geographical and climatal changes, which have certainly occurred within recent geological times, must have intercepted or rendered discontinuous the formerly continuous range of many species. So that we are reduced to consider whether the exceptions to continuity of range are so numerous and of so grave a nature, that we ought to give up the belief, rendered probable by general considerations, that each species has been produced within one area, and has migrated thence as far as it could. It would be hopelessly tedious to discuss all the exceptional cases of the same species, now living at distant and separated points; nor do I for a moment pretend that any explanation could be offered of many such cases. But after some preliminary remarks, I will discuss a few of the most striking classes of facts; namely, the existence of the same species on the summits of distant mountain-ranges, and at distant points in the arctic and antarctic regions; and secondly (in the following chapter), the wide distribution of fresh-water productions; and thirdly, the occurrence of the same terrestrial species on islands and on the mainland, though separated by hundreds of miles of open sea. If the existence of the same species at distant and isolated points of the earth's surface, can in many instances be explained on the view of each species having migrated from a single birthplace; then, considering our ignorance with respect to former climatal and geographical changes and various occasional means of transport, the belief that this has been the universal law, seems to me incomparably the safest.

In discussing this subject, we shall be enabled at the same time to consider a point equally important for us, namely, whether the several distinct species of a genus, which on my theory have all descended from a common progenitor, can have migrated (undergoing modification during some part of their migration) from the area inhabited by their progenitor. If it can be shown to

be almost invariably the case, that a region, of which
most of its inhabitants are closely related to, or belong
to the same genera with the species of a second region,
has probably received at some former period immigrants
from this other region, my theory will be strengthened;
for we can clearly understand, on the principle of
modification, why the inhabitants of a region should be
related to those of another region, whence it has been
stocked. A volcanic island, for instance, upheaved and
formed at the distance of a few hundreds of miles from a
continent, would probably receive from it in the course
of time a few colonists, and their descendants, though
modified, would still be plainly related by inheritance to
the inhabitants of that continent. Cases of this nature
are common, and are, as we shall hereafter more fully
see, inexplicable on the theory of independent creation.
This view of the relation of species in one region to
those in another, does not differ much (by substituting
the word variety for species) from that lately advanced
in an ingenious paper by Mr. Wallace, in which he con-
cludes, that "every species has come into existence
coincident both in space and time with a pre-existing
closely allied species." And I now know from correspondence, that this coincidence he attributes to generation with modification.

The previous remarks on "single and multiple centres
of creation" do not directly bear on another allied
question,—namely, whether all the individuals of the
same species have descended from a single pair, or
single hermaphrodite, or whether, as some authors
suppose, from many individuals simultaneously created.
With those organic beings which never intercross (if
such exist), the species, on my theory, must have descended from a succession of improved varieties, which
will never have blended with other individuals or varie-

5

ties, but will have supplanted each other ; so that, at each successive stage of modification and improvement, all the individuals of each variety will have descended from a single parent. But in the majority of cases, namely, with all organisms which habitually unite for each birth, or which often intercross, I believe that during the slow process of modification the individuals of the species will have been kept nearly uniform by intercrossing ; so that many individuals will have gone on simultaneously changing, and the whole amount of modification will not have been due, at each stage, to descent from a single parent. To illustrate what I mean : our English race-horses differ slightly from the horses of every other breed ; but they do not owe their difference and superiority to descent from any single pair, but to continued care in selecting and training many individuals during many generations.

Before discussing the three classes of facts, which I have selected as presenting the greatest ¦amount of difficulty on the theory of "single centres of creation," I must say a few words on the means of dispersal.

Means of Dispersal.—Sir C. Lyell and other authors have ably treated this subject. I can give here only the briefest abstract of the more important facts. Change of climate must have had a powerful influence on migration : a region when its climate was different may have been a high road for migration, but now be impassable ; I shall, however, presently have to discuss this branch of the subject in some detail. Changes of level in the land must also have been highly influ-ential : a narrow isthmus now separates two marine faunas ; submerge it, or let it formerly have been submerged, and the two faunas will now blend or may formerly have blended : where the sea now extends, land may at a former period have connected islands or possibly even continents together, and thus have allowed terrestrial productions to pass from one to the other. No geologist will dispute that great mutations of level have occurred within the period of existing organisms. Edward Forbes insisted that all the islands in the Atlantic must recently have been connected with Europe or Africa, and Europe likewise with America. Other authors have thus hypothetically bridged over every ocean, and have united almost every island to some mainland. If indeed the arguments used by Forbes are to be trusted, it must be admitted that scarcely a single island exists which has not recently been united to some continent. This view cuts the Gordian knot of the dispersal of the same species to the most distant points, and removes many a difficulty: but to the best of my judgment we are not authorized in admitting such enormous geographical changes within the period of existing species. It seems to me that we have abundant evidence of great oscillations of level in our continents; but not of such vast changes in their position and extension, as to have united them within the recent period to each other and to the several intervening oceanic islands. I freely admit the former existence of many islands, now buried beneath the sea, which may have served as halting places for plants and for many animals during their migration. In the coral-producing oceans such sunken islands are now marked by rings of coral or atolls standing over them. Whenever it is fully admitted, as I believe it will some day be, that each species has proceeded from a single birthplace, and when in the course of time we know something definite about the means of distribution, we shall be enabled to speculate with security on the former extension of the land. But I do not believe

c 2

that it will ever be proved that within the recent period continents which are now quite separate, have been continuously, or almost continuously, united with each other, and with the many existing oceanic islands. Several facts in distribution,—such as the great difference in the marine faunas on the opposite sides of almost every continent,—the close relation of the tertiary inhabitants of several lands and even seas to their present inhabitants,—a certain degree of relation (as we shall hereafter see) between the distribution of mammals and the depth of the sea,—these and other such facts seem to me opposed to the admission of such prodigious geographical revolutions within the recent period, as are necessitated on the view advanced by Forbes and admitted by his many followers. The nature and relative proportions of the inhabitants of oceanic islands likewise seem to me opposed to the belief of their former continuity with continents. Nor does their almost universally volcanic composition favour the admission that they are the wrecks of sunken continents;—if they had originally existed as mountain-ranges on the land, some at least of the islands would have been formed, like other mountain-summits, of granite, metamorphic schists, old fossiliferous or other such rocks, instead of consisting of mere piles of volcanic matter.

I must now say a few words on what are called accidental means, but which more properly might be called occasional means of distribution. I shall here confine myself to plants. In botanical works, this or that plant is stated to be ill adapted for wide dissemination; but for transport across the sea, the greater or less facilities may be said to be almost wholly unknown. Until I tried, with Mr. Berkeley's aid, a few experiments, it was not even known how far seeds could resist the inju-

rious action of sea-water. To my surprise I found that
out of 87 kinds, 64 germinated after an immersion of 28
days, and a few survived an immersion of 137 days. It
deserves notice that certain orders were far more affected
than others: nine Leguminosae were tried, and, with
one exception, they resisted the salt-water badly; seven
species of the allied orders, Hydrophyllaceae and Pole-
moniaceae, were all killed by a month's immersion. For
convenience' sake I chiefly tried small seeds, without
the capsule or fruit; and as all of these sank in a few
days, they could not be floated across wide spaces of
the sea, whether or not they were injured by the salt-
water. Afterwards I tried some larger fruits, capsules,
&c., and some of these floated for a long time. It is
well known what a difference there is in the buoyancy
of green and seasoned timber; and it occurred to me
that floods might wash down plants or branches, and
that these might be dried on the banks, and then by a
fresh rise in the stream be washed into the sea. Hence
I was led to dry stems and branches of 94 plants with
ripe fruit, and to place them on sea-water. The majority sank quickly, but some which whilst green floated
for a very short time, when dried floated much longer;
for instance, ripe hazel-nuts sank immediately, but when
dried they floated for 90 days, and afterwards when
planted they germinated; an asparagus plant with ripe
berries floated for 23 days, when dried it floated for
85 days, and the seeds afterwards germinated; the
ripe seeds of Helosciadium sank in two days, when dried
they floated for above 90 days, and afterwards germinated. Altogether, out of the 94 dried plants, 18 floated
for above 28 days, and some of the 18 floated for a
very much longer period. So that as 64 seeds germinated after an immersion of 28 days; and as 18 plants
with ripe fruit (but not all the same species as in the

foregoing experiment) floated, after being dried, for above 28 days, as far as we may infer anything from these scanty facts, we may conclude that the seeds of $\frac{14}{100}$ plants of any country might be floated by sea-currents during 28 days, and would retain their power of germination. In Johnston's Physical Atlas, the average rate of the several Atlantic currents is 33 miles per diem (some currents running at the rate of 60 miles per diem); on this average, the seeds of $\frac{14}{100}$ plants belonging to one country might be floated across 924 miles of sea to another country; and when stranded, if blown to a favourable spot by an inland gale, they would germinate.

Subsequently to my experiments, M. Martens tried similar ones, but in a much better manner, for he placed the seeds in a box in the actual sea, so that they were alternately wet and exposed to the air like really floating plants. He tried 98 seeds, mostly different from mine; but he chose many large fruits and likewise seeds from plants which live near the sea; and this would have favoured the average length of their flotation and of their resistance to the injurious action of the salt-water. On the other hand, he did not previously dry the plants or branches with the fruit; and this, as we have seen, would have caused some of them to have floated much longer. The result was that $\frac{18}{98}$ of his seeds floated for 42 days, and were then capable of germination. But I do not doubt that plants exposed to the waves would float for a less time than those protected from violent movement as in our experiments. Therefore it would perhaps be safer to assume that the seeds of about $\frac{10}{100}$ plants of a flora, after having been dried, could be floated across a space of sea 900 miles in width, and would then germinate. The fact of the larger fruits often floating longer than the small, is in-

terminating; as plants with large seeds or fruit could
hardly be transported by any other means; and A. de
de Candolle has shown that such plants generally have
restricted ranges.

But seeds may be occasionally transported in another
manner. Drift timber is thrown up on most islands,
even on those in the midst of the widest oceans; and
the natives of the coral-islands in the Pacific, procure
stones for their tools, solely from the roots of drifted
trees, these stones being a valuable royal tax. I find
on examination, that when irregularly shaped stones
are embedded in the roots of trees, small parcels of
earth are very frequently enclosed in their interstices
and behind them, so perfectly that not a particle
could be washed away in the longest transport: out of
one small portion of earth thus completely enclosed by
wood in an oak about 50 years old, three dicotyle-
donous plants germinated: I am certain of the accuracy
of this observation. Again, I can show that the car-
casses of birds, when floating on the sea, sometimes
escape being immediately devoured; and seeds of
many kinds in the crops of floating birds long retain
their vitality: peas and vetches, for instance, are killed
by even a few days' immersion in sea-water; but some
taken out of the crop of a pigeon, which had floated on
artificial salt-water for 30 days, to my surprise nearly all
germinated.

Living birds can hardly fail to be highly effective
agents in the transportation of seeds. I could give
many facts showing how frequently birds of many kinds
are blown by gales to vast distances across the ocean.
We may safely assume that under such circumstances
their rate of flight would often be 35 miles an hour;
and some authors have given a far higher estimate. I
have never seen an instance of nutritious seeds passing

through the intestines of a bird; but hard seeds of fruit
pass uninjured through even the digestive organs of a
turkey. In the course of two months, I picked up in
my garden 12 kinds of seeds, out of the excrement of
small birds, and these seemed perfect, and some of
them, which were tried, germinated. But the following
fact is more important: the crops of birds do not secrete
gastric juice, and do not in the least injure, as I know
by trial, the germination of seeds; now after a bird has
found and devoured a large supply of food, it is positively
asserted that all the grains do not pass into the gizzard
for twelve or even eighteen hours. A bird in this interval
might easily be blown to the distance of 500 miles, and
hawks are known to look out for tired birds, and the
contents of their torn crops might thus readily get scattered. Some hawks and owls bolt their prey whole,
and after an interval of from twelve to twenty hours,
disgorge pellets, which, as I know from experiments
made in the Zoological Gardens, include seeds capable
of germination. Some seeds of the oat, wheat, millet,
canary, hemp, clover, and beet germinated after having
been from twelve to twenty-one hours in the stomachs
of different birds of prey; and two seeds of beet grew
after having been thus retained for two days and fourteen hours. Fresh-water fish, I find, eat seeds of many
land and water plants: fish are frequently devoured by
birds, and thus the seeds might be transported from
place to place. I forced many kinds of seeds into the
stomachs of dead fish, and then gave their bodies to
fishing-eagles, storks, and pelicans; these birds after an
interval of many hours, either rejected the seeds in
pellets or passed them in their excrement; and several
of these seeds retained their power of germination.
Certain seeds, however, were always killed by this
process.

Although the beaks and feet of birds are generally quite clean, earth sometimes adheres to them: in one case I removed sixty-one grains, and in another case twenty-two grains of dry argillaceous earth from one foot of a partridge, and in the earth there was a pebble quite as large as the seed of a vetch. Thus seeds might occasionally be transported to great distances; for many facts could be given, showing that soil almost everywhere is charged with seeds. Reflect for a moment on the millions of quails which annually cross the Mediterranean; and can we doubt that the earth adhering to their feet would sometimes include a few minute seeds? But I shall presently have to recur to this subject.

As icebergs are known to be sometimes loaded with earth and stones, and have even carried brushwood, bones, and the nest of a land-bird, it can hardly be doubted that they must occasionally have transported seeds from one part to another of the arctic and antarctic regions, as suggested by Lyell; and during the Glacial period from one part of the now temperate regions to another. In the Azores, from the large number of the species of plants common to Europe, in comparison with the plants of other oceanic islands nearer to the mainland, and (as remarked by Mr. H. C. Watson) from the somewhat northern character of the flora in comparison with the latitude, I suspected that these islands had been partly stocked by ice-borne seeds, during the Glacial epoch. At my request Sir C. Lyell wrote to M. Hartung to inquire whether he had observed erratic boulders on these islands, and he answered that he had found large fragments of granite and other rocks, which do not occur in the archipelago. Hence we may safely infer that icebergs formerly landed their rocky burthens on the shores of these mid-ocean

islands, and it is at least possible that they may have
brought thither the seeds of northern plants.

. Considering that these several means of transport,
and that several other means, which without doubt
remain to be discovered, have been in action year
after year, for centuries and tens of thousands of years,
it would I think be a marvellous fact if many plants
had not thus become widely transported. These means
of transport are sometimes called accidental, but this
is not strictly correct: the currents of the sea are
not accidental, nor is the direction of prevalent gales
of wind. It should be observed that scarcely any
means of transport would carry seeds for very great
distances; for seeds do not retain their vitality when
exposed for a great length of time to the action of sea-
water; nor could they be long carried in the crops or
intestines of birds. These means, however, would suffice
for occasional transport across tracts of sea some hundred miles in breadth, or from island to island, or from
a continent to a neighbouring island, but not from one
distant continent to another. The floras of distant
continents would not by such means become mingled
in any great degree; but would remain as distinct
as we now see them to be. The currents, from their
course, would never bring seeds from North America
to Britain, though they might and do bring seeds
from the West Indies to our western shores, where,
if not killed by so long an immersion in salt-water,
they could not endure our climate. Almost every
year, one or two land-birds are blown across the
whole Atlantic Ocean, from North America to the
western shores of Ireland and England; but seeds
could be transported by these wanderers only by one
means, namely, in dirt sticking to their feet, which
is in itself a rare accident. Even in this case, how

small would the chances be of a seed falling on favourable soil, and coming to maturity! But it would be a great error to argue that because a well-stocked island, like Great Britain, has not, as far as is known (and it would be very difficult to prove this), received within the last few centuries, through occasional means of transport, immigrants from Europe or any other continent, that a poorly-stocked island, though standing more remote from the mainland, would not receive colonists by similar means. Out of twenty seeds or animals transported to an island, even if far less well-stocked than Britain, perhaps not more than one would be so well fitted to its new home, so as to become naturalised. But this, as it seems to me, is no valid argument against what would be effected by occasional means of transport, during the long lapse of geological time, whilst an island was being upheaved, and before it had become fully stocked with inhabitants. On almost bare land, with few or no destructive insects or birds living there, nearly every seed, which chanced to arrive, if fitted for the climate, would be sure to germinate and survive.

Dispersal during the Glacial period.—The identity of many plants and animals, on mountain-summits, separated from each other by hundreds of miles of lowlands, where the Alpine species could not possibly exist, is one of the most striking cases known of the same species living at distant points, without the apparent possibility of their having migrated from one to the other. It is indeed a remarkable fact to see so many of the same plants living on the snowy regions of the Alps or Pyrenees, and in the extreme northern parts of Europe; but it is far more remarkable that the plants on the White Mountains, in the United States of America, are all the same with those of Labrador, and nearly all the same, as we hear from Asa Gray, with those on the loftiest mountains of Europe. Even as long ago as 1747, such facts led Gmelin to conclude that the same species must have been independently created at several distinct points; and we might have remained in this same belief, had not Agassiz and others called vivid attention to the Glacial period, which, as we shall immediately see, affords a simple explanation of these facts. We have evidence of almost every conceivable kind, organic and inorganic, that within a very recent geological period, central Europe and North America suffered under an Arctic climate. The ruins of a house burnt by fire do not tell their tale more plainly, than do the mountains of Scotland and Wales, with their scored flanks, polished surfaces, and perched boulders, of the icy streams with which their valleys were lately filled. So greatly has the climate of Europe changed, that in Northern Italy, gigantic moraines, left by old glaciers, are now clothed by the vine and maize. Throughout a large part of the United States, erratic boulders, and rocks scored by drifted icebergs and coast-ice, plainly reveal a former cold period.

The former influence of the glacial climate on the distribution of the inhabitants of Europe, as explained with remarkable clearness by Edward Forbes, is substantially as follows. But we shall follow the changes more readily, by supposing a new glacial period to come slowly on, and then pass away, as formerly occurred. As the cold came on, and as each more southern zone became fitted for arctic beings and ill-fitted for their former more temperate inhabitants, the latter would be supplanted and arctic productions would take their places. The inhabitants of the more temperate regions would at the same time travel southward, unless they were stopped by barriers, in which case they would perish.

The mountains would become covered with snow and ice, and their former Alpine inhabitants would descend to the plains. By the time that the cold had reached its maximum, we should have a uniform arctic fauna and flora, covering the central parts of Europe, as far south as the Alps and Pyrenees, and even stretching into Spain. The now temperate regions of the United States would likewise be covered by arctic plants and animals, and these would be nearly the same with those of Europe; for the present circumpolar inhabitants, which we suppose to have everywhere travelled southward, are remarkably uniform round the world. We may suppose that the Glacial period came on a little earlier or later in North America than in Europe, so will the southern migration there have been a little earlier or later; but this will make no difference in the final result.

As the warmth returned, the arctic forms would retreat northward, closely followed up in their retreat by the productions of the more temperate regions. And as the snow melted from the bases of the mountains, the arctic forms would seize on the cleared and thawed ground, always ascending higher and higher, as the warmth increased, whilst their brethren were pursuing their northern journey. Hence, when the warmth had fully returned, the same arctic species, which had lately lived in a body together on the lowlands of the Old and New Worlds, would be left isolated on distant mountain summits (having been exterminated on all lower heights) and in the arctic regions of both hemispheres.

Thus we can understand the identity of many plants at points so immensely remote as on the mountains of the United States and of Europe. We can thus also understand the fact that the Alpine plants of each mountain-range are more especially related to the arctic forms living due north or nearly due north of them: for the migration as the cold came on, and the re-migration on the returning warmth, will generally have been due south and north. The Alpine plants, for example, of Scotland, as remarked by Mr. H. C. Watson, and those of the Pyrenees, as remarked by Ramond, are more especially allied to the plants of northern Scandinavia; those of the United States to Labrador; those of the mountains of Siberia to the arctic regions of that country. These views, grounded as they are on the perfectly well-ascertained occurrence of a former Glacial period, seem to me to explain in so satisfactory a manner the present distribution of the Alpine and Arctic productions of Europe and America, that when in other regions we find the same species on distant mountain-summits, we may almost conclude without other evidence, that a colder climate permitted their former migration across the low intervening tracts, since become too warm for their existence.

If the climate, since the Glacial period, has ever been in any degree warmer than at present (as some geologists in the United States believe to have been the case), then the arctic and temperate productions will at a very late period have marched a little farther north, and subsequently have retreated to their present homes; but I have met with no satisfactory evidence with respect to this intercalated slightly warmer period, since the Glacial period.

The arctic forms, during their long southern migration and re-migration northward, will have been exposed to nearly the same climate, and, as is especially to be noticed, they will have kept in a body together; consequently their mutual relations will not have been much disturbed, and, in accordance with the principles inculcated in this volume, they will not have been liable to much modification. But with our Alpine productions, left isolated from the moment of the returning warmth, first at the bases and ultimately on the summits of the mountains, the case will have been somewhat different; for it is not likely that all the same arctic species will have been left on mountain-ranges distant from each other, and have survived there ever since; they will, also, in all probability have become mingled with ancient Alpine species, which must have existed on the mountains before the commencement of the Glacial epoch, and which during its coldest period will have been temporarily driven down to the plains; they will, also, have been exposed to somewhat different climatal influences. Their mutual relations will thus have been in some degree disturbed; consequently they will have been liable to modification; and this we find has been the case; for if we compare the present Alpine plants and animals of the several great European mountain-ranges, though very many of the species are identically the same, some present varieties, some are ranked as doubtful forms, and many are distinct yet closely allied or representative species.

In illustrating what, as I believe, actually took place during the Glacial period, I assumed that at its commencement the arctic productions were as uniform round the polar regions as they are at the present day. But the foregoing remarks on distribution apply not only to strictly arctic forms, but also to many sub-arctic and to some few northern temperate forms, for some of these are the same on the lower mountains and on the plains of North America and Europe; and it may be reasonably asked how I account for the necessary degree of uniformity of the sub-arctic and northern temperate forms round the world, at the commencement of the Glacial period. At the present day, the sub-arctic

and northern temperate productions of the Old and
New Worlds are separated from each other by the
Atlantic Ocean and by the extreme northern part of
the Pacific. During the Glacial period, when the inhabitants of the Old and New Worlds lived farther
southwards than at present, they must have been still
more completely separated by wider spaces of ocean.
I believe the above difficulty may be surmounted by
looking to still earlier changes of climate of an opposite
nature. We have good reason to believe that during
the newer Pliocene period, before the Glacial epoch,
and whilst the majority of the inhabitants of the world
were specifically the same as now, the climate was
warmer than at the present day. Hence we may suppose that the organisms now living under the climate
of latitude 60°, during the Pliocene period lived farther
north under the Polar Circle, in latitude 66°-67°; and
that the strictly arctic productions then lived on the
broken land still nearer to the pole. Now if we look at
a globe, we shall see that under the Polar Circle there
is almost continuous land from western Europe, through
Siberia, to eastern America. And to this continuity of
the circumpolar land, and to the consequent freedom
for intermigration under a more favourable climate, I
attribute the necessary amount of uniformity in the
sub-arctic and northern temperate productions of the
Old and New Worlds, at a period anterior to the Glacial
epoch.

Believing, from reasons before alluded to, that our
continents have long remained in nearly the same relative position, though subjected to large, but partial
oscillations of level, I am strongly inclined to extend
the above view, and to infer that during some earlier
and still warmer period, such as the older Pliocene
period, a large number of the same plants and animals

inhabited the almost continuous circumpolar land; and that these plants and animals, both in the Old and New Worlds, began slowly to migrate southwards as the climate became less warm, long before the commencement of the Glacial period. We now see, as I believe, their descendants, mostly in a modified condition, in the central parts of Europe and the United States. On this view we can understand the relationship, with very little identity, between the productions of North America and Europe,—a relationship which is most remarkable, considering the distance of the two areas, and their separation by the Atlantic Ocean. We can further understand the singular fact remarked on by several observers, that the productions of Europe and America during the later tertiary stages were more closely related to each other than they are at the present time; for during these warmer periods the northern parts of the Old and New Worlds will have been almost continuously united by land, serving as a bridge, since rendered impassable by cold, for the intermigration of their inhabitants.

During the slowly decreasing warmth of the Pliocene period, as soon as the species in common, which inhabited the New and Old Worlds, migrated south of the Polar Circle, they must have been completely cut off from each other. This separation, as far as the more temperate productions are concerned, took place long ages ago. And as the plants and animals migrated southward, they will have become mingled in the one great region with the native American productions, and have had to compete with them; and in the other great region, with those of the Old World. Consequently we have here everything favourable for much modification,—for far more modification than with the Alpine productions, left isolated, within a much more recent period, on the several mountain-ranges and on the arctic lands of the two Worlds. Hence it has come, that when we compare the now living productions of the temperate regions of the New and Old Worlds, we find very few identical species (though Asa Gray has lately shown that more plants are identical than was formerly supposed), but we find in every great class many forms, which some naturalists rank as geographical races, and others as distinct species; and a host of closely allied or representative forms which are ranked by all naturalists as specifically distinct.

As on the land, so in the waters of the sea, a slow southern migration of a marine fauna, which during the Pliocene or even a somewhat earlier period, was nearly uniform along the continuous shores of the Polar Circle, will account, on the theory of modification, for many closely allied forms now living in areas completely sundered. Thus, I think, we can understand the presence of many existing and tertiary representative forms on the eastern and western shores of temperate North America; and the still more striking case of many closely allied crustaceans (as described in Dana's admirable work), of some fish and other marine animals, in the Mediterranean and in the seas of Japan,—areas now separated by a continent and by nearly a hemisphere of equatorial ocean.

These cases of relationship, without identity, of the inhabitants of seas now disjoined, and likewise of the past and present inhabitants of the temperate lands of North America and Europe, are inexplicable on the theory of creation. We cannot say that they have been created alike, in correspondence with the nearly similar physical conditions of the areas; for if we compare, for instance, certain parts of South America with the southern continents of the Old World, we see countries closely corresponding in all their physical conditions, but with their inhabitants utterly dissimilar.

But we must return to our more immediate subject, the Glacial period. I am convinced that Forbes's view may be largely extended. In Europe we have the plainest evidence of the cold period, from the western shores of Britain to the Oural range, and southward to the Pyrenees. We may infer from the frozen mammals and nature of the mountain vegetation, that Siberia was similarly affected. Along the Himalaya, at points 900 miles apart, glaciers have left the marks of their former low descent; and in Sikkim, Dr. Hooker saw maize growing on gigantic ancient moraines. South of the equator, we have some direct evidence of former glacial action in New Zealand; and the same plants, found on widely separated mountains in that island, tell the same story. If one account which has been published can be trusted, we have direct evidence of glacial action in the south-eastern corner of Australia.

Looking to America; in the northern half, ice-borne fragments of rock have been observed on the eastern side as far south as lat. 36°-37°, and on the shores of the Pacific, where the climate is now so different, as far south as lat. 46°; erratic boulders have, also, been noticed on the Rocky Mountains. In the Cordillera of equatorial South America, glaciers once extended far below their present level. In Central Chile I examined a vast mound of detritus, crossing the Portillo valley, which I now fully believe to have been due to ice-action; but we shall hereafter have valuable information on this subject from Mr. D. Forbes, who informs me that he found in the Cordillera, from lat. 13° to 30° S., at about the height of 12,000 feet, strongly-furrowed rocks, resembling those with which he was familiar in Norway, and likewise great masses of detritus, including grooved pebbles. Along this whole space of the Cordillera true glaciers do not now exist even at much more considerable heights. Farther south on both sides of the continent, from lat. 41° to the southernmost extremity, we have the clearest evidence of former glacial action, in huge boulders transported far from their parent source.

We do not know that the Glacial epoch was strictly simultaneous at these several far distant points on opposite sides of the world. But we have good evidence in almost every case, that the epoch was included within the latest geological period. We have, also, excellent evidence, that it endured for an enormous time, as measured by years, at each point. The cold may have come on, or have ceased, earlier at one point of the globe than at another, but seeing that it endured for a long time at each, and that it was contemporaneous in a geological sense, it seems probable that it was, during a part at least of the period, actually simultaneous throughout the world. Without some distinct evidence to the contrary, we may at least admit as probable that the glacial action was simultaneous on the eastern and western sides of North America, in the Cordillera under the equatorial, tropical, and warmer temperate zones, and on both sides of the southern portion of the continent. If this be admitted, it is difficult to avoid believing that the temperature of the whole world was at this period simultaneously cooler. But it would suffice for my purpose, if the temperature was at the same time lower along certain broad belts of longitude.

On this view of the whole world, or at least of broad longitudinal belts, having been simultaneously colder from pole to pole, much light can be thrown on the present distribution of identical and allied species. In America, Dr. Hooker has shown that between forty and

Many of the flowering plants of Tierra del Fuego, forming no inconsiderable part of its scanty flora, are common to Europe, enormously remote as these two points are; and there are many closely allied species. On the lofty mountains of equatorial America a host of peculiar species belonging to European genera occur. On the highest mountains of Brazil, some few European genera were found by Gardner, which do not exist in the wide intervening hot countries. So on the Silla of Caraccas the illustrious Humboldt long ago found species belonging to genera characteristic of the Cordillera. On the mountains of Abyssinia, several forms characteristic of Europe and some few representatives of the flora of the Cape of Good Hope occur. At the Cape of Good Hope itself a very few European species, believed not to have been introduced by man, and on the mountains, several representative European forms are found, which have not been discovered in the intertropical parts of Africa. On the Himalaya, and on the isolated mountain-ranges of the peninsula of India, on the heights of Ceylon, and on the volcanic cones of Java, many plants occur, either identically the same or representing each other, and at the same time representing plants of Europe not found in the intervening hot lowlands. A list of the genera collected on the loftier peaks of Java raises a picture of a collection made on a hill in Europe! Still more striking is the fact that southern Australian forms are clearly represented by plants growing on the summits of the mountains of Borneo. Some of these Australian forms, as I hear from Dr. Hooker, extend along the heights of the peninsula of Malacca, and are thinly scattered, on the one hand over India and on the other as far north as Japan.

On the southern mountains of Australia, Dr. F. Müller has discovered several European species; other species, are introduced by man, several on the lowlands; and a long list can be given, as I am informed by Dr. Hooker, of European genera, found in Australia, but not in the intermediate torrid regions. In the admirable 'Introduction to the Flora of New Zealand,' by Dr. Hooker, analogous and striking facts are given in regard to the plants of that large island. Hence we see that throughout the world, the plants growing on the more lofty mountains, and on the temperate lowlands of the northern and southern hemispheres, are sometimes identical; the same; but they are much oftener specifically distinct, though related to each other in a most remarkable manner.

This brief abstract applies to plants alone: some strictly analogous facts could be given on the distribution of terrestrial animals. In marine productions, similar cases occur; as an example, I may quote a remark by the highest authority, Prof. Dana, that "it is certainly a wonderful fact that New Zealand should have a closer resemblance in its crustacea to Great Britain, its antipode, than to any other part of the world." Sir J. Richardson, also, speaks of the reappearance on the shores of New Zealand, Tasmania, &c., of northern forms of fish. Dr. Hooker informs me that twenty-five species of Algæ are common to New Zealand and to Europe, but have not been found in the intermediate tropical seas.

It should be observed that the northern species and forms found in the southern parts of the southern hemisphere, and on the mountain-ranges of the intertropical regions, are not arctic, but belong to the northern temperate zones. As Mr. H. C. Watson has recently remarked, "In receding from polar towards equatorial latitudes, the Alpine or mountain floras really become less and less arctic." Many of the forms living on the mountains of the warmer regions of the earth and in the southern hemisphere are of doubtful value, being ranked by some naturalists as specifically distinct, by others as varieties; but some are certainly identical, and many, though closely related to northern forms, must be ranked as distinct species.

Now let us see what light can be thrown on the foregoing facts, on the belief, supported as it is by a large body of geological evidence, that the whole world, or a large part of it, was during the Glacial period simultaneously much colder than at present. The Glacial period, as measured by years, must have been very long; and when we remember over what vast spaces some naturalised plants and animals have spread within a few centuries, this period will have been ample for any amount of migration. As the cold came slowly on, all the tropical plants and other productions will have retreated from both sides towards the equator, followed in the rear by the temperate productions, and these by the arctic; but with the latter we are not now concerned. The tropical plants probably suffered much extinction; how much no one can say; perhaps formerly the tropics supported as many species as we see at the present day crowded together at the Cape of Good Hope, and in parts of temperate Australia. As we know that many tropical plants and animals can withstand a considerable amount of cold, many might have escaped extermination during a moderate fall of temperature, more especially by escaping into the lowest, most protected, and warmest districts. But the great fact to bear in mind is, that all tropical productions will have suffered to a certain extent. On the other hand, the temperate productions, after migrating nearer to the equator, though they will have been placed under somewhat new conditions, will have suffered less. And it is certain that many temperate plants, if protected from the inroads of competitors, can withstand a much warmer climate than their own. Hence, it seems to me possible, bearing in mind that the tropical productions were in a suffering state, and could not have presented a firm front against intruders, that a certain number of the more vigorous and dominant temperate forms might have penetrated the native ranks, and have reached or even crossed the equator. The invasion would, of course, have been greatly favoured by high land, and perhaps by a dry climate; for Dr. Falconer informs me that it is the damp with the heat of the tropics which is so destructive to perennial plants from a temperate climate. On the other hand, the most humid and hottest districts will have afforded an asylum to the tropical natives. The mountain-ranges north-west of the Himalaya, and the long line of the Cordillera, seem to have afforded two great lines of invasion: and it is a striking fact, lately communicated to me by Dr. Hooker, that all the flowering plants, about forty-six in number, common to Tierra del Fuego and to Europe, still exist in North America, which must have lain on the line of march. But I do not doubt that some temperate productions entered and crossed even the *lowlands* of the tropics at the period when the cold was most intense, when arctic forms had migrated some twenty-five degrees of latitude from their native country, and covered the land at the foot of the Pyrenees. At this period of extreme cold, I believe that the climate under the equator at the level of the sea was about the same with that now felt there at the height of six or seven thousand feet. During this the coldest period, the lowlands of the tropics were probably clothed with a mingled tropical and temperate vegetation, like that now growing with strange luxuriance at the base of the Himalaya, as graphically described by Hooker.

Thus, as I believe, a considerable number of plants, a few terrestrial animals, and some marine productions, migrated during the Glacial period from the northern and southern temperate zones into the intertropical regions, and some even crossed the equator. As the warmth returned, these temperate forms would naturally ascend the higher mountains, being exterminated on the lowlands; those which had not reached the equator would re-migrate northward or southward towards their former homes; but the forms, chiefly northern, which had crossed the equator, would travel still farther from their homes into the more temperate latitudes of the opposite hemisphere. Although we have reason to believe from geological evidence that the whole body of arctic shells underwent scarcely any modification during their long southern migration and re-migration northward, the case may have been wholly different with those intruding forms which settled themselves on the intertropical mountains, and in the southern hemisphere. These being surrounded by strangers will have had to compete with many new forms of life; and it is probable that selected modifications in their structure, habits, and constitutions will have profited them. Thus many of these wanderers, though still plainly related by inheritance to their brethren of the northern or southern hemispheres, now exist in their new homes as well-marked varieties, or as distinct species.

It is a remarkable fact, strongly insisted on by Hooker in regard to America, and by Alph. de Candolle in regard to Australia, that many more identical plants and allied forms have apparently migrated from the north to the south, than in a reversed direction. We see, however, a few southern vegetable forms on the mountains of Borneo and Abyssinia. I suspect that

T

this preponderant migration from north to south is due to the greater extent of land in the north, and to the northern forms having existed in their own homes in greater numbers, and having consequently been advanced through natural selection and competition to a higher stage of perfection or dominating power, than the southern forms. And thus, when they became commingled during the Glacial period, the northern forms were enabled to beat the less powerful southern forms. Just in the same manner as we see at the present day, that very many European productions cover the ground in La Plata, and in a lesser degree in Australia, and have to a certain extent beaten the natives; whereas extremely few southern forms have become naturalised in any part of Europe, though hides, wool, and other objects likely to carry seeds have been largely imported into Europe during the last two or three centuries from La Plata, and during the last thirty or forty years from Australia. Something of the same kind must have occurred on the intertropical mountains: no doubt before the Glacial period they were stocked with endemic Alpine forms; but these have almost everywhere largely yielded to the more dominant forms, generated in the larger areas and more efficient workshops of the north. In many islands the native productions are nearly equalled or even outnumbered by the naturalised; and if the natives have not been actually exterminated, their numbers have been greatly reduced, and this is the first stage towards extinction. A mountain is an island on the land; and the intertropical mountains before the Glacial period must have been completely isolated; and I believe that the productions of these islands on the land yielded to those produced within the larger areas of the north, just in

the same way as the productions of real islands have
everywhere lately yielded to continental forms, naturalised by man's agency.

I am far from supposing that all difficulties are removed on the view here given in regard to the range
and affinities of the allied species which live in the
northern and southern temperate zones and on the
mountains of the intertropical regions. Very many
difficulties remain to be solved. I do not pretend to
indicate the exact lines and means of migration, or the
reason why certain species and not others have migrated;
why certain species have been modified and have given
rise to new groups of forms, and others have remained
unaltered. We cannot hope to explain such facts,
until we can say why one species and not another becomes naturalised by man's agency in a foreign land;
why one ranges twice or thrice as far, and is twice or
thrice as common, as another species within their own
homes.

I have said that many difficulties remain to be solved:
some of the most remarkable are stated with admirable
clearness by Dr. Hooker in his botanical works on the
antarctic regions. These cannot be here discussed. I
will only say that as far as regards the occurrence of
identical species at points so enormously remote as
Kerguelen Land, New Zealand, and Fuegia, I believe
that towards the close of the Glacial period, icebergs,
as suggested by Lyell, have been largely concerned in
their dispersal. But the existence of several quite
distinct species, belonging to genera exclusively confined
to the south, at these and other distant points of the
southern hemisphere, is, on my theory of descent with
modification, a far more remarkable case of difficulty.
For some of these species are so distinct, that we cannot

T 2

suppose that there has been time since the commencement of the Glacial period for their migration, and for
their subsequent modification to the necessary degree.
The facts seem to me to indicate that peculiar and very
distinct species have migrated in radiating lines from
some common centre; and I am inclined to look in
the southern, as in the northern hemisphere, to a former
and warmer period, before the commencement of the
Glacial period, when the antarctic lands, now covered
with ice, supported a highly peculiar and isolated flora.
I suspect that before this flora was exterminated by the
Glacial epoch, a few forms were widely dispersed to
various points of the southern hemisphere by occasional
means of transport, and by the aid, as halting-places, of
existing and now sunken islands. By these means, as I
believe, the southern shores of America, Australia, New
Zealand, have become slightly tinted by the same
peculiar forms of vegetable life.

Sir C. Lyell in a striking passage has speculated, in
language almost identical with mine, on the effects of
great alternations of climate on geographical distribution. I believe that the world has recently felt one
of his great cycles of change; and that on this view,
combined with modification through natural selection,
a multitude of facts in the present distribution both
of the same and of allied forms of life can be explained. The living waters may be said to have flowed
during one short period from the north and from the
south, and to have crossed at the equator; but to
have flowed with greater force from the north so as to
have freely inundated the south. As the tide leaves
its drift in horizontal lines, though rising higher on
the shores where the tide rises highest, so have the
living waters left their living drift on our mountain-

summits, in a line gently rising from the arctic lowlands to a great height under the equator. The various beings thus left stranded may be compared with savage races of man, driven up and surviving in the mountain-fastnesses of almost every land, which serve as a record, full of interest to us, of the former inhabitants of the surrounding lowlands.

CHAPTER XII.

Geographical Distribution—continued.

Distribution of fresh-water productions — On the inhabitants of oceanic islands — Absence of Batrachians and of terrestrial Mammals — On the relation of the inhabitants of islands to those of the nearest mainland — On colonisation from the nearest source with subsequent modification — Summary of the last and present chapters.

As lakes and river-systems are separated from each other by barriers of land, it might have been thought that fresh-water productions would not have ranged widely within the same country, and as the sea is apparently a still more impassable barrier, that they never would have extended to distant countries. But the case is exactly the reverse. Not only have many fresh-water species, belonging to quite different classes, an enormous range, but allied species prevail in a remarkable manner throughout the world. I well remember, when first collecting in the fresh waters of Brazil, feeling much surprise at the similarity of the fresh-water insects, shells, &c., and at the dissimilarity of the surrounding terrestrial beings, compared with those of Britain.

But this power in fresh-water productions of ranging widely, though so unexpected, can, I think, in most cases be explained by their having become fitted, in a manner highly useful to them, for short and frequent migrations from pond to pond, or from stream to stream; and liability to wide dispersal would follow from this capacity as an almost necessary consequence.

We can here consider only a few cases. In regard to fish, I believe that the same species never occur in the fresh waters of distant continents. But on the same continent the species often range widely and almost capriciously; for two river-systems will have some fish in common and some different. A few facts seem to favour the possibility of their occasional transport by accidental means; like that of the live fish not rarely dropped by whirlwinds in India, and the vitality of their ova when removed from the water. But I am inclined to attribute the dispersal of fresh-water fish mainly to slight changes within the recent period in the level of the land, having caused rivers to flow into each other. Instances, also, could be given of this having occurred during floods, without any change of level. We have evidence in the loess of the Rhine of considerable changes of level in the land within a very recent geological period, and when the surface was peopled by existing land and fresh-water shells. The wide difference of the fish on opposite sides of continuous mountain-ranges, which from an early period must have parted river-systems and completely prevented their inosculation, seems to lead to this same conclusion. With respect to allied fresh-water fish occurring at very distant points of the world, no doubt there are many cases which cannot at present be explained: but some fresh-water fish belong to very ancient forms, and in such cases there will have been ample time for great geographical changes, and consequently time and means for much migration. In the second place, salt-water fish can with care be slowly accustomed to live in fresh water; and, according to Valenciennes, there is hardly a single group of fishes confined exclusively to fresh water, so that we may imagine that a marine member of a fresh-water group might travel far along the shores of the sea, and subsequently become modified and adapted to the fresh waters of a distant land.

Some species of fresh-water shells have a very wide range, and allied species, which, on my theory, are descended from a common parent, and must have proceeded from a single source, prevail throughout the world. Their distribution at first perplexed me much, as their ova are not likely to be transported by birds, and they are immediately killed by sea-water, as are the adults. I could not even understand how some naturalised species have rapidly spread throughout the same country. But two facts, which I have observed —and no doubt many others remain to be observed— throw some light on this subject. When a duck suddenly emerges from a pond covered with duck-weed, I have twice seen these little plants adhering to its back; and it has happened to me, in removing a little duckweed from one aquarium to another, that I have quite unintentionally stocked the one with fresh-water shells from the other. But another agency is perhaps more effectual: I suspended a duck's feet in an aquarium, where many ova of fresh-water shells were hatching; and I found that numbers of the extremely minute and just-hatched shells crawled on the feet, and clung to them so firmly that when taken out of the water they could not be jarred off, though at a somewhat more advanced age they would voluntarily drop off. These just-hatched molluscs, though aquatic in their nature, survived on the duck's feet, in damp air, from twelve to twenty hours; and in this length of time a duck or heron might fly at least six or seven hundred miles, and would be sure to alight on a pool or rivulet, if blown across sea to an oceanic island or to any other distant point. Sir Charles Lyell also informs me that a Dyticus has been caught with an Ancylus (a fresh-water shell like a limpet, firmly adhering to it; and a water-beetle of the same family, a Colymbetes, once flew on board the 'Beagle,' when forty-five miles distant from the nearest land: how much farther it might have flown with a favouring gale no one can tell.

With respect to plants, it has long been known what enormous ranges many fresh-water, and even marsh species, have, both over continents and to the most remote oceanic islands. This is strikingly shown, as remarked by Alph. de Candolle, in large groups of terrestrial plants, which have only a very few aquatic members; for these latter seem immediately to acquire, as if in consequence, a very wide range. I think favourable means of dispersal explain this fact. I have before mentioned that earth occasionally, though rarely, adheres in some quantity to the feet and beaks of birds. Wading birds, which frequent the muddy edges of ponds, if suddenly flushed, would be the most likely to have muddy feet. Birds of this order, I can show, are the greatest wanderers, and are occasionally found on the most remote and barren islands in the open ocean; they would not be likely to alight on the surface of the sea, so that the dirt would not be washed off their feet; when making land, they would be sure to fly to their natural fresh-water haunts. I do not believe that botanists are aware how charged the mud of ponds is with seeds: I have tried several little experiments, but will here give only the most striking case: I took in February three table-spoonfuls of mud from three different points, beneath water, on the edge of a little pond; this mud when dry weighed only 6¾ ounces; I kept it covered up in my study for six months, pulling up and counting each plant as it grew; the plants were of many kinds, and were altogether 537 in number; and yet the viscid mud was all contained in a breakfast cup? Considering these facts, I think it would be an inexplicable circumstance if water-birds did not transport the seeds of fresh-water plants to vast distances, and if consequently the range of these plants was not very great. The same agency may have come into play with the eggs of some of the smaller fresh-water animals.

Other and unknown agencies probably have also played a part. I have stated that fresh-water fish eat some kinds of seeds, though they reject many other kinds after having swallowed them; even small fish swallow seeds of moderate size, as of the yellow water-lily and Potamogeton. Herons and other birds, century after century, have gone on daily devouring fish; they then take flight and go to other waters, or are blown across the sea; and we have seen that seeds retain their power of germination, when rejected in pellets or in excrement, many hours afterwards. When I saw the great size of the seeds of that fine water-lily, the Nelumbium, and remembered Alph. de Candolle's remarks on this plant, I thought that its distribution must remain quite inexplicable; but Audubon states that he found the seeds of the great southern water-lily (probably, according to Dr. Hooker, the Nelumbium luteum) in a heron's stomach; although I do not know the fact, yet analogy makes me believe that a heron flying to another pond and getting a hearty meal of fish, would probably reject from its stomach a pellet containing the seeds of the Nelumbium undigested; or the seeds might be dropped by the bird whilst feeding its young, in the same way as fish are known sometimes to be dropped.

In considering these several means of distribution, it should be remembered that when a pond or stream is first formed, for instance, on a rising islet, it will be

unoccupied; and a single seed or egg will have a good
chance of succeeding. Although there will always be
a struggle for life between the individuals of the species,
however few, already occupying any pond, yet as the
number of kinds is small, compared with those on the
land, the competition will probably be less severe
between aquatic than between terrestrial species; consequently an intruder from the waters of a foreign
country, would have a better chance of seizing on a
place, than in the case of terrestrial colonists. We
should also remember that some, perhaps many, fresh-
water productions are low in the scale of nature, and
that we have reason to believe that such low beings
change or become modified less quickly than the high;
and this will give longer time than the average for the
migration of the same aquatic species. We should not
forget the probability of many species having formerly
ranged as continuously as fresh-water productions ever
can range, over immense areas, and having subsequently
become extinct in intermediate regions. But the wide
distribution of fresh-water plants and of the lower
animals, whether retaining the same identical form
or in some degree modified, I believe mainly depends
on the wide dispersal of their seeds and eggs by animals,
more especially by fresh-water birds, which have large
powers of flight, and naturally travel from one to another and often distant pieces of water. Nature, like
a careful gardener, thus takes her seeds from a bed of
a particular nature, and drops them in another equally
well fitted for them.

On the Inhabitants of Oceanic Islands. We now
come to the last of the three classes of facts, which I
have selected as presenting the greatest amount of
difficulty, on the view that all the individuals both of

the same and of allied species have descended from a
single parent; and therefore have all proceeded from a
common birthplace, notwithstanding that in the course
of time they have come to inhabit distant points of the
globe. I have already stated that I cannot honestly
admit Forbes's view on continental extensions, which,
if legitimately followed out, would lead to the belief
that within the recent period all existing islands have
been nearly or quite joined to some continent. This
view would remove many difficulties, but it would not
explain all the facts in regard to insular productions.
In the following remarks I shall not confine myself to
the mere question of dispersal; but shall consider some
other facts, which bear on the truth of the two theories
of independent creation and of descent with modification.

The species of all kinds which inhabit oceanic islands
are few in number compared with those on equal continental areas: Alph. de Candolle admits this for plants,
and Wollaston for insects. If we look to the large size
and varied stations of New Zealand, extending over
780 miles of latitude, and compare its flowering plants,
only 750 in number, with those on an equal area at the
Cape of Good Hope or in Australia, we must, I think,
admit that something quite independently of any difference in physical conditions has caused so great a
difference in number. Even the uniform county of
Cambridge has 847 plants, and the little island of
Anglesea 764, but a few ferns and a few introduced
plants are included in these numbers, and the comparison in some other respects is not quite fair. We
have evidence that the barren island of Ascension aboriginally possessed under half-a-dozen flowering plants;
yet many have become naturalised on it, as they have
on New Zealand and on every other oceanic island

which can be named. In St. Helena there is reason to
believe that the naturalised plants and animals have
nearly or quite exterminated many native productions.
He who admits the doctrine of the creation of each
separate species, will have to admit that a sufficient
number of the best adapted plants and animals have
not been created on oceanic islands; for man has unintentionally stocked them from various sources far more
fully and perfectly than has nature.

Although in oceanic islands the number of kinds of
inhabitants is scanty, the proportion of endemic species
(i.e. those found nowhere else in the world) is often
extremely large. If we compare, for instance, the
number of the endemic land-shells in Madeira, or of
the endemic birds in the Galapagos Archipelago, with
the number found on any continent, and then compare
the area of the islands with that of the continent, we
shall see that this is true. This fact might have been
expected on my theory, for, as already explained, species
occasionally arriving after long intervals in a new and
isolated district, and having to compete with new associates, will be eminently liable to modification, and will
often produce groups of modified descendants. But it
by no means follows that, because in an island nearly
all the species of one class are peculiar, those of another
class, or of another section of the same class, are peculiar; and this difference seems to depend partly on the
species which do not become modified having immigrated with facility and in a body, so that their mutual
relations have not been much disturbed; and partly on
the frequent arrival of unmodified immigrants from the
mother-country, and the consequent intercrossing with
them. With respect to the effects of this intercrossing,
it should be remembered that the offspring of such

crosses would almost certainly gain in vigour; so that even an occasional cross would produce more effect than might at first be anticipated. To give a few examples: in the Galapagos Islands there are 26 land-birds; of these 21 (or perhaps 23) are peculiar, whereas of the 11 marine birds only 2 are peculiar; and it is obvious that marine birds could arrive at these islands more easily than land-birds. Bermuda, on the other hand, which lies at about the same distance from North America as the Galapagos Islands do from South America, and which has a very peculiar soil, does not possess one endemic land-bird; and we know from Mr. J. M. Jones's admirable account of Bermuda, that very many North American birds, during their great annual migrations, visit either periodically or occasionally this island. Madeira does not possess one peculiar bird, and many European and African birds are almost every year blown there, as I am informed by Mr. E. V. Harcourt. So that these two islands of Bermuda and Madeira have been stocked by birds, which for long ages have struggled together in their former homes, and have become mutually adapted to each other; and when settled in their new homes, each bird will have been kept by the others to their proper places and habits, and will consequently have been little liable to modification. Any tendency to modification will also have been checked by intercrossing with the unmodified immigrants from the mother-country. Madeira again is inhabited by a wonderful number of peculiar land-shells, whereas not one species of sea-shell is confined to its shores: now, though we do not know how sea-shells are dispersed, yet we can see that their eggs or larvæ, perhaps attached to sea-weed or floating timber, or to the feet of wading-birds, might be transported far more

easily than land-shells, across three or four hundred
miles of open sea. The different orders of insects in
Madeira apparently present analogous facts.

Oceanic islands are sometimes deficient in certain
classes, and their places are apparently occupied by
the other inhabitants; in the Galapagos Islands reptiles,
and in New Zealand gigantic wingless birds, take the
place of mammals. In the plants of the Galapagos
Islands, Dr. Hooker has shown, that the proportional
numbers of the different orders are very different from
what they are elsewhere. Such cases are generally accounted for by the physical conditions of the islands;
but this explanation seems to me not a little doubtful.
Facility of immigration seems to have been at least as
important as the nature of the conditions.

Many remarkable little facts could be given with
respect to the inhabitants of remote islands. For instance, in certain islands not tenanted by mammals,
some of the endemic plants have beautifully hooked
seeds; yet few relations are more striking than the
adaptation of hooked seeds for transportal by the wool
and fur of quadrupeds. This case presents no difficulty
on my view, for a hooked seed might be transported to
an island by some other means; and the plant then
becoming slightly modified, but still retaining its hooked
seeds, would form an endemic species, having as useless
an appendage as any rudimentary organ,—for instance,
as the shrivelled wings under the soldered elytra of
many insular beetles. Again, islands often possess trees
or bushes belonging to orders which elsewhere include
only herbaceous species; now trees, as A. de Candolle has shown, generally have, whatever the cause
may be, confined ranges. Hence trees would be little
likely to reach distant oceanic islands; and an herbaceous plant, though it would have no chance of suc-

beautifully competing in stature with a fully developed tree, when established on an island and having to compete with herbaceous plants alone, might readily gain an advantage by growing taller and taller and overtopping the other plants. If so, natural selection would often tend to add to the stature of herbaceous plants when growing on an oceanic island, to whatever order they belonged, and thus convert them first into bushes and ultimately into trees.

With respect to the absence of whole orders on oceanic islands, Bory St. Vincent long ago remarked that Batrachians (frogs, toads, newts) have never been found on any of the many islands with which the great oceans are studded. I have taken pains to verify this assertion, and I have found it strictly true. I have, however, been assured that a frog exists on the mountains of the great island of New Zealand; but I suspect that this exception (if the information be correct) may be explained through glacial agency. This general absence of frogs, toads, and newts on so many oceanic islands cannot be accounted for by their physical conditions; indeed it seems that islands are peculiarly well fitted for these animals; for frogs have been introduced into Madeira, the Azores, and Mauritius, and have multiplied so as to become a nuisance. But as these animals and their spawn are known to be immediately killed by sea-water, there would be great difficulty in their transportal across the sea, and therefore on my view we can see why they do not exist on any oceanic island. But why, on the theory of creation, they should not have been created there, it would be very difficult to explain.

Mammals offer another and similar case. I have carefully searched the oldest voyages, but have not finished my search; as yet I have not found a single instance, free from doubt, of a terrestrial mammal (excluding domesticated animals kept by the natives) inhabiting an island situated above 300 miles from a continent or great continental island; and many islands situated at a much less distance are equally barren. The Falkland Islands, which are inhabited by a wolf-like fox, come nearest to an exception; but this group cannot be considered as oceanic, as it lies on a bank connected with the mainland, distant from it about 280 miles; moreover, icebergs formerly brought boulders to its western shores, and they may have formerly transported foxes, as so frequently now happens in the arctic regions. Yet it cannot be said that small islands will not support small mammals, for they occur in many parts of the world on very small islands, if close to a continent; and hardly an island can be named on which our smaller quadrupeds have not become naturalised and greatly multiplied. It cannot be said, on the ordinary view of creation, that there has not been time for the creation of mammals; many volcanic islands are sufficiently ancient, as shown by the stupendous degradation which they have suffered, and by their tertiary strata: there has also been time for the production of endemic species belonging to other classes; and on continents it is thought that mammals appear and disappear at a quicker rate than other and lower animals. Though terrestrial mammals do not occur on oceanic islands, aerial mammals do occur on almost every island. New Zealand possesses two bats found nowhere else in the world: Norfolk Island, the Viti Archipelago, the Bonin Islands, the Caroline and Marianne Archipelagoes, and Mauritius, all possess their peculiar bats. Why, it may be asked, has the supposed creative force produced bats and no other mammals on remote islands? On my view this question can easily be answered; for no terrestrial mammal can be transported across a wide space of sea, but bats can fly across. Bats have been seen wandering by day far over the Atlantic Ocean; and two North American species either regularly or occasionally visit Bermuda, at the distance of 600 miles from the mainland. I hear from Mr. Tomes, who has specially studied this family, that many of the same species have enormous ranges, and are found on continents and on far distant islands. Hence we have only to suppose that such wandering species have been modified through natural selection in their new homes in relation to their new position, and we can understand the presence of endemic bats on islands, with the absence of all terrestrial mammals.

Besides the absence of terrestrial mammals in relation to the remoteness of islands from continents, there is also a relation, to a certain extent independent of distance, between the depth of the sea separating an island from the neighbouring mainland, and the presence in both of the same or allied species or of allied species in a more or less modified condition. Mr. Windsor Earl has made some striking observations on this head in regard to the great Malay Archipelago, which is traversed near Celebes by a space of deep ocean; and this space separates two widely distinct mammalian faunas. On either side the islands are situated on moderately deep submarine banks, and they are inhabited by closely allied or identical quadrupeds. No doubt some few anomalies occur in this great archipelago, and there is much difficulty in forming a judgment in some cases owing to the probable naturalisation of certain mammals through man's agency; but we shall soon have much light thrown on the natural history of this archipelago by the admirable zeal and researches of Mr. Wallace. I have not as yet had time to

follow on this subject in all other quarters of the world;
but as far as I have gone, the relation generally holds
good. We see Britain separated by a shallow channel
from Europe, and the mammals are the same on both
sides: we meet with analogous facts on many islands
separated by similar channels from Australia. The
West Indian Islands stand on a deeply submerged bank,
nearly 1000 fathoms in depth, and here we find American
forms, but the species and even the genera are distinct.
As the amount of modification in all cases depends to
a certain degree on the lapse of time, and as during
changes of level it is obvious that islands separated by
shallow channels are more likely to have been continuously united within a recent period to the mainland than islands separated by deeper channels, we can
understand the frequent relation between the depth of
the sea and the degree of affinity of the mammalian
inhabitants of islands with those of a neighbouring continent,—an inexplicable relation on the view of independent acts of creation.

All the foregoing remarks on the inhabitants of
oceanic islands,—namely, the scarcity of kinds—the
richness in endemic forms in particular classes or sections of classes,—the absence of whole groups, as of
batrachians and of terrestrial mammals notwithstanding
the presence of aërial bats,—the singular proportions of
certain orders of plants,—herbaceous forms having been
developed into trees, &c.,—seem to me to accord better
with the view of occasional means of transport having
been largely efficient in the long course of time, than
with the view of all our oceanic islands having been
formerly connected by continuous land with the nearest
continent; for on this latter view the migration would
probably have been more complete; and if modification

be admitted, all the forms of life would have been more
equally modified, in accordance with the paramount importance of the relation of organism to organism.

I do not deny that there are many and grave difficulties in understanding how several of the inhabitants
of the more remote islands, whether still retaining the
same specific form or modified since their arrival, could
have reached their present homes. But the probability
of many islands having existed as halting-places, of
which not a wreck now remains, must not be overlooked. I will here give a single instance of one of
the cases of difficulty. Almost all oceanic islands,
even the most isolated and smallest, are inhabited by
land-shells, generally by endemic species, but sometimes
by species found elsewhere. Dr. Aug. A. Gould has
given several interesting cases in regard to the land-
snells of the islands of the Pacific. Now it is notorious
that land-shells are very easily killed by salt; their
eggs, at least such as I have tried, sink in sea-water,
and are killed by it. Yet there must be, on my view,
some unknown, but highly efficient means for their
transportal. Would the just-hatched young occasionally
crawl on and adhere to the feet of birds roosting on the
ground, and thus get transported? It occurred to
me that land-shells, when hybernating and having a
membranous diaphragm over the mouth of the shell,
might be floated in chinks of drifted timber across
moderately wide arms of the sea. And I found that
several species did in this state withstand uninjured an
immersion in sea-water during seven days: one of these
shells was the Helix pomatia, and after it had again
hybernated I put it in sea-water for twenty days, and
it perfectly recovered. As this species has a thick
calcareous operculum, I removed it, and when it had

formed a new membranous one. I immersed it for fourteen days in sea-water, and it recovered and crawled
away: but more experiments are wanted on this head.

The most striking and important fact for us in regard
to the inhabitants of islands, is their affinity to those of
the nearest mainland, without being actually the same
species. Numerous instances could be given of this
law. I will give only one, that of the Galapagos
Archipelago, situated under the equator, between 500
and 600 miles from the shores of South America. Here
almost every product of the land and water bears the
unmistakeable stamp of the American continent. There
are twenty-six land-birds, and twenty-one, or, perhaps,
twenty-three, of these are ranked as distinct species,
and are supposed to have been created here; yet the
close affinity of most of these birds to American species
in every character, in their habits, gestures, and tones
of voice, was manifest. So it is with the other animals,
and with nearly all the plants, as shown by Dr. Hooker
in his admirable memoir on the Flora of this archipelago.
The naturalist, looking at the inhabitants of these volcanic islands in the Pacific, distant several hundred
miles from the continent, yet feels that he is standing on
American land. Why should this be so? why should
the species which are supposed to have been created in
the Galapagos Archipelago, and nowhere else, bear so
plain a stamp of affinity to those created in America?
There is nothing in the conditions of life, in the geological nature of the islands, in their height or climate,
or in the proportions in which the several classes are
associated together, which resembles closely the conditions of the South American coast: in fact there is
a considerable dissimilarity in all these respects. On
the other hand, there is a considerable degree of resemblance in the volcanic nature of the soil, in climate,

height, and size of the islands, between the Galapagos and Cape de Verde Archipelagoes: but what an entire and absolute difference in their inhabitants! The inhabitants of the Cape de Verde Islands are related to those of Africa, like those of the Galapagos to America. I believe this grand fact can receive no sort of explanation on the ordinary view of independent creation; whereas on the view here maintained, it is obvious that the Galapagos Islands would be likely to receive colonists, whether by occasional means of transport or by formerly continuous land, from America; and the Cape de Verde Islands from Africa; and that such colonists would be liable to modification;—the principle of inheritance still betraying their original birthplace.

Many analogous facts could be given: indeed it is an almost universal rule that the endemic productions of islands are related to those of the nearest continent, or of other near islands. The exceptions are few, and most of them can be explained. Thus the plants of Kerguelen Land, though standing nearer to Africa than to America, are related, and that very closely, as we know from Dr. Hooker's account, to those of America: but on the view that this island has been mainly stocked by seeds brought with earth and stones on icebergs, drifted by the prevailing currents, this anomaly disappears. New Zealand in its endemic plants is much more closely related to Australia, the nearest mainland, than to any other region: and this is what might have been expected; but it is also plainly related to South America, which, although the next nearest continent, is so enormously remote, that the fact becomes an anomaly. But this difficulty almost disappears on the view that New Zealand, South America, and other southern lands were long ago partially stocked from a nearly intermediate though distant point, namely from the antarctic islands, when they were clothed with vegetation, before the commencement of the Glacial period. The affinity, which, though feeble, I am assured by Dr. Hooker is real, between the flora of the south-western corner of Australia and of the Cape of Good Hope, is a far more remarkable case, and is at present inexplicable: but this affinity is confined to the plants, and will, no doubt, be some day explained.

The law which causes the inhabitants of an archipelago, though specifically distinct, to be closely allied to those of the nearest continent, we sometimes see displayed on a small scale, yet in a most interesting manner, within the limits of the same archipelago. Thus the several islands of the Galapagos Archipelago are tenanted, as I have elsewhere shown, in a quite marvellous manner, by very closely related species; so that the inhabitants of each separate island, though mostly distinct, are related in an incomparably closer degree to each other than to the inhabitants of any other part of the world. And this is just what might have been expected on my view, for the islands are situated so near each other that they would almost certainly receive immigrants from the same original source, or from each other. But this dissimilarity between the endemic inhabitants of the islands may be used as an argument against my views; for it may be asked, how has it happened in the several islands situated within sight of each other, having the same geological nature, the same height, climate, &c., that many of the immigrants should have been differently modified, though only in a small degree? This long appeared to me a great difficulty: but it arises in chief part from the deeply-seated error of considering the physical conditions of a country as the most important for its inhabitants; whereas it cannot, I think, be disputed that the nature of the other inhabitants with which each has to compete, is at least as important, and generally a far more important element of success. Now if we look to those inhabitants of the Galapagos Archipelago which are found in other parts of the world (laying on one side for the moment the endemic species, which cannot be here fairly included, as we are considering how they have come to be modified since their arrival), we find a considerable amount of difference in the several islands. This difference might indeed have been expected on the view of the islands having been stocked by occasional means of transport—a seed, for instance, of one plant having been brought to one island, and that of another plant to another island. Hence when in former times an immigrant settled on any one or more of the islands, or when it subsequently spread from one island to another, it would undoubtedly be exposed to different conditions of life in the different islands, for it would have to compete with different sets of organisms: a plant for instance, would find the best-fitted ground more perfectly occupied by distinct plants in one island than in another, and it would be exposed to the attacks of somewhat different enemies. If then it varied, natural selection would probably favour different varieties in the different islands. Some species, however, might spread and yet retain the same character throughout the group, just as we see on continents some species spreading widely and remaining the same.

The really surprising fact in this case of the Galapagos Archipelago, and in a lesser degree in some analogous instances, is that the new species formed in the separate islands have not quickly spread to the other islands. But the islands, though in sight of each other, are separated by deep arms of the sea, in most cases

wider than the British Channel, and there is no reason
to suppose that they have at any former period been
continuously united. The currents of the sea are rapid
and sweep across the archipelago, and gales of wind
are extraordinarily rare; so that the islands are far
more effectually separated from each other than they
appear to be on a map. Nevertheless a good many
species, both those found in other parts of the world
and those confined to the archipelago, are common to
the several islands, and we may infer from certain facts
that these have probably spread from some one island
to the others. But we often take, I think, an erroneous view of the probability of closely-allied species
invading each other's territory, when put into free intercommunication. Undoubtedly if one species has any
advantage whatever over another, it will in a very brief
time wholly or in part supplant it; but if both are
equally well fitted for their own places in nature, both
probably will hold their own places and keep separate
for almost any length of time. Being familiar with the
fact that many species, naturalised through man's
agency, have spread with astonishing rapidity over new
countries, we are apt to infer that most species would
thus spread; but we should remember that the forms
which become naturalised in new countries are not
generally closely allied to the aboriginal inhabitants,
but are very distinct species, belonging in a large proportion of cases, as shown by Alph. de Candolle, to
distinct genera. In the Galapagos Archipelago, many
even of the birds, though so well adapted for flying
from island to island, are distinct on each; thus there
are three closely-allied species of mocking-thrush, each
confined to its own island. Now let us suppose the
mocking-thrush of Chatham Island to be blown to
Charles Island, which has its own mocking-thrush; why

should it succeed in establishing itself there? We may safely infer that Charles Island is well stocked with its own species, for annually more eggs are laid there than can possibly be reared; and we may infer that the mocking-thrush peculiar to Charles Island is at least as well fitted for its home as is the species peculiar to Chatham Island. Sir C. Lyell and Mr. Wollaston have communicated to me a remarkable fact bearing on this subject; namely, that Madeira and the adjoining islet of Porto Santo possess many distinct but representative land-shells, some of which live in crevices of stone; and although large quantities of stone are annually transported from Porto Santo to Madeira, yet this latter island has not become colonised by the Porto Santo species: nevertheless both islands have been colonised by some European land-shells, which no doubt had some advantage over the indigenous species. From these considerations I think we need not greatly marvel at the endemic and representative species, which inhabit the several islands of the Galapagos Archipelago, not having universally spread from island to island. In many other instances, as in the several districts of the same continent, pre-occupation has probably played an important part in checking the commingling of species under the same conditions of life. Thus, the south-east and south-west corners of Australia have nearly the same physical conditions, and are united by continuous land, yet they are inhabited by a vast number of distinct mammals, birds, and plants.

The principle which determines the general character of the fauna and flora of oceanic islands, namely, that the inhabitants, when not identically the same, yet are plainly related to the inhabitants of that region whence colonists could most readily have been derived,—the colonists having been subsequently modified and better fitted to their new homes,—is of the widest application throughout nature. We see this on every mountain, in every lake and marsh. For Alpine species, excepting in so far as the same forms, chiefly of plants, have spread widely throughout the world during the recent Glacial epoch, are related to those of the surrounding lowlands;—thus we have in South America, Alpine humming-birds, Alpine rodents, Alpine plants, &c., all of strictly American forms, and it is obvious that a mountain, as it became slowly upheaved, would naturally be colonised from the surrounding lowlands. So it is with the inhabitants of lakes and marshes, excepting in so far as great facility of transport has given the same general forms to the whole world. We see this same principle in some of the blind animals inhabiting the caves of America and of Europe. Other analogous facts could be given. And it will, I believe, be universally found to be true, that wherever in two regions, let them be ever so distant, many closely-allied or representative species occur, there will likewise be found some identical species, showing, in accordance with the foregoing view, that at some former period there has been intercommunication or migration between the two regions. And wherever many closely-allied species occur, there will be found many forms which some naturalists rank as distinct species, and some as varieties; these doubtful forms showing us the steps in the process of modification.

This relation between the power and extent of migration of a species, either at the present time or at some former period under different physical conditions, and the existence at remote points of the world of other species allied to it, is shown in another and more general way. Mr. Gould remarked to me long ago, that in those genera of birds which range over the world, many of the species have very wide ranges. I can hardly doubt that this rule is generally true, though it would be difficult to prove it. Amongst mammals, we see it strikingly displayed in Bats, and in a lesser degree in the Felidæ and Canidæ. We see it, if we compare the distribution of butterflies and beetles. So it is with most fresh-water productions, in which so many genera range over the world, and many individual species have enormous ranges. It is not meant that in world-ranging genera all the species have a wide range, or even that they have on an *average* a wide range; but only that some of the species range very widely; for the facility with which widely-ranging species vary and give rise to new forms will largely determine their average range. For instance, two varieties of the same species inhabit America and Europe, and the species thus has an immense range; but, if the variation had been a little greater, the two varieties would have been ranked as distinct species, and the common range would have been greatly reduced. Still less is it meant, that a species which apparently has the capacity of crossing barriers and ranging widely, as in the case of certain powerfully-winged birds, will necessarily range widely; for we should never forget that to range widely implies not only the power of crossing barriers, but the more important power of being victorious in distant lands in the struggle for life with foreign associates. But on the view of all the species of a genus having descended from a single parent, though now distributed to the most remote points of the world, we ought to find, and I believe as a general rule we do find, that some at least of the species range very widely; for it is necessary that the unmodified parent should range widely, undergoing modification during its diffusion, and should place itself under diverse conditions favourable for the conversion of its offspring, firstly into new varieties and ultimately into new species.

In considering the wide distribution of certain genera, we should bear in mind that some are extremely ancient, and that the species must have branched off from a common parent at a remote epoch; so that in such cases there will have been ample time for great climatal and geographical changes and for accidents of transport; and consequently for the migration of some of the species into all quarters of the world, where they may have become slightly modified in relation to their new conditions. There is, also, some reason to believe from geological evidence that organisms low in the scale within each great class, generally change at a slower rate than the higher forms; and consequently the lower forms will have had a better chance of ranging widely and of still retaining the same specific character. This fact, together with the seeds and eggs of many low forms being very minute and better fitted for distant transportation, probably accounts for a law which has long been observed, and which has lately been admirably discussed by Alph. de Candolle in regard to plants, namely, that the lower any group of organisms is, the more widely it is apt to range.

The relations just discussed,—namely, low and slowly-changing organisms ranging more widely than the high,—some of the species of widely-ranging genera themselves ranging widely,—such facts, as alpine, lacustrine, and marsh productions being related (with the exceptions before specified) to those on the surrounding low lands and dry lands, though these stations are so different,—the very close relation of the distinct species which inhabit the islets of the same archipelago,—and especially the striking relation of the inhabitants of

each whole archipelago or island to those of the nearest
mainland,—are, I think, utterly inexplicable on the
ordinary view of the independent creation of each species, but are explicable on the view of colonisation
from the nearest or readiest source, together with the
subsequent modification and better adaptation of the
colonists to their new homes.

Summary of last and present Chapter.—In these
chapters I have endeavoured to show, that if we make
due allowance for our ignorance of the full effects of all
the changes of climate and of the level of the land,
which have certainly occurred within the recent period,
and of other similar changes which may have occurred
within the same period; if we remember how profoundly ignorant we are with respect to the many
and curious means of occasional transport,—a subject
which has hardly ever been properly experimented on ;
if we bear in mind how often a species may have ranged
continuously over a wide area, and then have become
extinct in the intermediate tracts, the difficulties in
believing that all the individuals of the same species,
wherever located, have descended from the same parents,
are not insuperable. And we are led to this conclusion,
which has been arrived. at by many naturalists under
the designation of single centres of creation, by some
general considerations, more especially from the importance of barriers and from the analogical distribution
of sub-genera, genera, and families.

With respect to the distinct species of the same genus,
which on my theory must have spread from one parent-
source ; if we make the same allowances as before for
our ignorance, and remember that some forms of life
change most slowly, enormous periods of time being
thus granted for their migration, I do not think that the

difficulties are insuperable; though they often are in this case, and in that of the individuals of the same species, extremely great.

As exemplifying the effects of climatal changes on distribution, I have attempted to show how important has been the influence of the modern Glacial period, which I am fully convinced simultaneously affected the whole world, or at least great longitudinal belts. As showing how diversified are the means of occasional transport, I have discussed at some little length the means of dispersal of fresh-water productions.

If the difficulties be not insuperable in admitting that in the long course of time the individuals of the same species, and likewise of allied species, have proceeded from some one source; then all the grand leading facts of geographical distribution are explicable on the theory of migration (generally of the more dominant forms of life), together with subsequent modification and the multiplication of new forms. We can thus understand the high importance of barriers, whether of land or water, which separate our several zoological and botanical provinces. We can thus understand the localisation of sub-genera, genera, and families; and how it is that under different latitudes, for instance in South America, the inhabitants of the plains and mountains, of the forests, marshes, and deserts, are in so mysterious a manner linked together by affinity, and are likewise linked to the extinct beings which formerly inhabited the same continent. Bearing in mind that the mutual relation of organism to organism is of the highest importance, we can see why two areas having nearly the same physical conditions should often be inhabited by very different forms of life; for according to the length of time which has elapsed since new inhabitants entered one region; according to the nature of the communication

which allowed certain forms and not others to enter, either
in greater or lesser numbers; according or not, as those
which entered happened to come into more or less direct
competition with each other and with the aborigines;
and according as the immigrants were capable of varying more or less rapidly, there would ensue in different
regions, independently of their physical conditions, infinitely diversified conditions of life. There would be an
almost endless amount of organic action and reaction,—
and we should find, as we do find, some groups of beings
greatly, and some only slightly modified,—some developed in great force, some existing in scanty numbers—
in the different great geographical provinces of the
world.

On these same principles we can understand, as I
have endeavoured to show, why oceanic islands should
have few inhabitants, but of these a great number
should be endemic or peculiar; and why, in relation to
the means of migration, one group of beings, even within
the same class, should have all its species endemic, and
another group should have all its species common to
other quarters of the world. We can see why whole
groups of organisms, as batrachians and terrestrial mammals, should be absent from oceanic islands, whilst the
most isolated islands possess their own peculiar species of
aerial mammals or bats. We can see why there should
be some relation between the presence of mammals, in
a more or less modified condition, and the depth of the
sea between an island and the mainland. We can clearly
see why all the inhabitants of an archipelago, though
specifically distinct on the several islets, should be closely
related to each other, and likewise be related, but less
closely, to those of the nearest continent or other source
whence immigrants were probably derived. We can
see why in two areas, however distant from each other,

there should be a correlation, in the presence of identical species, of varieties, of doubtful species, and of distinct but representative species.

As the late Edward Forbes often insisted, there is a striking parallelism in the laws of life throughout time and space: the laws governing the succession of forms in past times being nearly the same with those governing at the present time the differences in different areas. We see this in many facts. The endurance of each species and group of species is continuous in time; for the exceptions to the rule are so few, that they may fairly be attributed to our not having as yet discovered in an intermediate deposit the forms which are therein absent, but which occur above and below: so in space, it certainly is the general rule that the area inhabited by a single species, or by a group of species, is continuous; and the exceptions, which are not rare, may, as I have attempted to show, be accounted for by migration at some former period under different conditions or by occasional means of transport, and by the species having become extinct in the intermediate tracts. Both in time and space, species and groups of species have their points of maximum development. Groups of species, belonging either to a certain period of time, or to a certain area, are often characterised by trifling characters in common, as of sculpture or colour. In looking to the long succession of ages, as in now looking to distant provinces throughout the world, we find that some organisms differ little, whilst others belonging to a different class, or to a different order, or even only to a different family of the same order, differ greatly. In both time and space the lower members of each class generally change less than the higher; but there are in both cases marked exceptions to the rule. On my theory these several relations throughout time and space are intelligible; for whether we look to the forms of life which have changed during successive ages within the same quarter of the world, or to those which have changed after having migrated into distant quarters, in both cases the forms within each class have been connected by the same bond of ordinary generation; and the more nearly any two forms are related in blood, the nearer they will generally stand to each other in time and space; in both cases the laws of variation have been the same, and modifications have been accumulated by the same power of natural selection.

CHAPTER XIII.

MUTUAL AFFINITIES OF ORGANIC BEINGS: MORPHOLOGY: EMBRYOLOGY: RUDIMENTARY ORGANS.

CLASSIFICATION, groups subordinate to groups — Natural system — Rules and difficulties in classification, explained on the theory of descent with modification — Classification of varieties — Descent always used in classification — Analogical or adaptive characters — Affinities, general, complex and radiating — Extinction separates and defines groups — Morphology, between members of the same class, between parts of the same individual — Embryology, laws of, explained by variations not supervening at an early age, and being inherited at a corresponding age — Rudimentary organs; their origin explained — Summary.

From the first dawn of life, all organic beings are found to resemble each other in descending degrees, so that they can be classed in groups under groups. This classification is evidently not arbitrary like the grouping of the stars in constellations. The existence of groups would have been of simple signification, if one group had been exclusively fitted to inhabit the land, and another the water; one to feed on flesh, another on vegetable matter, and so on; but the case is widely different in nature; for it is notorious how commonly members of even the same sub-group have different habits. In the second and fourth chapters, on Variation and on Natural Selection, I have attempted to show that within each country it is the widely ranging, the much diffused and common, that is the dominant species belonging to the larger genera in each class, which vary most. The varieties, or incipient species, thus produced, ultimately become converted, as I believe, into new and distinct species; and these, on the principle of inheritance, tend to produce other new and dominant species. Consequently the groups which are now large, and which generally include many dominant species, tend to go on increasing in size. I further attempted to show that from the varying descendants of each species trying to occupy as many and as different places as possible in the economy of nature, there is a constant tendency in their characters to diverge. This conclusion was supported by looking at the great diversity of the forms of life which, in any small area, come into the closest competition, and by looking to certain facts in naturalisation.

I attempted also to show that there is a constant tendency in the forms which are increasing in number and diverging in character, to supplant and exterminate the less divergent, the less improved and preceding forms. I request the reader to turn to the diagram illustrating the action, as formerly explained, of these several principles; and he will see that the inevitable result is that the modified descendants proceeding from one progenitor become broken up into groups subordinate to groups. In the diagram each letter on the uppermost line may represent a genus including several species; and all the genera on this upper line form together one class, for all have descended from one ancient parent, and, consequently, have inherited something in common. But the three genera on the left hand have, on this same principle, much in common, and form a sub-family, distinct from that including the next two genera on the right hand, which diverged from a common parent at the fifth stage of descent. These five genera have also much, though less, in common; and they form a family distinct from that including the three genera still farther to the right hand, which diverged at a still earlier period. And all these genera, descended from (A), form an order distinct from the genera descended from (I). So that we here have many species descended from a single progenitor grouped into genera; and the genera are included in, or subordinate to, sub-families, families, and orders, all united into one class. Thus, the grand fact in natural history of the subordination of group under group, which, from its familiarity, does not always sufficiently strike us, is in my judgment explained.

Naturalists try to arrange the species, genera, and families in each class, on what is called the Natural System. But what is meant by this system? Some authors look at it merely as a scheme for arranging together those living objects which are most alike, and for separating those which are most unlike; or as an artificial means for enunciating, as briefly as possible, general propositions,—that is, by one sentence to give the characters common, for instance, to all mammals, by another those common to all carnivora, by another those common to the dog-genus, and then by adding a single sentence, a full description is given of each kind of dog. The ingenuity and utility of this system are indisputable. But many naturalists think that something more is meant by the Natural System; they believe that it reveals the plan of the Creator; but unless it be specified whether order in time or space, or both, or what else is meant by the plan of the Creator, it seems to me that nothing is thus added to our knowledge. Such expressions as that famous one of Linnæus, and which we often meet with in a more or less concealed form, that the characters do not make the genus, but that the genus gives the characters, seem to imply that something more is included in our classification than mere resemblance. I believe that something more is included; and that propinquity of descent—the only known cause of the similarity of organic beings—is the bond, hidden as it is by various degrees of modification, which is partially revealed to us by our classifications.

Let us now consider the rules followed in classification, and the difficulties which are encountered on the view that classification either gives some unknown plan of creation, or is simply a scheme for enunciating general propositions and of placing together the forms most like each other. It might have been thought (and was in ancient times thought) that those parts of the structure which determined the habits of life, and the general place of each being in the economy of nature, would be of very high importance in classification. Nothing can be more false. No one regards the external similarity of a mouse to a shrew, of a dugong to a whale, of a whale to a fish, as of any importance. These resemblances, though so intimately connected with the whole life of the being, are ranked as merely "adaptive or analogical characters;" but to the consideration of these resemblances we shall have to recur. It may even be given as a general rule, that the less any part of the organisation is concerned with special habits, the more important it becomes for classification. As an instance: Owen, in speaking of the dugong, says, "The generative organs being those which are most remotely related to the habits and food of an animal, I have always regarded as affording very clear indications of its true affinities. We are least likely in the modifications of these organs to mistake a merely adaptive for an essential character." So with plants, how remarkable it is that the organs of vegetation, on which their whole life depends, are of little signification, excepting in the first main divisions; whereas the organs of reproduction, with their product the seed, are of paramount importance!

We must not, therefore, in classifying, trust to resemblances in parts of the organisation, however important

they may be for the welfare of the being in relation to
the outer world.   Perhaps from this cause it has partly
arisen, that almost all naturalists lay the greatest stress
on resemblances in organs of high vital or physiological
importance.   No doubt this view of the classificatory
importance of organs which are important is generally,
but by no means always, true.   But their importance
for classification, I believe, depends on their greater constancy throughout large groups of species; and this
constancy depends on such organs having generally been
subjected to less change in the adaptation of the species
to their conditions of life.   That the mere physiological
importance of an organ does not determine its classificatory value, is almost shown by the one fact, that in
allied groups, in which the same organ, as we have every
reason to suppose, has nearly the same physiological
value, its classificatory value is widely different.   No
naturalist can have worked at any group without being
struck with this fact; and it has been fully acknowledged in the writings of almost every author.   It will
suffice to quote the highest authority, Robert Brown,
who in speaking of certain organs in the Proteaceæ,
says their generic importance, "like that of all their
parts, not only in this, but, as I apprehend, in every
natural family, is very unequal, and in some cases seems
to be entirely lost."   Again in another work he says,
the genera of the Connaraceæ "differ in having one or
more ovaria, in the existence or absence of albumen, in
the imbricate or valvular aestivation.   Any one of these
characters singly is frequently of more than generic
importance, though here even when all taken together
they appear insufficient to separate Cnestis from Con-
narus."   To give an example amongst insects, in one
great division of the Hymenoptera, the antennæ, as
Westwood has remarked, are most constant in structure;

In another division they differ much, and the differences
are of quite subordinate value in classification; yet no
one probably will say that the antennæ in these two
divisions of the same order are of unequal physiological
importance. Any number of instances could be given
of the varying importance for classification of the same
important organ within the same group of beings.

Again, no one will say that rudimentary or atrophied
organs are of high physiological or vital importance;
yet, undoubtedly, organs in this condition are often of
high value in classification. No one will dispute that
the rudimentary teeth in the upper jaws of young ruminants, and certain rudimentary bones of the leg, are
highly serviceable in exhibiting the close affinity between
Ruminants and Pachyderms. Robert Brown has strongly
insisted on the fact that the position of the rudimentary
florets is of the highest importance in the classification
of the Grasses.

Numerous instances could be given of characters
derived from parts which must be considered of very
trifling physiological importance, but which are universally admitted as highly serviceable in the definition
of whole groups. For instance, whether or not there is
an open passage from the nostrils to the mouth, the
only character, according to Owen, which absolutely dis-
tinguishes fishes and reptiles—the inflection of the angle
of the jaws in Marsupials—the manner in which the
wings of insects are folded—mere colour in certain
Algæ—mere pubescence on parts of the flower in
grasses—the nature of the dermal covering, as hair or
feathers, in the Vertebrata. If the Ornithorhynchus had
been covered with feathers instead of hair, this external
and trifling character would have been considered by
naturalists as important an aid in determining the
degree of affinity of this strange creature to birds and

reptiles, as an approach in structure in any one internal organ.

The importance, for classification, of trifling characters, mainly depends on their being correlated with several other characters of more or less importance. The value indeed of an aggregate of characters is very evident in natural history. Hence, as has often been remarked, a species may depart from its allies in several characters, both of high physiological importance and of almost universal prevalence, and yet leave us in no doubt where it should be ranked. Hence, also, it has been found, that a classification founded on any single character, however important that may be, has always failed; for no part of the organisation is universally constant. The importance of an aggregate of characters, even when none are important, alone explains, I think, that saying of Linnæus, that the characters do not give the genus, but the genus gives the characters; for this saying seems founded on an appreciation of many trifling points of resemblance, too slight to be defined. Certain plants, belonging to the Malpighiaceæ, bear perfect and degraded flowers; in the latter, so A. de Jussieu has remarked, "the greater number of the characters proper to the species, to the genus, to the family, to the class, disappear, and thus laugh at our classification." But when Aspicarpa produced in France, during several years, only degraded flowers, departing so wonderfully in a number of the most important points of structure from the proper type of the order, yet M. Richard sagaciously saw, as Jussieu observes, that this genus should still be retained amongst the Malpighiaceæ. This case seems to me well to illustrate the spirit with which our classifications are sometimes necessarily founded.

Practically when naturalists are at work, they do

not trouble themselves about the physiological value
of the characters which they use in defining a group,
or in allocating any particular species. If they find
a character nearly uniform, and common to a great
number of forms, and not common to others, they use
it as one of high value: if common to some lesser
number, they use it as of subordinate value. This
principle has been broadly confessed by some naturalists
to be the true one; and by none more clearly than by
that excellent botanist, Aug. St. Hilaire. If certain
characters are always found correlated with others,
though no apparent bond of connexion can be discovered between them, especial value is set on them.
As in most groups of animals, important organs, such as
those for propelling the blood, or for aerating it, or those
for propagating the race, are found nearly uniform, they
are considered as highly serviceable in classification;
but in some groups of animals all these, the most important vital organs, are found to offer characters of quite
subordinate value.

We can see why characters derived from the embryo
should be of equal importance with those derived from
the adult, for our classifications of course include all
ages of each species. But it is by no means obvious,
on the ordinary view, why the structure of the embryo
should be more important for this purpose than that of
the adult, which alone plays its full part in the economy
of nature. Yet it has been strongly urged by those
great naturalists, Milne Edwards and Agassiz, that embryonic characters are the most important of any in the
classification of animals; and this doctrine has very
generally been admitted as true. The same fact holds
good with flowering plants, of which the two main divisions have been founded on characters derived from
the embryo,—on the number and position of the co-

bryonic leaves or cotyledons, and on the mode of development of the plumule and radicle. In our discussion on embryology, we shall see why such characters are so valuable, on the view of classification tacitly including the idea of descent.

Our classifications are often plainly influenced by chains of affinities. Nothing can be easier than to define a number of characters common to all birds; but in the case of crustaceans, such definition has hitherto been found impossible. There are crustaceans at the opposite ends of the series, which have hardly a character in common; yet the species at both ends, from being plainly allied to others, and these to others, and so onwards, can be recognised as unequivocally belonging to this, and to no other class of the Articulata.

Geographical distribution has often been used, though perhaps not quite logically, in classification, more especially in very large groups of closely allied forms. Temminck insists on the utility or even necessity of this practice in certain groups of birds; and it has been followed by several entomologists and botanists.

Finally, with respect to the comparative value of the various groups of species, such as orders, sub-orders, families, sub-families, and genera, they seem to be, at least at present, almost arbitrary. Several of the best botanists, such as Mr. Bentham and others, have strongly insisted on their arbitrary value. Instances could be given amongst plants and insects, of a group of forms, first ranked by practised naturalists as only a genus, and then raised to the rank of a sub-family or family; and this has been done, not because further research has detected important structural differences, at first overlooked, but because numerous allied species, with slightly different grades of difference, have been subsequently discovered.

All the foregoing rules and aids and difficulties in classification are explained, if I do not greatly deceive myself, on the view that the natural system is founded on descent with modification; that the characters which naturalists consider as showing true affinity between any two or more species, are those which have been inherited from a common parent, and, in so far, all true classification is genealogical; that community of descent is the hidden bond which naturalists have been unconsciously seeking, and not some unknown plan of creation, or the enunciation of general propositions, and the mere putting together and separating objects more or less alike.

But I must explain my meaning more fully. I believe that the arrangement of the groups within each class, in due subordination and relation to the other groups, must be strictly genealogical in order to be natural; but that the amount of difference in the several branches or groups, though allied in the same degree in blood to their common progenitor, may differ greatly, being due to the different degrees of modification which they have undergone; and this is expressed by the forms being ranked under different genera, families, sections, or orders. The reader will best understand what is meant, if he will take the trouble of referring to the diagram in the fourth chapter. We will suppose the letters A to L to represent allied genera, which lived during the Silurian epoch, and these have descended from a species which existed at an unknown anterior period. Species of three of these genera (A, F, and I) have transmitted modified descendants to the present day, represented by the fifteen genera (a[

to a¹⁴) on the uppermost horizontal line. Now all these modified descendants from a single species, are represented as related in blood or descent to the same degree; they may metaphorically be called cousins to the same millionth degree; yet they differ widely and in different degrees from each other. The forms descended from A, now broken up into two or three families, constitute a distinct order from those descended from I, also broken up into two families. Nor can the existing species, descended from A, be ranked in the same genus with the parent A; or those from I, with the parent I. But the existing genus F¹⁴ may be supposed to have been but slightly modified; and it will then rank with the parent-genus F; just as some few still living organic beings belong to Silurian genera. So that the amount or value of the differences between organic beings all related to each other in the same degree in blood, has come to be widely different. Nevertheless their genealogical arrangement remains strictly true, not only at the present time, but at each successive period of descent. All the modified descendants from A will have inherited something in common from their common parent, as will all the descendants from I; so will it be with each subordinate branch of descendants, at each successive period. If, however, we choose to suppose that any of the descendants of A or of I have been so much modified as to have more or less completely lost traces of their parentage, in this case, their places in a natural classification will have been more or less completely lost,—as sometimes seems to have occurred with existing organisms. All the descendants of the genus F, along its whole line of descent, are supposed to have been but little modified, and they yet form a single genus. But this genus, though much isolated, will still occupy its proper intermediate position; for F originally was intermediate in character between A and I, and the several genera descended from these two genera will have inherited to a certain extent their characters. This natural arrangement is shown, as far as is possible on paper, in the diagram, but in much too simple a manner. If a branching diagram had not been used, and only the names of the groups had been written in a linear series, it would have been still less possible to have given a natural arrangement; and it is notoriously not possible to represent in a series, on a flat surface, the affinities which we discover in nature amongst the beings of the same group. Thus, on the view which I hold, the natural system is genealogical in its arrangement, like a pedigree; but the degrees of modification which the different groups have undergone, have to be expressed by ranking them under different so-called genera, sub-families, families, sections, orders, and classes.

It may be worth while to illustrate this view of classification, by taking the case of languages. If we possessed a perfect pedigree of mankind, a genealogical arrangement of the races of man would afford the best classification of the various languages now spoken throughout the world; and if all extinct languages, and all intermediate and slowly changing dialects, had to be included, such an arrangement would be the only possible one. Yet it might be that some very ancient language had altered little, and had given rise to few new languages, whilst others owing to the spreading and subsequent isolation and states of civilisation of the several races, descended from a common race) had altered much, and had given rise to many new languages and dialects. The various degrees of difference in the languages from the same stock would have to be expressed by groups subordinate to groups; but the proper or even only possible arrangement would still be genealogical; and this would be strictly natural, as it would connect together all languages, extinct and modern, by the closest affinities, and would give the filiation and origin of each tongue.

In confirmation of this view, let us glance at the classification of varieties, which are believed or known to have descended from one species. These are grouped under species, with sub-varieties under varieties; and with our domestic productions, several other grades of difference are requisite, as we have seen with pigeons. The origin of the existence of groups subordinate to groups is the same with varieties as with species, namely, closeness of descent with various degrees of modification. Nearly the same rules are followed in classifying varieties, as with species. Authors have insisted on the necessity of classing varieties on a natural instead of an artificial system; we are cautioned, for instance, not to class two varieties of the pine-apple together, merely because their fruit, though the most important part, happens to be nearly identical; no one puts the Swedish and common turnips together, though the esculent and thickened stems are so similar. Whatever part is found to be most constant, is used in classing varieties: thus the great agriculturist Marshall says the horns are very useful for this purpose with cattle, because they are less variable than the shape or colour of the body, &c.; whereas with sheep the horns are much less serviceable, because less constant. In classing varieties, I apprehend if we had a real pedigree, a genealogical classification would be universally preferred; and it has been attempted by some authors. For we might feel sure, whether there had been more or less modification, the principle of inheritance would keep the forms together which were allied in the greatest number of points. In tumbler pigeons, though some sub-varieties differ from the others in the important character of having a longer beak, yet all are kept together from having the common habit of tumbling; but the short-faced breed has nearly or quite lost this habit; nevertheless, without any reasoning or thinking on the subject, these tumblers are kept in the same group, because allied in blood and alike in some other respects. If it could be proved that the Hottentot had descended from the Negro, I think he would be classed under the Negro group, however much he might differ in colour and other important characters from negroes.

With species in a state of nature, every naturalist has in fact brought descent into his classification; for he includes in his lowest grade, or that of a species, the two sexes; and how enormously these sometimes differ in the most important characters, is known to every naturalist: scarcely a single fact can be predicated in common of the males and hermaphrodites of certain cirripedes, when adult, and yet no one dreams of separating them. The naturalist includes as one species the several larval stages of the same individual, however much they may differ from each other and from the adult; as he likewise includes the so-called alternate generations of Steenstrup, which can only in a technical sense be considered as the same individual. He includes monsters; he includes varieties, not solely because they closely resemble the parent-form, but because they are descended from it. He who believes that the cowslip is descended from the primrose, or conversely, ranks them together as a single species, and gives a single definition. As soon as three Orchidean forms (Monochanthus, Myanthus, and Catasetum), which had previously been ranked as three distinct genera, were known to be sometimes produced on the same spike, they were immediately included as a single species.

As descent has universally been used in classing together the individuals of the same species, though the males and females and larvæ are sometimes extremely different; and as it has been used in classing varieties which have undergone a certain, and sometimes a considerable amount of modification, may not this same element of descent have been unconsciously used in grouping species under genera, and genera under higher groups, though in these cases the modification has been greater in degree, and has taken a longer time to complete? I believe it has thus been unconsciously used; and only thus can I understand the several rules and guides which have been followed by our best systematists. We have no written pedigrees; we have to make out community of descent by resemblances of any kind. Therefore we choose those characters which, as far as we can judge, are the least likely to have been modified in relation to the conditions of life to which each species has been recently exposed. Rudimentary structures on this view are as good as, or even sometimes better than, other parts of the organisation. We care not how trifling a character may be—let it be the mere inflection of the angle of the jaw, the manner in which an insect's wing is folded, whether the skin be covered by hair or feathers—if it prevail throughout many and different species, especially those having very different habits of life, it assumes high value; for we can account for its presence in so many forms with such different habits, only by its inheritance from a common parent. We may err in this respect in regard to single points of structure, but when several characters, let them be ever so trifling, occur together throughout a

X

large group of beings having different habits, we may
feel almost sure, on the theory of descent, that these
characters have been inherited from a common ancestor.
And we know that such correlated or aggregated characters have especial value in classification.

We can understand why a species or a group of
species may depart, in several of its most important
characteristics, from its allies, and yet be safely classed
with them. This may be safely done, and is often done,
as long as a sufficient number of characters, let them
be ever so unimportant, betrays the hidden bond of community of descent. Let two forms have not a single
character in common, yet if these extreme forms are
connected together by a chain of intermediate groups,
we may at once infer their community of descent, and
we put them all into the same class. As we find organs
of high physiological importance—those which serve to
preserve life under the most diverse conditions of existence—are generally the most constant, we attach especial value to them; but if these same organs, in another
group or section of a group, are found to differ much, we
at once value them less in our classification. We shall
hereafter, I think, clearly see why embryological characters are of such high classificatory importance. Geographical distribution may sometimes be brought use-
fully into play in classing large and widely-distributed
genera, because all the species of the same genus, inhabiting any distinct and isolated region, have in all
probability descended from the same parents.

We can understand, on these views, the very important distinction between real affinities and analogical
or adaptive resemblances. Lamarck first called attention to this distinction, and he has been ably followed
by Macleay and others. The resemblances, in the shape
of the body and in the fin-like anterior limbs, between

the dugong, which is a pachydermatous animal, and the whale, and between both these mammals and fishes, is analogical. Amongst insects there are innumerable instances : thus Linnæus, misled by external appearances, actually classed an homopterous insect as a moth. We see something of the same kind even in our domestic varieties, as in the thickened stems of the common and Swedish turnip. The resemblance of the greyhound and racehorse is hardly more fanciful than the analogies which have been drawn by some authors between very distinct animals. On my view of characters being of real importance for classification, only in so far as they reveal descent, we can clearly understand why analogical or adaptive characters, although of the utmost importance to the welfare of the being, are almost valueless to the systematist. For animals, belonging to two most distinct lines of descent, may readily become adapted to similar conditions, and thus assume a close external resemblance; but such resemblances will not reveal—will rather tend to conceal their blood-relationship to their proper lines of descent. We can also understand the apparent paradox, that the very same characters are analogical when one class or order is compared with another, but give true affinities when the members of the same class or order are compared one with another : thus, the shape of the body and fin-like limbs are only analogical when whales are compared with fishes, being adaptations in both classes for swimming through the water; but the shape of the body and fin-like limbs serve as characters exhibiting true affinity between the several members of the whale family; for these cetaceans agree in so many characters, great and small, that we cannot doubt that they have inherited their general shape of body and structure of limbs from a common ancestor. So it is with fishes.

x 2

As members of distinct classes have often been adapted by successive slight modifications to live under nearly similar circumstances,—to inhabit, for instance, the three elements of land, air, and water,—we can perhaps understand how it is that a numerical parallellism has sometimes been observed between the subgroups in distinct classes. A naturalist, struck by a parallelism of this nature in any one class, by arbitrarily raising or sinking the value of the groups in other classes (and all our experience shows that this valuation has hitherto been arbitrary), could easily extend the parallelism over a wide range; and thus the septenary, quinary, quaternary, and ternary classifications have probably arisen.

As the modified descendants of dominant species belonging to the larger genera, tend to inherit the advantages which made the groups to which they belong large and their parents dominant, they are almost sure to spread widely, and to seize on more and more places in the economy of nature. The larger and more dominant groups within each class thus tend to go on increasing in size; and they consequently supplant many smaller and feebler groups. Thus we can account for the fact that all organisms, recent and extinct, are included under a few great orders, under still fewer classes, and all in one great natural system. As showing how few the higher groups are in number, and how widely spread they are throughout the world, the fact is striking, that the discovery of Australia has not added a single insect belonging to a new class; and that in the vegetable kingdom, as I learn from Dr. Hooker, it has added only two or three families of small size.

In the chapter on Geological Succession I attempted to show, on the principle of each group having generally diverged much in character during the long-continued process of modification, how it is that the more ancient forms of life often present characters in some slight degree intermediate between existing groups. A few old and intermediate parent-forms having occasionally transmitted to the present day descendants but little modified, will give to us our so-called osculant or aberrant groups. The more aberrant any form is, the greater must be the number of connecting forms which on my theory have been exterminated and utterly lost. And we have some evidence of aberrant forms having suffered severely from extinction, for they are generally represented by extremely few species; and such species as do occur are generally very distinct from each other, which again implies extinction. The genera Ornithorhynchus and Lepidosiren, for example, would not have been less aberrant had each been represented by a dozen species instead of by a single one; but such richness in species, as I find after some investigation, does not commonly fall to the lot of aberrant genera. We can, I think, account for this fact only by looking at aberrant forms as failing groups conquered by more successful competitors, with a few members preserved by some unusual coincidence of favourable circumstances.

Mr. Waterhouse has remarked that, when a member belonging to one group of animals exhibits an affinity to a quite distinct group, this affinity in most cases is general and not special: thus, according to Mr. Waterhouse, of all Rodents, the bizcacha is most nearly related to Marsupials; but in the points in which it approaches this order, its relations are general, and not to any one marsupial species more than to another. As the points of affinity of the bizcacha to Marsupials are believed to be real and not merely adaptive, they are due on my theory to inheritance in common. Therefore we must suppose either that all Rodents, including the biz-

cies, branched off from some very ancient Marsupial,
which will have had a character in some degree intermediate with respect to all existing Marsupials; or
that both Rodents and Marsupials branched off from a
common progenitor, and that both groups have since
undergone much modification in divergent directions.
On either view we may suppose that the bizcacha has
retained, by inheritance, more of the character of its
ancient progenitor than have other Rodents; and
therefore it will not be specially related to any one
existing Marsupial, but indirectly to all or nearly all
Marsupials, from having partially retained the character
of their common progenitor, or of an early member of
the group. On the other hand, of all Marsupials, as
Mr. Waterhouse has remarked, the phascolomys resembles most nearly, not any one species, but the
general order of Rodents. In this case, however, it
may be strongly suspected that the resemblance is only
analogical, owing to the phascolomys having become
adapted to habits like those of a Rodent. The elder
De Candolle has made nearly similar observations on
the general nature of the affinities of distinct families of
plants.

On the principle of the multiplication and gradual
divergence in character of the species descended from
a common parent, together with their retention by
inheritance of some characters in common, we can
understand the excessively complex and radiating
affinities by which all the members of the same family
or higher group are connected together. For the common parent of a whole family of species, now broken
up by extinction into distinct groups and sub-groups,
will have transmitted some of its characters, modified
in various ways and degrees, to all; and the several
species will consequently be related to each other by

circuitous lines of affinity of various lengths (as may be seen in the diagram so often referred to), mounting up through many predecessors. As it is difficult to show the blood-relationship between the numerous kindred of any ancient and noble family, even by the aid of a genealogical tree, and almost impossible to do this without this aid, we can understand the extraordinary difficulty which naturalists have experienced in describing, without the aid of a diagram, the various affinities which they perceive between the many living and extinct members of the same great natural class.

Extinction, as we have seen in the fourth chapter, has played an important part in defining and widening the intervals between the several groups in each class. We may thus account even for the distinctness of whole classes from each other—for instance, of birds from all other vertebrate animals—by the belief that many ancient forms of life have been utterly lost, through which the early progenitors of birds were formerly connected with the early progenitors of the other vertebrate classes. There has been much less entire extinction of the forms of life which once connected fishes with batrachians. There has been still less in some other classes, as in that of the Crustacea, for here the most wonderfully diverse forms are still tied together by long, but broken chains of affinities. Extinction has only separated groups: it has by no means made them; for if every form which has ever lived on this earth were suddenly to reappear, though it would be quite impossible to give definitions by which each group could be distinguished from other groups, as all would blend together by steps as fine as those between existing varieties, nevertheless a natural classification, or at least a natural arrangement, would be possible. We shall see this by turning to the diagram: the letters A to L, may represent eleven Silurian genera, some of which have produced large groups of modified descendants. Every intermediate link between these eleven genera and their primordial parent, and every intermediate link in each branch and sub-branch of their descendants, may be supposed to be still alive; and the links to be as fine as those between the finest varieties. In this case it would be quite impossible to give any definition by which the several members of the several groups could be distinguished from their more immediate parents; or these parents from their ancient and unknown progenitor. Yet the natural arrangement in the diagram would still hold good; and, on the principle of inheritance, all the forms descended from A, or from I, would have something in common. In a tree we can specify this or that branch, though at the actual fork the two unite and blend together. We could not, as I have said, define the several groups; but we could pick out types, or forms, representing most of the characters of each group, whether large or small, and thus give a general idea of the value of the differences between them. This is what we should be driven to, if we were ever to succeed in collecting all the forms in any class which have lived throughout all time and space. We shall assuredly never succeed in making so perfect a collection: nevertheless, in certain classes, we are tending in this direction; and Milne Edwards has lately insisted, in an able paper, on the high importance of looking to types, whether or not we can separate and define the groups to which such types belong.

Finally, we have seen that natural selection, which results from the struggle for existence, and which almost inevitably induces extinction and divergence of character in the many descendants from one dominant parent-species, explains that great and universal feature in the affinities of all organic beings, namely, their subordination in group under group. We use the element of descent in classing the individuals of both sexes and of all ages, although having few characters in common, under one species; we use descent in classing acknowledged varieties, however different they may be from their parent; and I believe this element of descent is the hidden bond of connexion which naturalists have sought under the term of the Natural System. On this idea of the natural system being, in so far as it has been perfected, genealogical in its arrangement, with the grades of difference between the descendants from a common parent, expressed by the terms genera, families, orders, &c., we can understand the rules which we are compelled to follow in our classification. We can understand why we value certain resemblances far more than others; why we are permitted to use rudimentary and useless organs, or others of trifling physiological importance; why, in comparing one group with a distinct group, we summarily reject analogical or adaptive characters, and yet use these same characters within the limits of the same group. We can clearly see how it is that all living and extinct forms can be grouped together in one great system; and how the several members of each class are connected together by the most complex and radiating lines of affinities. We shall never, probably, disentangle the inextricable web of affinities between the members of any one class; but when we have a distinct object in view, and do not look to some unknown plan of creation, we may hope to make sure but slow progress.

Morphology.—We have seen that the members of the same class, independently of their habits of life, resemble each other in the general plan of their organ-

x 3

lisation. This resemblance is often expressed by the
term "unity of type;" or by saying that the several
parts and organs in the different species of the class are
homologous. The whole subject is included under the
general name of Morphology. This is the most interesting department of natural history, and may be said to
be its very soul. What can be more curious than that
the hand of a man, formed for grasping, that of a mole
for digging, the leg of the horse, the paddle of the porpoise, and the wing of the bat, should all be constructed
on the same pattern, and should include similar bones,
in the same relative positions? Geoffroy St. Hilaire
has insisted strongly on the high importance of relative
connexion in homologous organs: the parts may change
to almost any extent in form and size, and yet they
always remain connected together in the same order.
We never find, for instance, the bones of the arm and
forearm, or of the thigh and leg, transposed. Hence
the same names can be given to the homologous bones
in widely different animals. We see the same great
law in the construction of the mouths of insects: what
can be more different than the immensely long spiral
proboscis of a sphinx-moth, the curious folded one of a
bee or bug, and the great jaws of a beetle?—yet all
these organs, serving for such different purposes, are
formed by infinitely numerous modifications of an upper
lip, mandibles, and two pairs of maxillae. Analogous
laws govern the construction of the mouths and limbs
of crustaceans. So it is with the flowers of plants.

Nothing can be more hopeless than to attempt to
explain this similarity of pattern in members of the
same class, by utility or by the doctrine of final causes.
The hopelessness of the attempt has been expressly
admitted by Owen in his most interesting work on the
'Nature of Limbs.' On the ordinary view of the inde-

pendent creation of each being, we can only say that so
it is;—that it has so pleased the Creator to construct
each animal and plant.

The explanation is manifest on the theory of the
natural selection of successive slight modifications,—
each modification being profitable in some way to the
modified form, but often affecting by correlation of
growth other parts of the organisation. In changes
of this nature, there will be little or no tendency to
modify the original pattern, or to transpose parts. The
bones of a limb might be shortened and widened to any
extent, and become gradually enveloped in thick membrane, so as to serve as a fin; or a webbed foot might
have all its bones, or certain bones lengthened to any
extent, and the membrane connecting them increased
to any extent, so as to serve as a wing: yet in all this
great amount of modification there will be no tendency
to alter the framework of bones or the relative connexion of the several parts. If we suppose that the
ancient progenitor, the archetype as it may be called, of
all mammals, had its limbs constructed on the existing
general pattern, for whatever purpose they served, we
can at once perceive the plain signification of the homologous construction of the limbs throughout the whole
class. So with the mouths of insects, we have only to
suppose that their common progenitor had an upper
lip, mandibles, and two pair of maxillæ, these parts
being perhaps very simple in form; and then natural
selection will account for the infinite diversity in structure and function of the mouths of insects. Nevertheless,
it is conceivable that the general pattern of an organ
might become so much obscured as to be finally lost, by
the atrophy and ultimately by the complete abortion of
certain parts, by the soldering together of other parts,
and by the doubling or multiplication of others,—varia-

tions which we know to be within the limits of possibility. In the paddles of the extinct gigantic sea-lizards, and in the mouths of certain suctorial crustaceans, the general pattern seems to have been thus to a certain extent conserved.

There is another and equally curious branch of the present subject; namely, the comparison not of the same part in different members of a class, but of the different parts or organs in the same individual. Most physiologists believe that the bones of the skull are homologous with—that is correspond in number and in relative connexion with—the elemental parts of a certain number of vertebræ. The anterior and posterior limbs in each member of the vertebrate and articulate classes are plainly homologous. We see the same law in comparing the wonderfully complex jaws and legs in crustaceans. It is familiar to almost every one, that in a flower the relative position of the sepals, petals, stamens, and pistils, as well as their intimate structure, are intelligible on the view that they consist of metamorphosed leaves, arranged in a spire. In monstrous plants, we often get direct evidence of the possibility of one organ being transformed into another; and we can actually see in embryonic crustaceans and in many other animals, and in flowers, that organs, which when mature become extremely different, are at an early stage of growth exactly alike.

How inexplicable are these facts on the ordinary view of creation! Why should the brain be enclosed in a box composed of such numerous and such extraordinarily shaped pieces of bone? As Owen has remarked, the benefit derived from the yielding of the separate pieces in the act of parturition of mammals, will by no means explain the same construction in the skulls of birds. Why should similar bones have been created

in the formation of the wing and leg of a bat, used as
they are for such totally different purposes? Why
should one crustacean, which has an extremely complex
mouth formed of many parts, consequently always have
fewer legs; or conversely, those with many legs have
simpler mouths? Why should the sepals, petals, stamens and pistils in any individual flower, though fitted
for such widely different purposes, be all constructed
on the same pattern?

On the theory of natural selection, we can satisfactorily
answer these questions. In the vertebrata, we see a series
of internal vertebræ bearing certain processes and appendages; in the articulata, we see the body divided into a
series of segments, bearing external appendages; and in
flowering plants, we see a series of successive spiral
whorls of leaves. An indefinite repetition of the same
part or organ is the common characteristic (as Owen
has observed) of all low or little modified forms; therefore we may readily believe that the unknown progenitor
of the vertebrata possessed many vertebræ; the unknown
progenitor of the articulata, many segments; and the
unknown progenitor of flowering plants, many spiral
whorls of leaves. We have formerly seen that parts many
times repeated are eminently liable to vary in number
and structure; consequently it is quite probable that
natural selection, during a long-continued course of modification, should have seized on a certain number of the
primordially similar elements, many times repeated, and
have adapted them to the most diverse purposes. And
as the whole amount of modification will have been
effected by slight successive steps, we need not wonder
at discovering in such parts or organs a certain degree
of fundamental resemblance, retained by the strong
principle of inheritance.

In the great class of molluscs, though we can homo-

logise the parts of one species with those of other and
distinct species, we can indicate but few serial homologies; that is, we are seldom enabled to say that one
part or organ is homologous with another in the same
individual. And we can understand this fact; for in
molluscs, even in the lowest members of the class, we
do not find nearly so much indefinite repetition of any
one part, as we find in the other great classes of the
animal and vegetable kingdoms.

Naturalists frequently speak of the skull as formed of
metamorphosed vertebræ; the jaws of crabs as metamorphosed legs; the stamens and pistils of flowers as
metamorphosed leaves; but it would in these cases probably be more correct, as Professor Huxley has remarked,
to speak of both skull and vertebræ, both jaws and legs,
&c., as having been metamorphosed, not one from the
other, but from some common element. Naturalists,
however, use such language only in a metaphorical
sense: they are far from meaning that during a long
course of descent, primordial organs of any kind—vertebræ in the one case and legs in the other—have actually
been modified into skulls or jaws. Yet so strong is the
appearance of a modification of this nature having occurred, that naturalists can hardly avoid employing
language having this plain signification. On my view
these terms may be used literally; and the wonderful
fact of the jaws, for instance, of a crab retaining numerous characters, which they would probably have retained
through inheritance, if they had really been metamorphosed during a long course of descent from true legs,
or from some simple appendage, is explained.

Embryology.—It has already been casually remarked
that certain organs in the individual, which when mature
become widely different and serve for different purposes,

are in the embryo exactly alike. The embryos, also, of distinct animals within the same class are often strikingly similar: a better proof of this cannot be given, than a statement made by Von Baer, namely, that "the embryos of mammalia, of birds, lizards, and snakes, probably also of chelonia, are in their earliest states exceedingly like one another, both as a whole and in the mode of development of their parts: so much so, in fact, that we can often distinguish the embryos only by their size. In my possession are two little embryos in spirit, whose names I have omitted to attach, and at present I am quite unable to say to what class they belong. They may be lizards or small birds, or very young mammalia, so complete is the similarity in the mode of formation of the head and trunk in these animals. The extremities, however, are still absent in these embryos. But even if they had existed in the earliest stage of their development we should learn nothing, for the feet of lizards and mammals, the wings and feet of birds, no less than the hands and feet of man, all arise from the same fundamental form." The vermiform larvæ of moths, flies, beetles, &c., resemble each other much more closely than do the mature insects; but in the case of larvæ, the embryos are active, and have been adapted for special lines of life. A trace of the law of embryonic resemblance, sometimes lasts till a rather late age: thus birds of the same genus, and of closely allied genera, often resemble each other in their first and second plumage; as we see in the spotted feathers in the thrush group. In the cat tribe, most of the species are striped or spotted in lines; and stripes or spots can be plainly distinguished in the whelp of the lion and the puma. We occasionally though rarely see something of this kind in plants: thus the first leaves of the ulex or furze, and the first leaves of the phyllodinoses acarina are pinnate or divided like the ordinary leaves of their respective.

The points of structure, in which the embryos of widely different animals of the same class resemble each other, often have no direct relation to their conditions of existence. We cannot, for instance, suppose that in the embryos of the vertebrata the peculiar loop-like course of the arteries near the branchial slits are related to similar conditions,—in the young mammal which is nourished in the womb of its mother, in the egg of the bird which is hatched in a nest, and in the spawn of a frog under water. We have no more reason to believe in such a relation, than we have to believe that the similar bones in the hand of a man, wing of a bat, and fin of a porpoise, are related to similar conditions of life. No good observer will suppose that the stripes on the whelp of a lion, or the spots on the young blackbird, are of any use to these animals, or are related to the conditions to which they are exposed.

The case, however, is different when an animal during any part of its embryonic career is active, and has to provide for itself. The period of activity may come on earlier or later in life; but whenever it comes on, the adaptation of the larva to its conditions of life is just as perfect and as beautiful as in the adult animal. From such special adaptations, the similarity of the larvae or active embryos of allied animals is sometimes much obscured; and cases could be given of the larvae of two species, or of two groups of species, differing quite as much, or even more, from each other than do their adult parents. In most cases, however, the larvae, though active, still obey, more or less closely, the law of common embryonic resemblance. Cirripedes afford a good instance of this: even the illustrious Cuvier did not perceive that a barnacle was, as it certainly is, a crustacean;

but a glance at the larva shows this to be the case in an unmistakeable manner. So again the two main divisions of cirripedes, the pedunculated and sessile, which differ widely in external appearance, have larvæ in all their stages barely distinguishable.

The embryo in the course of development generally rises in organisation: I use this expression, though I am aware that it is hardly possible to define clearly what is meant by the organisation being higher or lower. But no one probably will dispute that the butterfly is higher than the caterpillar. In some cases, however, the mature animal is generally considered as lower in the scale than the larva, as with certain parasitic crustaceans. To refer once again to cirripedes: the larvæ in the first stage have three pairs of legs, a very simple single eye, and a probosciformed mouth, with which they feed largely, for they increase much in size. In the second stage, answering to the chrysalis stage of butterflies, they have six pairs of beautifully constructed natatory legs, a pair of magnificent compound eyes, and extremely complex antennæ; but they have a closed and imperfect mouth, and cannot feed: their function at this stage is, to search by their well-developed organs of sense, and to reach by their active powers of swimming, a proper place on which to become attached and to undergo their final metamorphosis. When this is completed they are fixed for life: their legs are now converted into prehensile organs; they again obtain a well-constructed mouth; but they have no antennæ, and their two eyes are now reconverted into a minute, single, and very simple eye-spot. In this last and complete state, cirripedes may be considered as either more highly or more lowly organised than they were in the larval condition. But in some genera the larvæ become developed either into hermaphrodites having the ordinary structure, or into what I have called complemental males; and in the latter, the development has assuredly been retrograde; for the male is a mere sack, which lives for a short time, and is destitute of mouth, stomach, or other organ of importance, excepting for reproduction.

We are so much accustomed to see differences in structure between the embryo and the adult, and likewise a close similarity in the embryos of widely different animals within the same class, that we might be led to look at these facts as necessarily contingent in some manner on growth. But there is no obvious reason why, for instance, the wing of a bat, or the fin of a porpoise, should not have been sketched out with all the parts in proper proportion, as soon as any structure became visible in the embryo. And in some whole groups of animals and in certain members of other groups, the embryo does not at any period differ widely from the adult: thus Owen has remarked in regard to cuttle-fish, "there is no metamorphosis; the cephalopodic character is manifested long before the parts of the embryo are completed;" and again in spiders, "there is nothing worthy to be called a metamorphosis." The larvæ of insects, whether adapted to the most diverse and active habits, or quite inactive, being fed by their parents or placed in the midst of proper nutriment, yet nearly all pass through a similar worm-like stage of development; but in some few cases, as in that of Aphis, if we look to the admirable drawings by Professor Huxley of the development of this insect, we see no trace of the vermiform stage.

How, then, can we explain these several facts in embryology,—namely the very general, but not universal difference in structure between the embryo and the adult; of parts in the same individual embryo,

which ultimately become very unlike and serve for
diverse purposes, being at this early period of growth
alike;—of embryos of different species within the same
class, generally, but not universally, resembling each
other;—of the structure of the embryo not being closely
related to its conditions of existence, except when the
embryo becomes at any period of life active and has to
provide for itself;—of the embryo apparently having
sometimes a higher organisation than the mature
animal, into which it is developed? I believe that all
these facts can be explained, as follows, on the view of
descent with modification.

It is commonly assumed, perhaps from monstrosities
often affecting the embryo at a very early period,
that slight variations necessarily appear at an equally
early period. But we have little evidence on this head—
indeed the evidence rather points the other way; for it
is notorious that breeders of cattle, horses, and various
fancy animals, cannot positively tell, until some time
after the animal has been born, what its merits or form
will ultimately turn out. We see this plainly in our own
children; we cannot always tell whether the child will
be tall or short, or what its precise features will be. The
question is not, at what period of life any variation has
been caused, but at what period it is fully displayed.
The cause may have acted, and I believe generally has
acted, even before the embryo is formed; and the variation may be due to the male and female sexual elements
having been affected by the conditions to which either
parent, or their ancestors, have been exposed. Nevertheless an effect thus caused at a very early period, even
before the formation of the embryo, may appear late in
life; as when an hereditary disease, which appears in
old age alone, has been communicated to the offspring
from the reproductive element of one parent. Or

again, as when the horns of cross-bred cattle have been affected by the shape of the horns of either parent. For the welfare of a very young animal, as long as it remains in its mother's womb, or in the egg, or as long as it is nourished and protected by its parent, it must be quite unimportant whether most of its characters are fully acquired a little earlier or later in life. It would not signify, for instance, to a bird which obtained its food best by having a long beak, whether or not it acquired a beak of this particular length, as long as it was fed by its parents. Hence, I conclude, that it is quite possible, that each of the many successive modifications, by which each species has acquired its present structure, may have supervened at a not very early period of life; and some direct evidence from our domestic animals supports this view. But in other cases it is quite possible that each successive modification, or most of them, may have appeared at an extremely early period.

I have stated in the first chapter, that there is some evidence to render it probable, that at whatever age any variation first appears in the parent, it tends to reappear at a corresponding age in the offspring. Certain variations can only appear at corresponding ages, for instance, peculiarities in the caterpillar, cocoon, or imago states of the silk-moth; or, again, in the horns of almost full-grown cattle. But further than this, variations which, for all that we can see, might have appeared earlier or later in life, tend to appear at a corresponding age in the offspring and parent. I am far from meaning that this is invariably the case; and I could give a good many cases of variations (taking the word in the largest sense) which have supervened at an earlier age in the child than in the parent.

These two principles, if their truth be admitted, will, I believe, explain all the above specified leading facts in embryology. But first let us look at a few analogous cases in domestic varieties. Some authors who have written on Dogs, maintain that the greyhound and bull-dog, though appearing so different, are really varieties closely allied, and have probably descended from the same wild stock; hence I was curious to see how far their puppies differed from each other: I was told by breeders that they differed just as much as their parents, and this, judging by the eye, seemed almost to be the case; but on actually measuring the old dogs and their six-days old puppies, I found that the puppies had not nearly acquired their full amount of proportional difference. So, again, I was told that the foals of cart and race-horses differed as much as the full-grown animals; and this surprised me greatly, as I think it probable that the difference between these two breeds has been wholly caused by selection under domestication: but having had careful measurements made of the dam and of a three-days old colt of a race and heavy cart-horses, I find that the colts have by no means acquired their full amount of proportional difference.

As the evidence appears to me conclusive, that the several domestic breeds of Pigeon have descended from one wild species, I compared young pigeons of various breeds, within twelve hours after being hatched; I carefully measured the proportions (but will not here give details) of the beak, width of mouth, length of nostril and of eyelid, size of foot and length of leg, in the wild stock, in pouters, fantails, runts, barbs, dragons, carriers, and tumblers. Now some of these birds, when mature, differ so extraordinarily in length and form of beak, that they would, I cannot doubt, be ranked in distinct genera, had they been natural productions. But when the nestling birds of these several breeds were placed in a row, though most of them could be distinguished from each other, yet their proportional differences in the above specified several points were incomparably less than in the full-grown birds. Some characteristic points of difference—for instance, that of the width of mouth—could hardly be detected in the young. But there was one remarkable exception to this rule, for the young of the short-faced tumbler differed from the young of the wild rock-pigeon and of the other breeds, in all its proportions, almost exactly as much as in the adult state.

The two principles above given seem to me to explain these facts in regard to the later embryonic stages of our domestic varieties. Fanciers select their horses, dogs, and pigeons for breeding, when they are nearly grown up: they are indifferent whether the desired qualities and structures have been acquired earlier or later in life, if the full-grown animal possesses them. And the cases just given, more especially that of pigeons, seem to show that the characteristic differences which give value to each breed, and which have been accumulated by man's selection, have not generally first appeared at an early period of life, and have been inherited by the offspring at a corresponding not early period. But the case of the short-faced tumbler, which when twelve hours old had acquired its proper proportions, proves that this is not the universal rule; for here the characteristic differences must either have appeared at an earlier period than usual, or, if not so, the differences must have been inherited, not at the corresponding, but at an earlier age.

Now let us apply these facts and the above two principles—which latter, though not proved true, can be shown to be in some degree probable—to species in a state of nature. Let us take a genus of birds, descended on my theory from some one parent-species,

ard of which the several new species have become
modified through natural selection in accordance with
their diverse habits. Then, from the many slight successive steps of variation having supervened at a rather
late age, and having been inherited at a corresponding
age, the young of the new species of our supposed genus
will manifestly tend to resemble each other much more
closely than do the adults, just as we have seen in the
case of pigeons. We may extend this view to whole
families or even classes. The fore-limbs, for instance,
which served as legs in the parent-species, may have
become, by a long course of modification, adapted in
one descendant to act as hands, in another as paddles,
in another as wings; and on the above two principles—
namely of each successive modification supervening at
a rather late age, and being inherited at a corresponding late age—the fore-limbs in the embryos of
the several descendants of the parent species will still
resemble each other closely, for they will not have been
modified. But in each of our new species, the embryonic
fore-limbs will differ greatly from the fore-limbs in the
mature animal; the limbs in the latter having undergone much modification at a rather late period of life,
and having thus been converted into hands, or paddles,
or wings. Whatever influence long-continued exercise
or use on the one hand, and disuse on the other, may
have in modifying an organ, such influence will mainly
affect the mature animal, which has come to its full powers
of activity and has to gain its own living; and the
effects thus produced will be inherited at a corresponding mature age. Whereas the young will remain unmodified, or be modified in a lesser degree, by the effects
of use and disuse.

In certain cases the successive steps of variation might
supervene, from causes of which we are wholly ignorant,

480 EMBRYOLOGY. CHAP. XIII.

at a very early period of life, or each step might be
inherited at an earlier period than that at which it first
appeared. In either case (as with the short-faced
tumbler) the young or embryo would closely resemble
the mature parent-form. We have seen that this is the
rule of development in certain whole groups of animals,
as with cuttle-fish and spiders, and with a few members
of the great class of insects, as with Aphis. With
respect to the final cause of the young in these cases
not undergoing any metamorphosis, or closely resembling
their parents from their earliest age, we can see that
this would result from the two following contingencies:
firstly, from the young, during a course of modification
carried on for many generations, having to provide for
their own wants at a very early stage of development,
and secondly, from their following exactly the same
habits of life with their parents; for in this case, it
would be indispensable for the existence of the species,
that the child should be modified at a very early age in
the same manner with its parents, in accordance with
their similar habits. Some further explanation, however, of the embryo not undergoing any metamorphosis
is perhaps requisite. If, on the other hand, it profited
the young to follow habits of life in any degree different
from those of their parent, and consequently to be constructed in a slightly different manner, then, on the
principle of inheritance at corresponding ages, the active
young or larvæ might be rendered by natural selection
different to any conceivable extent from their parents.
Such differences might, also, become correlated with
successive stages of development; so that the larvæ in
the first stage, might differ greatly from the larvæ in
the second stage, as we have seen to be the case with
cirripedes. The adult might become fitted for sites or
habits, in which organs of locomotion or of the senses,

&c., would be useless; and in this case the final metamorphosis would be said to be retrograde.

As all the organic beings, extinct and recent, which have ever lived on this earth have to be classed together, and as all have been connected by the finest gradations, the best, or indeed, if our collections were nearly perfect, the only possible arrangement, would be genealogical; descent being on my view the hidden bond of connexion which naturalists have been seeking under the term of the natural system. On this view we can understand how it is that, in the eyes of most naturalists, the structure of the embryo is even more important for classification than that of the adult. For the embryo is the animal in its less modified state; and in so far it reveals the structure of its progenitor. In two groups of animals, however much they may at present differ from each other in structure and habits, if they pass through the same or similar embryonic stages, we may feel assured that they have both descended from the same or nearly similar parents, and are therefore in that degree closely related. Thus, community in embryonic structure reveals community of descent. It will reveal this community of descent, however much the structure of the adult may have been modified and obscured; we have seen, for instance, that cirripedes can at once be recognised by their larvæ as belonging to the great class of crustaceans. As the embryonic state of each species and group of species partially shows us the structure of their less modified ancient progenitors, we can clearly see why ancient and extinct forms of life should resemble the embryos of their descendants,—our existing species. Agassiz believes this to be a law of nature; but I am bound to confess that I only hope to see the law hereafter proved true. It can be proved true in those cases alone in which the ancient state, now supposed to be represented in existing embryos, has not been obliterated, either by the successive variations in a long course of modification having supervened at a very early period of growth, or by the variations having been inherited at an earlier period than that at which they first appeared. It should also be borne in mind, that the supposed law of resemblance of ancient forms of life to the embryonic stages of recent forms, may be true, but yet, owing to the geological record not extending far enough back in time, may remain for a long period, or for ever, incapable of demonstration.

Thus, as it seems to me, the leading facts in embryology, which are second in importance to none in natural history, are explained on the principle of slight modifications not appearing, in the many descendants from some one ancient progenitor, at a very early period in the life of each, though perhaps caused at the earliest, and being inherited at a corresponding not early period. Embryology rises greatly in interest, when we thus look at the embryo as a picture, more or less obscured, of the common parent-form of each great class of animals.

Rudimentary, atrophied, or aborted Organs.—Organs or parts in this strange condition, bearing the stamp of inutility, are extremely common throughout nature. For instance, rudimentary mammæ are very general in the males of mammals : I presume that the " bastard-wing " in birds may be safely considered as a digit in a rudimentary state : in very many snakes one lobe of the lungs is rudimentary ; in other snakes there are rudiments of the pelvis and hind limbs. Some of the cases of rudimentary organs are extremely curious ; for instance, the presence of teeth in fœtal whales, which when grown up have not a tooth in their heads ; and the presence of

teeth, which never cut through the gums, in the upper
jaws of our unborn calves. It has even been stated on
good authority that rudiments of teeth can be detected
in the beaks of certain embryonic birds. Nothing can
be plainer than that wings are formed for flight, yet in
how many insects do we see wings so reduced in size as
to be utterly incapable of flight, and not rarely lying
under wing-cases, firmly soldered together!

The meaning of rudimentary organs is often quite
unmistakeable : for instance there are beetles of the
same genus (and even of the same species) resembling
each other most closely in all respects, one of which will
have full-sized wings, and another mere rudiments of
membrane ; and here it is impossible to doubt, that the
rudiments represent wings. Rudimentary organs sometimes retain their potentiality, and are merely not developed : this seems to be the case with the mammæ of
male mammals, for many instances are on record of
these organs having become well developed in full-grown
males, and having secreted milk. So again there are
normally four developed and two rudimentary teats in
the udders of the genus Bos; but in our domestic
cows the two sometimes become developed and give
milk. In plants of the same species the petals sometimes occur as mere rudiments, and sometimes in a well-
developed state. In plants with separated sexes, the
male flowers often have a rudiment of a pistil ; and
Kölreuter found that by crossing such male plants with
an hermaphrodite species, the rudiment of the pistil in
the hybrid offspring was much increased in size ; and
this shows that the rudiment and the perfect pistil are
essentially alike in nature.

An organ serving for two purposes, may become rudimentary or utterly aborted for one, even the more
important purpose, and remain perfectly efficient for

the other. Thus in plants, the office of the pistil is to
allow the pollen-tubes to reach the ovules protected in
the ovarium at its base. The pistil consists of a stigma
supported on the style; but in some compositæ, the
male florets, which of course cannot be fecundated, have
a pistil, which is in a rudimentary state, for it is not
crowned with a stigma; but the style remains well
developed, and is clothed with hairs as in other compositæ, for the purpose of brushing the pollen out of the
surrounding anthers. Again, an organ may become
rudimentary for its proper purpose, and be used for a
distinct object: in certain fish the swim-bladder seems
to be nearly rudimentary for its proper function of giving
buoyancy, but has become converted into a nascent
breathing organ or lung. Other similar instances could
be given.

Organs, however little developed, if of use, should not
be called rudimentary: they may be called nascent, and
may hereafter be developed by natural selection to any
further extent. Rudimentary organs, on the other hand,
are essentially useless, as teeth which never cut through
the gums. As when in a still less developed condition
they would be of still less use, they cannot under the
present state of things have been formed by natural
selection, which acts solely by the preservation of useful
modifications. They relate to a former condition of their
possessor, and have been retained, as we shall see, by
inheritance. It is difficult to know what organs are
nascent; looking to the future, we cannot of course tell
how any part will be developed, and whether it is now
nascent; looking to the past, creatures with an organ
in a nascent condition will generally have been supplanted and exterminated by their successors with the
organ in a more perfect and developed condition. The
wing of the penguin is of high service, and acts as a fin;

it may, therefore, represent the nascent state of the wings
of birds; not that I believe this to be the case; it is more
probably a reduced organ, modified for a new function:
the wing of the Apteryx is useless, and is truly rudimentary. The mammary glands of the Ornithorhynchus
may, perhaps, be considered, in comparison with the
udder of a cow, as in a nascent state. The ovigerous
frena of certain cirripedes, which are only slightly developed and which have ceased to give attachment to the
ova, are nascent branchiæ.

Rudimentary organs in the individuals of the same
species are very liable to vary in degree of development
and in other respects. Moreover, in closely allied
species, the degree to which the same organ has been
rendered rudimentary occasionally differs much. This
latter fact is well exemplified in the state of the wings
of the female moths in certain groups. Rudimentary
organs may be utterly aborted; and this implies, that
we find in an animal or plant no trace of an organ,
which analogy would lead us to expect to find, and
which is occasionally found in monstrous individuals of
the species. Thus in the snapdragon (antirrhinum)
we generally do not find a rudiment of a fifth stamen;
but this may sometimes be seen. In tracing the homologies of the same part in different members of a class,
nothing is more common, or more necessary, than the
use and discovery of rudiments. This is well shown in
the drawings given by Owen of the bones of the leg of
the horse, ox, and rhinoceros.

It is an important fact that rudimentary organs, such
as teeth in the upper jaws of whales and ruminants,
can often be detected in the embryo, but afterwards
wholly disappear. It is also, I believe, a universal
rule, that a rudimentary part or organ is of greater

size relatively to the adjoining parts in the embryo, than in the adult; so that the organ at this early age is less rudimentary, or even cannot be said to be in any degree rudimentary. Hence, also, a rudimentary organ in the adult is often said to have retained its embryonic condition.

I have now given the leading facts with respect to rudimentary organs. In reflecting on them, every one must be struck with astonishment : for the same reasoning power which tells us plainly that most parts and organs are exquisitely adapted for certain purposes, tells us with equal plainness that these rudimentary or atrophied organs are imperfect and useless. In works on natural history rudimentary organs are generally said to have been created " for the sake of symmetry," or in order " to complete the scheme of nature ; " but this seems to me no explanation, merely a re-statement of the fact. Would it be thought sufficient to say that because planets revolve in elliptic courses round the sun, satellites follow the same course round the planets, for the sake of symmetry, and to complete the scheme of nature ? An eminent physiologist accounts for the presence of rudimentary organs, by supposing that they serve to excrete matter in excess, or injurious to the system ; but can we suppose that the minute papilla, which often represents the pistil in male flowers, and which is formed merely of cellular tissue, can thus act ? Can we suppose that the formation of rudimentary teeth, which are subsequently absorbed, can be of any service to the rapidly growing embryonic calf by the excretion of precious phosphate of lime ? When a man's fingers have been amputated, imperfect nails sometimes appear on the stumps : I could as soon believe that these vestiges of nails have appeared, not from unknown laws of growth, but in order to excrete horny matter, as that the rudimentary teeth on the fin of the manatee were formed for this purpose.

On my view of descent with modification, the origin of rudimentary organs is simple. We have plenty of cases of rudimentary organs in our domestic productions,—as the stump of a tail in tailless breeds,—the vestige of an ear in earless breeds,—the reappearance of minute dangling horns in hornless breeds of cattle, more especially, according to Youatt, in young animals, —and the state of the whole flower in the cauliflower. We often see rudiments of various parts in monsters. But I doubt whether any of these cases throw light on the origin of rudimentary organs in a state of nature, further than by showing that rudiments can be produced; for I doubt whether species under nature ever undergo abrupt changes. I believe that disuse has been the main agency; that it has led in successive generations to the gradual reduction of various organs, until they have become rudimentary,—as in the case of the eyes of animals inhabiting dark caverns, and of the wings of birds inhabiting oceanic islands, which have seldom been forced to take flight, and have ultimately lost the power of flying. Again, an organ useful under certain conditions, might become injurious under others, as with the wings of beetles living on small and exposed islands; and in this case natural selection would continue slowly to reduce the organ, until it was rendered harmless and rudimentary.

Any change in function, which can be effected by insensibly small steps, is within the power of natural selection; so that an organ rendered, during changed habits of life, useless or injurious for one purpose, might be modified and used for another purpose. Or an organ might be retained for one alone of its former functions.

An organ, when rendered useless, may well be variable, for its variations cannot be checked by natural selection. At whatever period of life disuse or selection reduces an organ, and this will generally be when the being has come to maturity and to its full powers of action, the principle of inheritance at corresponding ages will reproduce the organ in its reduced state at the same age, and consequently will seldom affect or reduce it in the embryo. Thus we can understand the greater relative size of rudimentary organs in the embryo, and their lesser relative size in the adult. But if each step of the process of reduction were to be inherited, not at the corresponding age, but at an extremely early period of life (as we have good reason to believe to be possible), the rudimentary part would tend to be wholly lost, and we should have a case of complete abortion. The principle, also, of economy, explained in a former chapter, by which the materials forming any part or structure, if not useful to the possessor, will be saved as far as is possible, will probably often come into play; and this will tend to cause the entire obliteration of a rudimentary organ.

As the presence of rudimentary organs is thus due to the tendency in every part of the organisation, which has long existed, to be inherited—we can understand, on the genealogical view of classification, how it is that systematists have found rudimentary parts as useful as, or even sometimes more useful than, parts of high physiological importance. Rudimentary organs may be compared with the letters in a word, still retained in the spelling, but become useless in the pronunciation, but which serve as a clue in seeking for its derivation. On the view of descent with modification, we may conclude that the existence of organs in a rudimentary, imperfect, and useless condition, or quite aborted, far from presenting a strange difficulty, as they assuredly do on the ordinary doctrine of creation, might even have been anticipated, and can be accounted for by the laws of inheritance.

Summary.—In this chapter I have attempted to show, that the subordination of group to group in all organisms throughout all time; that the nature of the relationship, by which all living and extinct beings are united by complex, radiating, and circuitous lines of affinities into one grand system; the rules followed and the difficulties encountered by naturalists in their classifications; the value set upon characters, if constant and prevalent, whether of high vital importance, or of the most trifling importance, or, as in rudimentary organs, of no importance; the wide opposition in value between analogical or adaptive characters, and characters of true affinity; and other such rules;—all naturally follow on the view of the common parentage of those forms which are considered by naturalists as allied, together with their modification through natural selection, with its contingencies of extinction and divergence of character. In considering this view of classification, it should be borne in mind that the element of descent has been universally used in ranking together the sexes, ages, and acknowledged varieties of the same species, however different they may be in structure. If we extend the use of this element of descent,—the only certainly known cause of similarity in organic beings,—we shall understand what is meant by the natural system: it is genealogical in its attempted arrangement, with the grades of acquired difference marked by the terms varieties, species, genera, families, orders, and classes.

On this same view of descent with modification, all

Y 8

the great facts in Morphology become intelligible,
whether we look to the same pattern displayed in the
homologous organs, to whatever purpose applied, of the
different species of a class; or to the homologous parts
constructed on the same pattern in each individual
animal and plant.

On the principle of successive slight variations, not
necessarily or generally supervening at a very early
period of life, and being inherited at a corresponding
period, we can understand the great leading facts in
Embryology; namely, the resemblance in an individual
embryo of the homologous parts, which when matured
become widely different from each other in structure
and function; and the resemblance in different species
of a class of the homologous parts or organs, though
fitted in the adult members for purposes as different as
possible. Larvæ are active embryos, which have become
specially modified in relation to their habits of life,
through the principle of modifications being inherited
at corresponding ages. On this same principle—and
bearing in mind, that when organs are reduced in size,
either from disuse or selection, it will generally be at
that period of life when the being has to provide for its
own wants—and bearing in mind how strong is the principle of inheritance—the occurrence of rudimentary
organs and their final abortion, present to us no inexplicable difficulties; on the contrary, their presence
might have been even anticipated. The importance of
embryological characters and of rudimentary organs in
classification is intelligible, on the view that an arrangement is only so far natural as it is genealogical.

Finally, the several classes of facts which have been
considered in this chapter, seem to me to proclaim
so plainly, that the innumerable species, genera, and

families of organic beings, with which this world is peopled, have all descended, each within its own class or group, from common parents, and have all been modified in the course of descent, that I should without hesitation adopt this view, even if it were unsupported by other facts or arguments.

CHAPTER XIV.

RECAPITULATION AND CONCLUSION.

Recapitulation of the difficulties on the theory of Natural Selection
—Recapitulation of the general and special circumstances in its
favour — Causes of the general belief in the immutability of
species — How far the theory of natural selection may be
extended — Effects of its adoption on the study of Natural
history — Concluding remarks.

As this whole volume is one long argument, it may be
convenient to the reader to have the leading facts and
inferences briefly recapitulated.

That many and serious objections may be advanced
against the theory of descent with modification through
natural selection, I do not deny. I have endeavoured
to give to them their full force. Nothing at first can
appear more difficult to believe than that the more
complex organs and instincts should have been perfected, not by means superior to, though analogous with,
human reason, but by the accumulation of innumerable
slight variations, each good for the individual possessor.
Nevertheless, this difficulty, though appearing to our
imagination insuperably great, cannot be considered real
if we admit the following propositions, namely, that all
organs and instincts are, in ever so slight a degree,
variable—that there is a struggle for existence leading
to the preservation of each profitable deviation of structure or instinct—and, lastly, that gradations in the perfection of every organ may have existed, each good of
its kind. The truth of these propositions cannot, I think,
be disputed.

It is, no doubt, extremely difficult even to conjecture by what gradations many structures have been perfected, more especially amongst broken and failing groups of organic beings, which have suffered much extinction; but we see so many strange gradations in nature, that we ought to be extremely cautious in saying that any organ or instinct, or any whole being, could not have arrived at its present state by many graduated steps. There are, it must be admitted, cases of special difficulty on the theory of natural selection; and one of the most curious of these is the existence of two or three defined castes of workers or sterile females in the same community of ants; but I have attempted to show how these difficulties can be mastered.

With respect to the almost universal sterility of species when first crossed, which forms so remarkable a contrast with the almost universal fertility of varieties when crossed, I must refer the reader to the recapitulation of the facts given at the end of the eighth chapter, which seem to me conclusively to show that this sterility is no more a special endowment than is the incapacity of two trees to be grafted together; but that it is incidental on constitutional differences in the reproductive systems of the intercrossed species. We see the truth of this conclusion in the vast difference in the result, when the same two species are crossed reciprocally; that is, when one species is first used as the father and then as the mother.

Although the fertility of varieties when intercrossed and of their mongrel offspring has been asserted by so many authors to be universal, this cannot be considered correct after the facts given on the authority of Gärtner and Kölreuter. Nor is the very general fertility of varieties, when crossed, surprising, when we remember that it is not likely that either their constitutions or their reproductive systems should have been profoundly modified. Moreover, most of the varieties which have been experimented on have been produced under domestication; and as domestication (I do not mean mere confinement) apparently tends to eliminate sterility, we ought not to expect it also to produce sterility.

The sterility of hybrids is a very different case from that of a first cross, for their reproductive organs are more or less functionally impotent: whereas in first crosses, the organs of both species are in a perfect condition. As we continually see that organisms of all kinds are rendered in some degree sterile from their constitutions having been disturbed by slightly different and new conditions of life, we need not feel surprise at hybrids being in some degree sterile, for their constitutions can hardly fail to have been disturbed from being compounded of two distinct organisations. This parallelism is supported by another parallel, but directly opposite, class of facts, namely, that the vigour and fertility of all organic beings are increased by slight changes in their conditions of life, and that the offspring of slightly modified forms or varieties acquire from being crossed increased vigour and fertility. So that, on the one hand, considerable changes in the conditions of life and crosses between greatly modified forms, lessen fertility; and on the other hand, lesser changes in the conditions of life and crosses between less modified forms, increase fertility.

Turning to geographical distribution, the difficulties encountered on the theory of descent with modification are serious enough. All the individuals of the same species, and all the species of the same genus, or even higher group, must have descended from common parents; and therefore, in however distant and isolated parts of the world they are now found, they must in the course of successive generations have passed from some one part to the others. We are often wholly unable even to conjecture how this could have been effected. Yet, as we have reason to believe that some species have retained the same specific form for very long periods, enormously long as measured by years, too much stress ought not to be laid on the occasional wide diffusion of the same species; for during very long periods of time there will always have been a good chance for wide migration by many means. A broken or interrupted range may often be accounted for by the extinction of the species in the intermediate regions. It cannot be denied that we are as yet very ignorant of the full extent of the various climatal and geographical changes which have affected the earth during modern periods; and such changes will obviously have greatly facilitated migration. As an example, I have attempted to show how potent has been the influence of the Glacial period on the distribution both of the same and of representative species throughout the world. We are as yet profoundly ignorant of the many occasional means of transport. With respect to distinct species of the same genus inhabiting very distant and isolated regions, as the process of modification has necessarily been slow, all the means of migration will have been possible during a very long period; and consequently the difficulty of the wide diffusion of species of the same genus is in some degree lessened.

As on the theory of natural selection an interminable number of intermediate forms must have existed, linking together all the species in each group by gradations as fine as our present varieties, it may be asked, Why do we not see these linking forms all around us? Why are not all organic beings blended together in an inextricable chaos? With respect to existing forms, we should remember that we have no right to expect (excepting in rare cases) to discover directly connecting links between them, but only between each and some extinct and supplanted form. Even on a wide area, which has during a long period remained continuous, and of which the climate and other conditions of life change insensibly in going from a district occupied by one species into another district occupied by a closely allied species, we have no just right to expect often to find intermediate varieties in the intermediate zone. For we have reason to believe that only a few species of a genus ever undergo change; the other species becoming utterly extinct and leaving no modified progeny. Of the species which do change, only a few within the same country change at the same time; and all modifications are slowly effected. I have also shown that the intermediate varieties which will at first probably exist in the intermediate zones, will be liable to be supplanted by the allied forms on either hand; and the latter, from existing in greater numbers, will generally be modified and improved at a quicker rate than the intermediate varieties, which exist in lesser numbers; so that the intermediate varieties will, in the long run, be supplanted and exterminated.

On this doctrine of the extermination of an infinitude of connecting links, between the living and extinct inhabitants of the world, and at each successive period between the extinct and still older species, why is not every geological formation charged with such links? Why does not every collection of fossil remains afford plain evidence of the gradation and mutation of the forms of life? Although geological research has undoubtedly revealed the former existence of many links, bringing numerous forms of life much closer together, it does not yield the infinitely many fine gradations between past and present species required on my theory; and this is the most obvious and forcible of the many objections which may be urged against it. Why, again, do whole groups of allied species appear, though certainly they often falsely appear, to have come in suddenly on the several geological stages? Why do we not find great piles of strata beneath the Silurian system, stored with the remains of the progenitors of the Silurian groups of fossils? For on my theory such strata must somewhere have been deposited at these ancient and utterly unknown epochs in the world's history.

I can answer these questions and objections only on the supposition that the geological record is far more imperfect than most geologists believe. It cannot be objected that there has not been time sufficient for any amount of organic change; for the lapse of time has been so great as to be utterly inappreciable by the human intellect. The number of specimens in all our museums is absolutely as nothing compared with the countless generations of countless species which certainly have existed. The parent form of any two or more species would not be in all its characters directly intermediate between its modified offspring, any more than the rock-pigeon is directly intermediate in crop and tail between its descendants the pouter and fantail pigeons. We should not be able to recognise a species as the parent of another species if we were to examine but ever so closely, unless we likewise possessed many of the intermediate links between their past and present states; and these many links we could hardly ever expect to discover, owing to the imperfection of the geological record. If two or three, or even more linking forms were discovered, they would simply be ranked as so many new species, more especially if found in different geological substages, let their differences be ever so

slight. Numerous existing doubtful forms could be
named which are probably varieties; but who will
pretend that in future ages so many fossil links will be
discovered, that naturalists will be able to decide, on
the common view, whether or not these doubtful forms
are varieties? Only a small portion of the world has
been geologically explored. Only organic beings of
certain classes can be preserved in a fossil condition, at
least in any great number. Widely ranging species vary
most, and varieties are often at first local,—both causes
rendering the discovery of intermediate links less likely.
Local varieties will not spread into other and distant
regions until they are considerably modified and improved; and when they do spread, if discovered in a
geological formation, they will appear as if suddenly
created there, and will be simply classed as new species.
Most formations have been intermittent in their accumulation; and their duration, I am inclined to believe,
has been shorter than the average duration of specific
forms. Successive formations are in most cases separated
from each other by enormous blank intervals of time;
for fossiliferous formations thick enough to resist future
degradation can generally be accumulated only where
much sediment is deposited on the subsiding bed of the
sea. During the alternate periods of elevation and of
stationary level the record will be generally blank.
During these latter periods there will probably be more
variability in the forms of life; during periods of subsidence, more extinction.

With respect to the absence of fossiliferous formations
beneath the lowest Silurian strata, I can only recur to
the hypothesis given in the ninth chapter. That the
geological record is imperfect all will admit; but that
it is imperfect to the degree which I require, few will
be inclined to admit. If we look to long enough in-

tervals of time, geology plainly declares that all species have changed; and they have changed in the manner which my theory requires, for they have changed slowly and in a graduated manner. We clearly see this in the fossil remains from consecutive formations invariably being much more closely related to each other, than are the fossils from formations distant from each other in time.

Such is the sum of the several chief objections and difficulties which may justly be urged against my theory; and I have now briefly recapitulated the answers and explanations which can be given to them. I have felt these difficulties far too heavily during many years to doubt their weight. But it deserves especial notice that the more important objections relate to questions on which we are confessedly ignorant; nor do we know how ignorant we are. We do not know all the possible transitional gradations between the simplest and the most perfect organs; it cannot be pretended that we know all the varied means of Distribution during the long lapse of years, or that we know how imperfect the Geological Record is. Grave as these several difficulties are, in my judgment they do not overthrow the theory of descent from a few primordial forms with subsequent modification.

Now let us turn to the other side of the argument. Under domestication we see much variability. This seems to be mainly due to the reproductive system being eminently susceptible to changes in the conditions of life; so that this system, when not rendered impotent, fails to reproduce offspring exactly like the parent-form. Variability is governed by many complex laws, by correlation of growth, by use and disuse, and by the direct action of the physical conditions of life. There is much difficulty in ascertaining how much modification our domestic productions have undergone; but we may safely infer that the amount has been large, and that modifications can be inherited for long periods. As long as the conditions of life remain the same, we have reason to believe that a modification, which has already been inherited for many generations, may continue to be inherited for an almost infinite number of generations. On the other hand we have evidence that variability, when it has once come into play, does not wholly cease; for new varieties are still occasionally produced by our most anciently domesticated productions.

Man does not actually produce variability; he only unintentionally exposes organic beings to new conditions of life, and then nature acts on the organisation, and causes variability. But man can and does select the variations given to him by nature, and thus accumulate them in any desired manner. He thus adapts animals and plants for his own benefit or pleasure. He may do this methodically, or he may do it unconsciously by preserving the individuals most useful to him at the time without any thought of altering the breed. It is certain that he can largely influence the character of a breed by selecting, in each successive generation, individual differences so slight as to be inappreciable by an uneducated eye. This process of selection has been the great agency in the production of the most distinct and useful domestic breeds. That many of the breeds produced by man have to a large extent the character of natural species, is shown by the inextricable doubts whether very many of them are varieties or aboriginally distinct species.

There is no obvious reason why the principles which have acted so efficiently under domestication should not have acted under nature. In the preservation of favoured individuals and races, during the constantly recurrent struggle for existence, we see the most powerful and ever-acting means of selection. The struggle for existence inevitably follows from the high geometrical ratio of increase which is common to all organic beings. This high rate of increase is proved by calculation,—by the rapid increase of many animals and plants during a succession of peculiar seasons, or when naturalised in a new country. More individuals are born than can possibly survive. A grain in the balance will determine which individual shall live and which shall die,—which variety or species shall increase in number, and which shall decrease, or finally become extinct. As the individuals of the same species come in all respects into the closest competition with each other, the struggle will generally be most severe between them; it will be almost equally severe between the varieties of the same species, and next in severity between the species of the same genus. But the struggle will often be very severe between beings most remote in the scale of nature. The slightest advantage in one being, at any age or during any season, over those with which it comes into competition, or better adaptation in however slight a degree to the surrounding physical conditions, will turn the balance.

With animals having separated sexes there will in most cases be a struggle between the males for possession of the females. The most vigorous individuals, or those which have most successfully struggled with their conditions of life, will generally leave most progeny. But success will often depend on having special weapons or means of defence, or on the charms of the males; and the slightest advantage will lead to victory.

As geology plainly proclaims that each land has undergone great physical changes, we might have expected that organic beings would have varied under nature, in the same way as they generally have varied under the changed conditions of domestication. And if there be any variability under nature, it would be an unaccountable fact if natural selection had not come into play. It has often been asserted, but the assertion is quite incapable of proof, that the amount of variation under nature is a strictly limited quantity. Man, though acting on external characters alone and often capriciously, can produce within a short period a great result by adding up mere individual differences in his domestic productions; and every one admits that there are at least individual differences in species under nature. But, besides such differences, all naturalists have admitted the existence of varieties, which they have considered sufficiently distinct to be worthy of record in their systematic works. No one can draw any clear distinction between individual differences and slight varieties; or between more plainly marked varieties and sub-species, and species. Let it be observed how naturalists differ in the rank which they assign to the many representative forms in Europe and North America.

If then we have variability as well as a powerful agent always ready to act, why should we doubt that variations in any way useful to beings, under their excessively complex relations of life, would be preserved, accumulated, and inherited? Why, if man can by patience select variations most useful to himself, should nature fail in selecting variations useful, under changing conditions of life, to her living products? What limit can be put to this power, acting during long ages and rigidly scrutinising the whole constitution, structure, and habits of each creature,—favouring the good and rejecting the bad? I can see no limit to this power, in slowly and beautifully adapting each form to the most complex relations of life. The theory of natural

selection, even if we looked no farther than this, seems
to me to be in itself probable. I have already recapitulated, as fairly as I could, the opposed difficulties and
objections: now let us turn to the special facts and
arguments in favour of the theory.

On the view that species are only strongly marked
and permanent varieties, and that each species first
existed as a variety, we can see why it is that no line
of demarcation can be drawn between species, commonly supposed to have been produced by special acts
of creation, and varieties which are acknowledged to
have been produced by secondary laws. On this same
view we can understand how it is that in each region
where many species of a genus have been produced,
and where they now flourish, these same species should
present many varieties; for where the manufactory of
species has been active, we might expect, as a general
rule, to find it still in action; and this is the case if
varieties be incipient species. Moreover, the species of
the larger genera, which afford the greater number of
varieties or incipient species, retain to a certain degree
the character of varieties; for they differ from each
other by a less amount of difference than do the species
of smaller genera. The closely allied species also of the
larger genera apparently have restricted ranges, and in
their affinities they are clustered in little groups round
other species—in which respects they resemble varieties.
These are strange relations on the view of each species
having been independently created, but are intelligible
if all species first existed as varieties.

As each species tends by its geometrical ratio of
reproduction to increase inordinately in number; and
as the modified descendants of each species will be
enabled to increase by so much the more as they

become diversified in habits and structure, so as to be enabled to seize on many and widely different places in the economy of nature, there will be a constant tendency in natural selection to preserve the most divergent offspring of any one species. Hence during a long-continued course of modification, the slight differences, characteristic of varieties of the same species, tend to be augmented into the greater differences characteristic of species of the same genus. New and improved varieties will inevitably supplant and exterminate the older, less improved and intermediate varieties ; and thus species are rendered to a large extent defined and distinct objects. Dominant species belonging to the larger groups within each class tend to give birth to new and dominant forms ; so that each large group tends to become still larger, and at the same time more divergent in character. But as all groups cannot thus succeed in increasing in size, for the world would not hold them, the more dominant groups beat the less dominant. This tendency in the large groups to go on increasing in size and diverging in character, together with the almost inevitable contingency of much extinction, explains the arrangement of all the forms of life, in groups subordinate to groups, all within a few great classes, which has prevailed throughout all time. This grand fact of the grouping of all organic beings is utterly inexplicable on the theory of creation.

As natural selection acts solely by accumulating slight, successive, favourable variations, it can produce no great or sudden modification ; it can act only by very short and slow steps. Hence the canon of " Natura non facit saltum," which every fresh addition to our knowledge tends to make truer, is on this theory simply intelligible. We can plainly see why nature is prodigal in variety, though niggard in innovation. But why this should be a law of nature if each species has been independently created, no man can explain.

Many other facts are, as it seems to me, explicable on this theory. How strange it is that a bird, under the form of woodpecker, should have been created to prey on insects on the ground; that upland geese, which never or rarely swim, should have been created with webbed feet; that a thrush should have been created to dive and feed on sub-aquatic insects; and that a petrel should have been created with habits and structure fitting it for the life of an auk! and so on in endless other cases. But on the view of each species constantly trying to increase in number, with natural selection always ready to adapt the slowly varying descendants of each to any unoccupied or ill-occupied place in nature, these facts cease to be strange, or perhaps might even have been anticipated.

As natural selection acts by competition, it adapts the inhabitants of each country only in relation to the degree of perfection of their associates; so that we need feel no surprise at the inhabitants of any one country, although on the ordinary view supposed to have been specially created and adapted for that country, being beaten and supplanted by the naturalised productions from another land. Nor ought we to marvel if all the contrivances in nature be not, as far as we can judge, absolutely perfect; and if some of them be abhorrent to our ideas of fitness. We need not marvel at the sting of the bee causing the bee's own death; at drones being produced in such vast numbers for one single act, with the great majority slaughtered by their sterile sisters; at the astonishing waste of pollen by our fir-trees; at the instinctive hatred of the queen bee for her own fertile daughters; at ichneumonidae feeding

Z

within the live bodies of caterpillars ; and at other such cases. The wonder indeed is, on the theory of natural selection, that more cases of the want of absolute perfection have not been observed.

The complex and little known laws governing variation are the same, as far as we can see, with the laws which have governed the production of so-called specific forms. In both cases physical conditions seem to have produced but little direct effect; yet when varieties enter any zone, they occasionally assume some of the characters of the species proper to that zone. In both varieties and species, use and disuse seem to have produced some effect; for it is difficult to resist this conclusion when we look, for instance, at the logger-headed duck, which has wings incapable of flight, in nearly the same condition as in the domestic duck ; or when we look at the burrowing tucutucu, which is occasionally blind, and then at certain moles, which are habitually blind and have their eyes covered with skin ; or when we look at the blind animals inhabiting the dark caves of America and Europe. In both varieties and species correlation of growth seems to have played a most important part, so that when one part has been modified other parts are necessarily modified. In both varieties and species reversions to long-lost characters occur. How inexplicable on the theory of creation is the variable appearance of stripes on the shoulder and legs of the several species of the horse-genus and in their hybrids! How simply is this fact explained if we believe that these species have descended from a striped progenitor, in the same manner as the several domestic breeds of pigeon have descended from the blue and barred rock-pigeon !

On the ordinary view of each species having been independently created, why should the specific characters, or those by which the species of the same genus differ from each other, be more variable than the generic characters in which they all agree? Why, for instance, should the colour of a flower be more likely to vary in any one species of a genus, if the other species, supposed to have been created independently, have differently coloured flowers, than if all the species of the genus have the same coloured flowers? If species are only well-marked varieties, of which the characters have become in a high degree permanent, we can understand this fact; for they have already varied since they branched off from a common progenitor in certain characters, by which they have come to be specifically distinct from each other; and therefore these same characters would be more likely still to be variable than the generic characters which have been inherited without change for an enormous period. It is inexplicable on the theory of creation why a part developed in a very unusual manner in any one species of a genus, and therefore, as we may naturally infer, of great importance to the species, should be eminently liable to variation; but, on my view, this part has undergone, since the several species branched off from a common progenitor, an unusual amount of variability and modification, and therefore we might expect this part generally to be still variable. But a part may be developed in the most unusual manner, like the wing of a bat, and yet not be more variable than any other structure, if the part be common to many subordinate forms, that is, if it has been inherited for a very long period; for in this case it will have been rendered constant by long-continued natural selection.

Glancing at instincts, marvellous as some are, they offer no greater difficulty than does corporeal structure on the theory of the natural selection of successive, slight, but profitable modifications. We can thus understand

why nature moves by graduated steps in endowing
different animals of the same class with their several
instincts. I have attempted to show how much light the
principle of gradation throws on the admirable architectural powers of the hive-bee. Habit no doubt sometimes comes into play in modifying instincts; but it
certainly is not indispensable, as we see, in the case of
neuter insects, which leave no progeny to inherit the
effects of long-continued habit. On the view of all the
species of the same genus having descended from a
common parent, and having inherited much in common,
we can understand how it is that allied species, when
placed under considerably different conditions of life,
yet should follow nearly the same instincts; why the
thrush of South America, for instance, lines her nest
with mud like our British species. On the view of
instincts having been slowly acquired through natural
selection we need not marvel at some instincts being
apparently not perfect and liable to mistakes, and at
many instincts causing other animals to suffer.

If species be only well-marked and permanent varieties, we can at once see why their crossed offspring
should follow the same complex laws in their degrees
and kinds of resemblance to their parents,—in being
absorbed into each other by successive crosses, and in
other such points,—as do the crossed offspring of
acknowledged varieties. On the other hand, these
would be strange facts if species have been independently created, and varieties have been produced by
secondary laws.

If we admit that the geological record is imperfect in
an extreme degree, then such facts as the record gives
support the theory of descent with modification. New
species have come on the stage slowly and at successive
intervals; and the amount of change, after equal

intervals of time, is widely different in different groups. The extinction of species and of whole groups of species, which has played so conspicuous a part in the history of the organic world, almost inevitably follows on the principle of natural selection; for old forms will be supplanted by new and improved forms. Neither single species nor groups of species reappear when the chain of ordinary generation has once been broken. The gradual diffusion of dominant forms, with the slow modification of their descendants, causes the forms of life, after long intervals of time, to appear as if they had changed simultaneously throughout the world. The fact of the fossil remains of each formation being in some degree intermediate in character between the fossils in the formations above and below, is simply explained by their intermediate position in the chain of descent. The grand fact that all extinct organic beings belong to the same system with recent beings, falling either into the same or into intermediate groups, follows from the living and the extinct being the offspring of common parents. As the groups which have descended from an ancient progenitor have generally diverged in character, the progenitor with its early descendants will often be intermediate in character in comparison with its later descendants; and thus we can see why the more ancient a fossil is, the oftener it stands in some degree intermediate between existing and allied groups. Recent forms are generally looked at as being, on the whole, higher than ancient and extinct forms; and they are in so far higher as the later and more improved forms have conquered the older and less improved organic beings in the struggle for life; they will also generally have had their organs more specialised for different functions. This fact is perfectly compatible with numerous beings still retaining a simple and little improved organisation fitted for simple conditions of life; it is likewise compatible with some forms having retrograded in organisation, though becoming under each grade of descent better fitted for their changed and degraded habits of life. Lastly, the law of the long endurance of allied forms on the same continent,—of marsupials in Australia, of edentata in America, and other such cases—is intelligible, for within a confined country, the recent and the extinct will naturally be allied by descent.

Looking to geographical distribution, if we admit that there has been during the long course of ages much migration from one part of the world to another, owing to former climatal and geographical changes and to the many occasional and unknown means of dispersal, then we can understand, on the theory of descent with modification, most of the great leading facts in Distribution. We can see why there should be so striking a parallelism in the distribution of organic beings throughout space, and in their geological succession throughout time; for in both cases the beings have been connected by the bond of ordinary generation, and the means of modification have been the same. We see the full meaning of the wonderful fact, which must have struck every traveller, namely, that on the same continent, under the most diverse conditions, under heat and cold, on mountain and lowland, on deserts and marshes, most of the inhabitants within each great class are plainly related; for they will generally be descendants of the same progenitors and early colonists. On this same principle of former migration, combined in most cases with modification, we can understand, by the aid of the Glacial period, the identity of some few plants, and the close alliance of many others, on the most distant mountains, under the most different climates; and likewise

the close alliance of some of the inhabitants of the sea
in the northern and southern temperate zones, though
separated by the whole intertropical ocean.   Although
two countries may present physical conditions as closely
similar as the same species ever require, we need feel
no surprise at their inhabitants being widely different, if
they have been for a long period completely sundered from
each other; for as the relation of organism to organism
is the most important of all relations, and as the two
countries will have received colonists from some third
source or from each other, at various periods and in
different proportions, the course of modification in the
two areas will inevitably have been different.

On this view of migration, with subsequent modification, we can see why oceanic islands should be inhabited
by few species, but of these, that many should be
peculiar.   We can clearly see why those animals which
cannot cross wide spaces of ocean, as frogs and terrestrial
mammals, should not inhabit oceanic islands; and why,
on the other hand, new and peculiar species of bats,
animals which can traverse the ocean, should so often
be found on islands far distant from any continent.
Such facts as the presence of peculiar species of bats,
and the absence of all other mammals, on oceanic islands,
are utterly inexplicable on the theory of independent
acts of creation.

The existence of closely allied or representative species in any two areas, implies, on the theory of descent
with modification, that the same parents formerly inhabited both areas; and we almost invariably find that
wherever many closely allied species inhabit two areas,
some identical species common to both still exist there.
Wherever many closely allied yet distinct species occur,
many doubtful forms and varieties of the same species
likewise occur.   It is a rule of high generality that the

inhabitants of each area are related to the inhabitants of the nearest source whence immigrants might have been derived. We see this in nearly all the plants and animals of the Galapagos archipelago, of Juan Fernandez, and of the other American islands being related in the most striking manner to the plants and animals of the neighbouring American mainland; and those of the Cape de Verde archipelago and other African islands to the African mainland. It must be admitted that these facts receive no explanation on the theory of creation.

The fact, as we have seen, that all past and present organic beings constitute one grand natural system, with group subordinate to group, and with extinct groups often falling in between recent groups, is intelligible on the theory of natural selection with its contingencies of extinction and divergence of character. On these same principles we see how it is that the mutual affinities of the species and genera within each class are so complex and circuitous. We see why certain characters are far more serviceable than others for classification; why adaptive characters, though of paramount importance to the being, are of hardly any importance in classification; why characters derived from rudimentary parts, though of no service to the being, are often of high classificatory value; and why embryological characters are the most valuable of all. The real affinities of all organic beings are due to inheritance or community of descent. The natural system is a genealogical arrangement, in which we have to discover the lines of descent by the most permanent characters, however slight their vital importance may be.

The framework of bones being similar in the hand of a man, wing of a bat, fin of the porpoise, and leg of

the horse,—the same number of vertebræ forming the
neck of the giraffe and of the elephant, and innumerable other such facts, at once explain themselves on
the theory of descent with slow and slight successive
modifications. The similarity of pattern in the wing and
leg of a bat, though used for such different purposes,—in
the jaws and legs of a crab,—in the petals, stamens, and
pistils of a flower, is likewise intelligible on the view of
the gradual modification of parts or organs, which were
alike in the early progenitor of each class. On the principle of successive variations not always supervening
at an early age, and being inherited at a corresponding
not early period of life, we can clearly see why the embryos of mammals, birds, reptiles, and fishes should be so
closely alike, and should be so unlike the adult forms.
We may cease marvelling at the embryo of an air-
breathing mammal or bird having branchial slits and
arteries running in loops, like those in a fish which has
to breathe the air dissolved in water by the aid of well-
developed branchiæ.

Disuse, aided sometimes by natural selection, will
often tend to reduce an organ, when it has become
useless by changed habits or under changed conditions
of life; and we can clearly understand on this view the
meaning of rudimentary organs. But disuse and selection will generally act on each creature, when it has
come to maturity and has to play its full part in the
struggle for existence, and will thus have little power
of acting on an organ during early life; hence the organ
will not be much reduced or rendered rudimentary at
this early age. The calf, for instance, has inherited
teeth, which never cut through the gums of the upper
jaw, from an early progenitor having well-developed
teeth; and we may believe, that the teeth in the
mature animal were reduced, during successive genera-

z 3

tions by disuse or by the tongue and palate, or lips, having become better fitted by natural selection to browse without their aid; whereas in the calf, the teeth have been left untouched by selection or disuse, and on the principle of inheritance at corresponding ages have been inherited from a remote period to the present day. On the view of each organic being and each separate organ having been specially created, how utterly inexplicable it is that parts, like the teeth in the embryonic calf or like the shrivelled wings under the soldered wing-covers of some beetles, should thus so frequently bear the plain stamp of inutility! Nature may be said to have taken pains to reveal, by rudimentary organs and by homologous structures, her scheme of modification, which it seems that we wilfully will not understand.

I have now recapitulated the chief facts and considerations which have thoroughly convinced me that species have been modified, during a long course of descent, by the preservation or the natural selection of many successive slight favourable variations. I cannot believe that a false theory would explain, as it seems to me that the theory of natural selection does explain, the several large classes of facts above specified. It is no valid objection that science as yet throws no light on the far higher problem of the essence or origin of life. Who can explain what is the essence of the attraction of gravity? No one now objects to following out the results consequent on this unknown element of attraction; notwithstanding that Leibnitz formerly accused Newton of introducing "occult qualities and miracles into philosophy."

I see no good reason why the views given in this volume should shock the religious feelings of any one.

It is satisfactory, as showing how transient such impressions are, to remember that the greatest discovery ever made by man, namely, the law of the attraction of gravity, was also attacked by Leibnitz, "as subversive of natural, and inferentially of revealed, religion." A celebrated author and divine has written to me that "he has gradually learnt to see that it is just as noble a conception of the Deity to believe that He created a few original forms capable of self-development into other and needful forms, as to believe that He required a fresh act of creation to supply the voids caused by the action of His laws."

Why, it may be asked, have nearly all the most eminent living naturalists and geologists rejected this view of the mutability of species? It cannot be asserted that organic beings in a state of nature are subject to no variation; it cannot be proved that the amount of variation in the course of long ages is a limited quantity; no clear distinction has been, or can be, drawn between species and well-marked varieties. It cannot be maintained that species when intercrossed are invariably sterile, and varieties invariably fertile; or that sterility is a special endowment and sign of creation. The belief that species were immutable productions was almost unavoidable as long as the history of the world was thought to be of short duration; and now that we have acquired some idea of the lapse of time, we are too apt to assume, without proof, that the geological record is so perfect that it would have afforded us plain evidence of the mutation of species, if they had undergone mutation.

But the chief cause of our natural unwillingness to admit that one species has given birth to other and distinct species, is that we are always slow in admitting any great change of which we do not see the intermed-

diate steps. The difficulty is the same as that felt by
so many geologists, when Lyell first insisted that long
lines of inland cliffs had been formed, and great valleys
excavated, by the slow action of the coast-waves. The
mind cannot possibly grasp the full meaning of the
term of a hundred million years; it cannot add up
and perceive the full effects of many slight variations,
accumulated during an almost infinite number of generations.

Although I am fully convinced of the truth of the
views given in this volume under the form of an
abstract, I by no means expect to convince experienced
naturalists whose minds are stocked with a multitude
of facts all viewed, during a long course of years, from
a point of view directly opposite to mine. It is so easy
to hide our ignorance under such expressions as the
"plan of creation," "unity of design," &c., and to think
that we give an explanation when we only restate a fact.
Any one whose disposition leads him to attach more
weight to unexplained difficulties than to the explanation of a certain number of facts will certainly reject
my theory. A few naturalists, endowed with much
flexibility of mind, and who have already begun to
doubt on the immutability of species, may be influenced
by this volume; but I look with confidence to the future,
to young and rising naturalists, who will be able to
view both sides of the question with impartiality. Who-
ever is led to believe that species are mutable will do
good service by conscientiously expressing his conviction; for only thus can the load of prejudice by which
this subject is overwhelmed be removed.

Several eminent naturalists have of late published
their belief that a multitude of reputed species in each
genus are not real species; but that other species are
real, that is, have been independently created. This

seems to me a strange conclusion to arrive at. They admit that a multitude of forms, which till lately they themselves thought were special creations, and which are still thus looked at by the majority of naturalists, and which consequently have every external characteristic feature of true species,—they admit that these have been produced by variation, but they refuse to extend the same view to other and very slightly different forms. Nevertheless they do not pretend that they can define, or even conjecture, which are the created forms of life, and which are those produced by secondary laws. They admit variation as a vera causa in one case, they arbitrarily reject it in another, without assigning any distinction in the two cases. The day will come when this will be given as a curious illustration of the blindness of preconceived opinion. These authors seem no more startled at a miraculous act of creation than at an ordinary birth. But do they really believe that at innumerable periods in the earth's history certain elemental atoms have been commanded suddenly to flash into living tissues? Do they believe that at each supposed act of creation one individual or many were produced? Were all the infinitely numerous kinds of animals and plants created as eggs or seed, or as full grown? and in the case of mammals, were they created bearing the false marks of nourishment from the mother's womb? Undoubtedly these same questions cannot be answered by those who, under the present state of science, believe in the creation of a few aboriginal forms, or of some one form of life. It has been asserted by several authors that it is as easy to believe in the creation of a hundred million beings as of one; but Maupertuis' philosophical axiom " of least action " leads the mind more willingly to admit the smaller number; and certainly we ought not to believe that innumerable beings within each great class have been created with plain, but deceptive, marks of descent from a single parent.

It may be asked how far I extend the doctrine of the modification of species. The question is difficult to answer, because the more distinct the forms are which we may consider, by so much the arguments fall away in force. But some arguments of the greatest weight extend very far. All the members of whole classes can be connected together by chains of affinities, and all can be classified on the same principle, in groups subordinate to groups. Fossil remains sometimes tend to fill up very wide intervals between existing orders. Organs in a rudimentary condition plainly show that an early progenitor had the organ in a fully developed state; and this in some instances necessarily implies an enormous amount of modification in the descendants. Throughout whole classes various structures are formed on the same pattern, and at an embryonic age the species closely resemble each other. Therefore I cannot doubt that the theory of descent with modification embraces all the members of the same class. I believe that animals have descended from at most only four or five progenitors, and plants from an equal or lesser number.

Analogy would lead me one step farther, namely, to the belief that all animals and plants have descended from some one prototype. But analogy may be a deceitful guide. Nevertheless all living things have much in common,—in their chemical composition, their cellular structure, their laws of growth, and their liability to injurious influences. We see this even in so trifling a circumstance as that the same poison often similarly affects plants and animals; or that the poison secreted by the gall-fly produces monstrous growths on the wild rose or oak tree. In all organic beings the union of a male and female element seems occasionally to be necessary for the production of a new being. In all, as far as is at present known, the germinal vesicle is the same. So that every individual organic being starts from a common origin. If we look even to the two main divisions—namely, to the animal and vegetable kingdoms—certain low forms are so far intermediate in character that naturalists have disputed to which kingdom they should be referred; and, as Professor Asa Gray has remarked, "the spores and other reproductive bodies of many of the lower algæ may claim to have first a characteristically animal, and then an unequivocally vegetable existence." Therefore, on the principle of natural selection with divergence of character, it does not seem incredible that, from some such low and intermediate form, both animals and plants may have been developed; and, if we admit this, we must admit that all the organic beings which have ever lived on this earth may have descended from some one primordial form. But this inference is chiefly grounded on analogy, and it is immaterial whether or not it be accepted. The case is different with the members of each great class, as the Vertebrata, the Articulata, &c.; for here, as has just been remarked, we have in the laws of homology and embryology, &c., distinct evidence that all have descended from a single parent.

When the views advanced by me in this volume, and by Mr. Wallace in the Linnean Journal, or when analogous views on the origin of species are generally admitted, we can dimly foresee that there will be a considerable revolution in natural history. Systematists will be able to pursue their labours as at present; but they will not be incessantly haunted by the shadowy doubt whether this or that form be in essence a species.

This I feel sure, and I speak after experience, will be no slight relief. The endless disputes whether or not some fifty species of British brambles are true species will cease. Systematists will have only to decide (not that this will be easy) whether any form be sufficiently constant and distinct from other forms, to be capable of definition; and if definable, whether the differences be sufficiently important to deserve a specific name. This latter point will become a far more essential consideration than it is at present; for differences, however slight, between any two forms, if not blended by intermediate gradations, are looked at by most naturalists as sufficient to raise both forms to the rank of species. Hereafter we shall be compelled to acknowledge that the only distinction between species and well-marked varieties is, that the latter are known, or believed, to be connected at the present day by intermediate gradations, whereas species were formerly thus connected. Hence, without rejecting the consideration of the present existence of intermediate gradations between any two forms, we shall be led to weigh more carefully and to value higher the actual amount of difference between them. It is quite possible that forms now generally acknowledged to be merely varieties may hereafter be thought worthy of specific names; and in this case scientific and common language will come into accordance. In short, we shall have to treat species in the same manner as those naturalists treat genera, who admit that genera are merely artificial combinations made for convenience. This may not be a cheering prospect; but we shall at least be freed from the vain search for the undiscovered and undiscoverable essence of the term species.

The other and more general departments of natural

history will rise greatly in interest. The terms used by naturalists of affinity, relationship, community of type, paternity, morphology, adaptive characters, rudimentary and aborted organs, &c., will cease to be metaphorical, and will have a plain signification. When we no longer look at an organic being as a savage looks at a ship, as at something wholly beyond his comprehension; when we regard every production of nature as one which has had a history; when we contemplate every complex structure and instinct as the summing up of many contrivances, each useful to the possessor, nearly in the same way as when we look at any great mechanical invention as the summing up of the labour, the experience, the reason, and even the blunders of numerous workmen; when we thus view each organic being, how far more interesting, I speak from experience, will the study of natural history become!

A grand and almost untrodden field of inquiry will be opened, on the causes and laws of variation, on correlation of growth, on the effects of use and disuse, on the direct action of external conditions, and so forth. The study of domestic productions will rise immensely in value. A new variety raised by man will be a more important and interesting subject for study than one more species added to the infinitude of already recorded species. Our classifications will come to be, as far as they can be so made, genealogies; and will then truly give what may be called the plan of creation. The rules for classifying will no doubt become simpler when we have a definite object in view. We possess no pedigrees or armorial bearings; and we have to discover and trace the many diverging lines of descent in our natural genealogies, by characters of any kind which have long been inherited. Rudimentary organs will speak infallibly with respect to the nature of long-lost structures. Species and groups of species, which are called aberrant, and which may fancifully be called living fossils, will aid us in forming a picture of the ancient forms of life. Embryology will reveal to us the structure, in some degree obscured, of the prototypes of each great class.

When we can feel assured that all the individuals of the same species, and all the closely allied species of most genera, have within a not very remote period descended from one parent, and have migrated from some one birthplace; and when we better know the many means of migration, then, by the light which geology now throws, and will continue to throw, on former changes of climate and of the level of the land, we shall surely be enabled to trace in an admirable manner the former migrations of the inhabitants of the whole world. Even at present, by comparing the differences of the inhabitants of the sea on the opposite sides of a continent, and the nature of the various inhabitants of that continent in relation to their apparent means of immigration, some light can be thrown on ancient geography.

The noble science of Geology loses glory from the extreme imperfection of the record. The crust of the earth with its embedded remains must not be looked at as a well-filled museum, but as a poor collection made at hazard and at rare intervals. The accumulation of each great fossiliferous formation will be recognised as having depended on an unusual concurrence of circumstances, and the blank intervals between the successive stages as having been of vast duration. But we shall be able to gauge with some security the duration of these intervals by a comparison of the preceding and succeeding organic forms. We must be cautious in attempting to correlate as strictly contemporaneous two formations, which include few identical species, by the general succession of their forms of life. As species are produced and exterminated by slowly acting and still existing causes, and not by miraculous acts of creation and by catastrophes; and as the most important of all causes of organic change is one which is almost independent of altered and perhaps suddenly altered physical conditions, namely, the mutual relation of organism to organism,—the improvement of one being entailing the improvement or the extermination of others; it follows, that the amount of organic change in the fossils of consecutive formations probably serves as a fair measure of the lapse of actual time. A number of species, however, keeping in a body might remain for a long period unchanged, whilst within this same period, several of these species, by migrating into new countries and coming into competition with foreign associates, might become modified; so that we must not overrate the accuracy of organic change as a measure of time. During early periods of the earth's history, when the forms of life were probably fewer and simpler, the rate of change was probably slower; and at the first dawn of life, when very few forms of the simplest structure existed, the rate of change may have been slow in an extreme degree. The whole history of the world, as at present known, although of a length quite incomprehensible by us, will hereafter be recognised as a mere fragment of time, compared with the ages which have elapsed since the first creature, the progenitor of innumerable extinct and living descendants, was created.

In the distant future I see open fields for far more important researches. Psychology will be based on a new foundation, that of the necessary acquirement of each mental power and capacity by gradation. Light will be thrown on the origin of man and his history.

Authors of the highest eminence seem to be fully

satisfied with the view that each species has been independently created. To my mind it accords better with what we know of the laws impressed on matter by the Creator, that the production and extinction of the past and present inhabitants of the world should have been due to secondary causes, like those determining the birth and death of the individual. When I view all beings not as special creations, but as the lineal descendants of some few beings which lived long before the first bed of the Silurian system was deposited, they seem to me to become ennobled. Judging from the past, we may safely infer that not one living species will transmit its unaltered likeness to a distant futurity. And of the species now living very few will transmit progeny of any kind to a far distant futurity; for the manner in which all organic beings are grouped, shews that the greater number of species of each genus, and all the species of many genera, have left no descendants, but have become utterly extinct. We can so far take a prophetic glance into futurity as to foretel that it will be the common and widely spread species, belonging to the larger and dominant groups within each class, which will ultimately prevail and procreate new and dominant species. As all the living forms of life are the lineal descendants of those which lived long before the Silurian epoch, we may feel certain that the ordinary succession by generation has never once been broken, and that no cataclysm has desolated the whole world. Hence we may look with some confidence to a secure future of equally inappreciable length. And as natural selection works solely by and for the good of each being, all corporeal and mental endowments will tend to progress towards perfection.

It is interesting to contemplate an entangled bank, clothed with many plants of many kinds, with birds singing on the bushes, with various insects flitting about, and with worms crawling through the damp earth, and to reflect that these elaborately constructed forms, so different from each other, and dependent on each other in so complex a manner, have all been produced by laws acting around us. These laws, taken in the largest sense, being Growth with Reproduction; Inheritance which is almost implied by reproduction; Variability from the indirect and direct action of the external conditions of life, and from use and disuse; a Ratio of Increase so high as to lead to a Struggle for Life, and as a consequence to Natural Selection, entailing Divergence of Character and the Extinction of less-improved forms. Thus, from the war of nature, from famine and death, the most exalted object which we are capable of conceiving, namely, the production of the higher animals, directly follows. There is grandeur in this view of life, with its several powers, having been originally breathed by the Creator into a few forms or into one; and that, whilst this planet has gone cycling on according to the fixed law of gravity, from so simple a beginning endless forms most beautiful and most wonderful have been, and are being, evolved.

INDEX.

2 A

2 A 3

LONDON: PRINTED BY W. CLOWES AND SONS, STAMFORD STREET,
AND CHARING CROSS.

www.ingramcontent.com/pod-product-compliance
Lightning Source LLC
Chambersburg PA
CBHW020854210326
41598CB00018B/1665